AF576058

Long-Term Monitoring and Research in Forest Hydrology: Towards Integrated Watershed Management

Long-Term Monitoring and Research in Forest Hydrology: Towards Integrated Watershed Management

Editor

Koichiro Kuraji

MDPI • Basel • Beijing • Wuhan • Barcelona • Belgrade • Manchester • Tokyo • Cluj • Tianjin

Editor
Koichiro Kuraji
Graduate School of
Agricultural and Life Sciences
The University of Tokyo
Tokyo
Japan

Editorial Office
MDPI
St. Alban-Anlage 66
4052 Basel, Switzerland

This is a reprint of articles from the Special Issue published online in the open access journal *Water* (ISSN 2073-4441) (available at: www.mdpi.com/journal/water/special_issues/Forest_Hydrology_Integrated_Watershed_Management).

For citation purposes, cite each article independently as indicated on the article page online and as indicated below:

LastName, A.A.; LastName, B.B.; LastName, C.C. Article Title. *Journal Name* **Year**, *Volume Number*, Page Range.

ISBN 978-3-0365-5208-8 (Hbk)
ISBN 978-3-0365-5207-1 (PDF)

Cover image courtesy of Koichiro Kuraji

Contents

About the Editor

Koichiro Kuraji

Professor Dr. Koichiro Kuraji is currently a Professor of the Laboratory of Forest and Water Resources Management, the University of Tokyo Forests. He obtained his Doctoral Degree in Forest Science from the University of Tokyo. His past working experience includes serving as a Research Officer in Forest Research Centre Sepilok, Sabah Forestry Department; and as an academician at the University of Tokyo Chiba Forest, Tokyo Institute of Technology, the University of Tokyo Aichi Forest, and Ecohydrology Research Institute. His research expertise is forest hydrology and integrated watershed management. His recent professional work includes serving as the Director of the Executive Office at the University of Tokyo Forests, Graduate School of Agricultural and Life Sciences, the University of Tokyo; Chairman of the Follow-up Committee of the Basic Act on Water Cycle (Act No. 16 of 2014).

Preface to "Long-Term Monitoring and Research in Forest Hydrology: Towards Integrated Watershed Management"

Forest hydrology, as a discipline, was designed to address fundamental questions regarding the impact of deforestation on floods and droughts. Recently, forest hydrology has become a primary discipline in the biophysical sciences to clarify how forests and water interact. Despite the remarkable and detailed progress of research on forest hydrology, the original questions have not yet been fully answered. Additionally, the knowledge gained through this research has not yet been integrated into real-world forest and water management. Payment for environmental services (PES) schemes have recently become available as a new tool for forest and water management; however, most of these schemes fail to consider recent advances in forest hydrology.

The influence of global warming continues to grow, and extreme weather events are increasing in frequency, posing a threat to people and property. To sustain and manage forests and water resources, and to avoid and mitigate disasters, it is important and urgent to understand long-term hydrological changes in forests and provide robust scientific knowledge on the response of forest and water resources to those changes. The detection of environmental changes and ecosystem responses requires baseline datasets based on long-term hydrological observations of forests. In recent years, the number of long-term forest hydrological observation sites has increased.

This reprint aims to gather both recent scientific research on forest hydrology based on long-term data, and integrated watershed management based on current research in forest hydrology. Ten original contributions from China, Japan, United States, Korea, Brazil, Vietnam, Indonesia, and Slovakia were included.

Koichiro Kuraji
Editor

 *water*

Editorial

Long-Term Monitoring and Research in Forest Hydrology: Towards Integrated Watershed Management

Koichiro Kuraji

Graduate School of Agricultural and Life Sciences, The University of Tokyo, Bunkyo-ku, Tokyo 113-8657, Japan; kuraji_koichiro@uf.a.u-tokyo.ac.jp

Citation: Kuraji, K. Long-Term Monitoring and Research in Forest Hydrology: Towards Integrated Watershed Management. *Water* **2022**, *14*, 2556. https://doi.org/10.3390/w14162556

Received: 15 August 2022
Accepted: 17 August 2022
Published: 19 August 2022

Publisher's Note: MDPI stays neutral with regard to jurisdictional claims in published maps and institutional affiliations.

1. Introduction

Forest hydrology, as a discipline, was designed to address fundamental questions regarding the impact of deforestation on floods and droughts. Recently, forest hydrology has become a primary discipline in the biophysical sciences to clarify how forests and water interact. Despite the remarkable and detailed progress of research on forest hydrology, the original questions have not yet been fully answered. Additionally, the knowledge gained through this research has not yet been integrated into real-world forest and water management. Payment for environmental services (PES) schemes have recently become available as a new tool for forest and water management; however, most of these schemes fail to consider recent advances in forest hydrology.

The influence of global warming continues to grow, and extreme weather events are increasing in frequency, posing a threat to people and property. To sustain and manage forests and water resources and avoid and mitigate disasters, it is important and urgent to understand long-term hydrological changes in forests and provide robust scientific knowledge on the response of forest and water resources to those changes. The detection of environmental changes and ecosystem responses requires baseline datasets based on long-term hydrological observations of forests. In recent years, the number of long-term forest hydrological observation sites has increased.

This Special Issue aims to gather both recent scientific research on forest hydrology based on long-term data, and integrated watershed management based on current research in forest hydrology.

2. Overview of This Special Issue

This Special Issue collected ten original contributions focused on forest hydrology based on long-term data and integrated watershed management. Two of them were developed in China [1,2] and Japan [3,4], while the rest were from the United States [5], Korea [6], Brazil [7], Vietnam [8], Indonesia [9], and Slovakia [10].

The publications are grouped by general themes: (1) hydrology, (2) sediment yield, and (3) payment for ecosystem services.

Topic 1 comprises eight publications, including various scales, such as global [2], large catchment [8], natural lake [4], experimental catchments [5,9], experimental plots [1,3], and simulations [6].

Hong et al. [1] investigated the negative hydraulic response to seasonal drought by mono-planting fast-growing species. They tested this hypothesis in a setting involving (a) a reforestation project, in which they mono-planted eight fast-growing tree species to successfully restore a 0.2 km^2 extremely degraded tropical rainforest, and (b) its adjacent undisturbed tropical rainforest in Sanya City, Hainan, China. They found that very high water demand from the wet to dry seasons for the mono-planted fast-growing species makes recovering the soil water content difficult.

Li et al. [6] investigated the maximum and minimum interception storage of litter layers using rainfall simulation experiments and examined the effects of litter type and rainfall characteristics on the rainfall retention and drainage processes that occur in the litter layer. Their results indicated that an increase in the intensity or duration of rainfall events led to an increase in the water retention storage of the litter. However, these factors do not influence the litter drainage capacity, which depends primarily on the force of gravity.

Wang and Zhang [2] reported research results for trend estimation of river discharge using a recently developed wavelet-based method, ensemble empirical mode decomposition (EEMD), which can separate nonstationary variations from the long-term nonlinear trend. Applying EEMD to annual discharge data of the world's 925 largest rivers from 1948 to 2004, they found that the global discharge decreased before 1978 and increased after that year, which contrasts the non-significant trend estimated by the linear method over the same period. They showed that precipitation had a consistent and dominant influence on the interannual variation of discharge on all six continents and globally, but the influences of precipitation and surface air temperature on the trend of discharge varied regionally.

Nainar et al. [3] investigated the impacts of ground litter removal and forest clearing on surface runoff using a paired runoff plot approach in the Ananomiya Experimental Watershed, Aichi, Japan. They found that the surface runoff increased four times when moving from the no-treatment to litter removed before the clearcutting phase, and 4.4 times when moving from the litter removed before clearcutting to after the clearcutting phase. The antecedent precipitation index had a significant influence on surface runoff in the litter removed before the clearcutting phase but not in the no-treatment and after the clearcutting phases.

Suryatmojo and Kosugi [9] investigated the impact of an intensive forest management system on soil hydraulic conductivity and the generation of surface runoff in different river buffer scenarios. Soil hydraulic properties were investigated in 11 plots, including one virgin forest plot and ten plots at different operational periods of the intensive forest management system in the headwater region of the Katingan watershed in Central Kalimantan, Indonesia. A two-dimensional saturated soil water flow simulation was applied to generate surface runoff from different periods of intensive forest management. The results showed that fundamental intensive forest management system activities associated with mechanized selective logging and intensive line planting reduced soil hydraulic conductivity within the near-surface profile. The recovery time for near-surface saturated hydraulic conductivity on non-skidder tracks was between 10 and 15 years, whereas on the skidder tracks it was more than 20 years.

Amatya et al. [5] tested pre-treatment hydrologic calibration relationships between paired headwater watersheds and explained the difference in flow compared to previously published data, using daily rainfall, runoff, and water table in the Santee Experimental Forest in coastal South Carolina, USA. The objective of this study was to re-evaluate and re-establish the paired calibration relationship of watersheds recovered since the 1989 hurricane, using climatic data for 2011–2019, which includes large rainfall and dry events. The results revealed that the historical pattern of runoff difference between the paired watersheds was maintained in the current baseline assessment. The difference in the mean monthly runoff between the two watersheds did not vary significantly between the pre- and post-hurricane periods, indicating complete runoff recovery.

Truong et al. [8] quantified the impact on the water cycle caused by the conversion of forests to coffee plantations in a tropical humid climate region by the application of a soil and water assessment tool (SWAT) hydrological model in the Dong Nai River Basin, Vietnam. They indicated that forest conversion into agriculture significantly increased surface runoff, while actual evapotranspiration, soil water content, and groundwater discharge decreased. These changes were mainly related to the decrease in infiltration and leaf area index after land cover changes. However, the soil was not completely destroyed after deforestation because the lost forest was replaced with crops and vegetation. Therefore, changes in infiltration were marginal and insufficient to cause substantial changes in annual flow.

Kuraji and Saito [4] identified changes in the relationship between water level and precipitation in Lake Yamanaka, Japan, by analyzing 93 years of precipitation, lake water level, and outflow data from 1928 to 2020. They found that the six-day maximum rise in the water level for the same six-day maximum precipitation was significantly greater in the latter than in the earlier period, and the difference increased with increasing precipitation. In particular, large increases in precipitation were sometimes caused by a single event or multiple events occurring in succession.

Topic 2 comprises one study. Sotiri et al. [7] validated sediment input modelling by measuring the sediment stock from the long-term siltation estimate in the Passaúna Reservoir catchment near the Metropolitan Region of Curitiba, Brazil. The sediment yield was calculated by combining a revised universal soil loss equation (RUSLE)-based model with a sediment delivery ratio model based on the connectivity approach. For RUSLE factors, a combination of remote sensing, literature review, and conventional sampling was used. They showed that the principal factors that create discrepancies in the case of the sediment budget are mostly associated with the sediment yield model. However, when including the errors due to the interpolation technique, the underestimation of sediment yield from the model may become even greater. Although they fully agree that a RUSLE-based model can reproduce the spatial and temporal patterns of sediment yield from a catchment, a comparison of the approaches in this study shows that there are clear limitations in using modelling approaches for reservoir sediment stock or reservoir lifetime assessment.

Finally, Topic 3 comprises one study. Gallay et al. [10] examined the monetary value of the ecosystem service provided by the ecosystem corresponding to its actual share in flood regulating processes, and the value of the property protected by this service was developed and demonstrated based on an example of the Cierny Hron River Basin, central Slovakia. The cost of the flood protection ecosystem service was assessed by the method of non-market monetary value to estimate the avoided damage costs of endangered infrastructure and calculated both for current and hypothetical land use. They identified areas that are crucial for water retention and deserve greater attention in management. Additionally, the monetary valuation of flood protection provided by current and hypothetical land uses enables competent and well-formulated decision-making processes.

Funding: Part of this work was supported by the Japan Society for the Promotion of Science [JPJSCCB20190007].

Acknowledgments: I acknowledge all authors of the ten papers in this SI for their contributions.

Conflicts of Interest: The author declares no conflict of interest.

References

1. Hong, W.J.; Yang, J.D.; Luo, J.H.; Jiang, K.; Xu, J.Z.; Zhang, H. Reforestation Based on Mono-Plantation of Fast-Growing Tree Species Make It Difficult to Maintain (High) Soil Water Content in Tropics, a Case Study in Hainan Island, China. *Water* **2020**, *12*, 3077. [CrossRef]
2. Wang, C.; Zhang, H. Trend and Variance of Continental Fresh Water Discharge over the Last Six Decades. *Water* **2020**, *12*, 3556. [CrossRef]
3. Nainar, A.; Kishimoto, K.; Takahashi, K.; Gomyo, M.; Kuraji, K. How Do Ground Litter and Canopy Regulate Surface Runoff? —A Paired-Plot Investigation after 80 Years of Broadleaf Forest Regeneration. *Water* **2021**, *13*, 1205. [CrossRef]
4. Kuraji, K.; Saito, H. Long-Term Changes in Relationship between Water Level and Precipitation in Lake Yamanaka. *Water* **2022**, *14*, 2232. [CrossRef]
5. Amatya, D.M.; Herbert, S.; Trettin, C.C.; Hamidi, M.D. Evaluation of Paired Watershed Runoff Relationships since Recovery from a Major Hurricane on a Coastal Forest—A Basis for Examining Effects of Pinus palustris Restoration on Water Yield. *Water* **2021**, *13*, 3121. [CrossRef]
6. Li, Q.; Lee, Y.E.; Im, S. Characterizing the Interception Capacity of Floor Litter with Rainfall Simulation Experiments. *Water* **2020**, *12*, 3145. [CrossRef]
7. Sotiri, K.; Hilgert, S.; Duraes, M.; Armindo, R.A.; Wolf, N.; Scheer, M.B.; Kishi, R.; Pakzad, K.; Fuchs, S. To What Extent Can a Sediment Yield Model Be Trusted? A Case Study from the Passaúna Catchment, Brazil. *Water* **2021**, *13*, 1045. [CrossRef]

8. Truong, N.C.Q.; Khoi, D.N.; Nguyen, H.Q.; Kondoh, A. Impact of Forest Conversion to Agriculture on Hydrologic Regime in the Large Basin in Vietnam. *Water* **2022**, *14*, 854. [CrossRef]
9. Suryatmojo, H.; Kosugi, K. River Buffer Effectiveness in Controlling Surface Runoff Based on Saturated Soil Hydraulic Conductivity. *Water* **2021**, *13*, 2383. [CrossRef]
10. Gallay, I.; Olah, B.; Gallayová, Z.; Lepeška, T. Monetary Valuation of Flood Protection Ecosystem Service Based on Hydrological Modelling and Avoided Damage Costs. An Example from the Čierny Hron River Basin, Slovakia. *Water* **2021**, *13*, 198. [CrossRef]

Article

Reforestation Based on Mono-Plantation of Fast-Growing Tree Species Make It Difficult to Maintain (High) Soil Water Content in Tropics, a Case Study in Hainan Island, China

Wenjun Hong [1,†], Jindian Yang [1,†], Jinhuan Luo [1,†], Kai Jiang [2,3], Junze Xu [1] and Hui Zhang [2,3,*]

1 Sanya Academy of Forestry, Sanya 572000, China; hongwenjun0827@126.com (W.H.); zhanghui1985052785@126.com (J.Y.); ljh3779@163.com (J.L.); zhanghuitianxia@163.com (J.X.)
2 College of Forestry/Wuzhishan National Long Term Forest Ecosystem Monitoring Research Station, Hainan University, Haikou 570228, China; y_feng000@163.com
3 Key Laboratory of Genetics and Germplasm Innovation of Tropical Special Forest Trees and Ornamental Plants (Hainan University), Ministry of Education, College of Forestry, Hainan University, Haikou 570228, China
* Correspondence: 993781@hainu.edu.cn
† These three authors contributed equally.

Received: 20 September 2020; Accepted: 28 October 2020; Published: 3 November 2020

Abstract: Reforestation has been assumed as a natural solution to recover soil water content, thereby increasing freshwater supply. Mono-plantation of fast-growing species is the first step for performing reforestation to prevent frequent and heavy rain-induced landslide in tropics. However, fast-growing species may have negative hydraulic response to seasonal drought to maintain high growth rate and, thus, may make it difficult for reforestation in tropics to recover soil water content. We tested this hypothesis in a setting involving (a) a reforestation project, which mono-planted eight fast-growing tree species to successfully restore a 0.2-km^2 extremely degraded tropical rainforest, and (b) its adjacent undisturbed tropical rainforest in Sanya City, Hainan, China. We found that, for maintaining invariably high growth rates across wet to dry seasons, the eight mono-planted fast-growing tree species had comparable transpiration rates and very high soil water uptake, which in turn led to a large (3 times) reduction in soil water content from the wet to dry seasons in this reforested area. Moreover, soil water content for the adjacent undisturbed tropical rainforest was much higher (1.5 to 5 times) than that for the reforested area in both wet and dry seasons. Thus, the invariably very high water demand from the wet to dry seasons for the mono-planted fast-growing species possesses difficulty in the recovery of soil water content. We suggest, in the next step, to mix many native-species along with the currently planted fast-growing nonnative species in this reforestation project to recover soil water content.

Keywords: deforestation; freshwater scarcity; hydraulic response to seasonal drought; limited leaf water supply; recovery of soil water content; tropical rainforest reforestation

---

1. Introduction

Human beings face a freshwater scarcity problem on account of the steadily increasing freshwater demand [1]. Currently, soil water content is one of the main freshwater resources [2], and globally, forests play a key role in maintaining them [3]. Historic human disturbance (e.g., ore mining and unreasonable agricultural use) have resulted in very high deforestation and degradation in tropical rainforest worldwide, which in turn has led to a large amount of global freshwater loss [4–7]. Thus, a number of reforestation projects have been performed worldwide to alleviate global water loss [8–11]. However, relatively few studies have evaluated the influence of large spatial or temporal reforestation projects on soil water content, especially in the humid subtropics and the tropics [12].

Forest evapotranspiration and enhanced soil infiltration can increase rainfall and soil water content [13], which are two main resources of freshwater [2]. Since the tropics would potentially witness high amounts of deforestation in the near future [14], this in turn may result in a large amount of freshwater loss. Indeed, many studies have found that the high deforestation in tropics have led to large scale freshwater loss [15–19]. A previous meta-analysis has found that reforestation-induced changes of landscape composition and configuration may be an effective way to increase freshwater supply in the tropics [20]. However, nearly no study has evaluated the influence of large spatial or temporal reforestation projects on soil water content in humid tropics [8].

Frequent typhoon and heavy rainfall during the monsoon season could easily cause landslides and tree lodging in tropical rainforest [16–19,21]. This make the reforestation of large areas of highly degraded tropical rainforests very difficult. Performing reforestation while using mono-plantation of nonnative fast-growing tree species with high survival rate could help in preventing landslide [22–24]. However, mono-plantation of nonnative fast-growing species may also lead to fast reductions in soil water content. That is because forest transpiration and its enhanced soil water infiltration are the key determinants of the final soil water content [9,10]. As shown in Figure 1a, when precipitation is absorbed into the soil, forest transpiration acts as a pump, as one part of the water is absorbed and finally lost to the atmosphere. Soil water infiltration, however, is like a sponge that permeates the rest of the precipitation and finally keeps them in the deep soil layers. Fast-growing trees usually have high transpiration [11], which may further trigger soil water uptake, reduce soil water infiltration [12], and thereby result in a low soil water content (Figure 1b). Thus, it may be difficult to recover soil water content by reforestation using mono-planted fast-growing species.

To investigate this, since 2016, we have performed a reforestation project in Baopoling mountains (BPL) in Sanya, Hainan island, China, which involves separately mono-planting eight fast-growing tree species to restore a 0.2 km^2 highly degraded tropical forest after ore mining. Specifically, we used information on the topographic and soil environments in an adjacent undisturbed tropical rainforest as a reference for this reforestation project. By this way, comparing the differences in transpiration rates between these eight fast-growing tree species and eight dominant tree species in the adjacent undisturbed tropical rainforest could reveal whether reforestation based on monoculture of fast-growing tree species might indeed recover soil water content.

Tree transpiration is also determined by some key tree hydraulic responses including photosynthesis rate, stomatal conductance, leaf hydraulic conductivity, and drought stress tolerance [13,25–29]. Functional traits (maximum photosynthesis rate, transpiration rate stomatal conductance, leaf hydraulic conductivity, and leaf turgor loss point) can directly capture these hydraulic responses [30,31]. It has been found that there is a seasonal drought in BPL [32], which should result in different hydraulic responses by the fast-growing "introduced" tree species and the slow-growing dominant tree species in the wet and the dry seasons. Thus, here, we compared the differences in these functional traits between the introduced fast-growing tree species and the slow-growing dominant tree species in the adjacent undisturbed tropical rainforest. We also tested how soil water content in both the reforestation project and the undisturbed tropical rainforest vary from the wet to dry seasons. We hypothesized that fast-growing species would develop negative hydraulic responses (e.g., having much higher transpiration and leaf hydraulic conductivity but lower drought stress tolerance than the slow-growing dominant tree species) between the wet and the dry seasons, thereby limiting recovery of soil water content during this reforestation project.

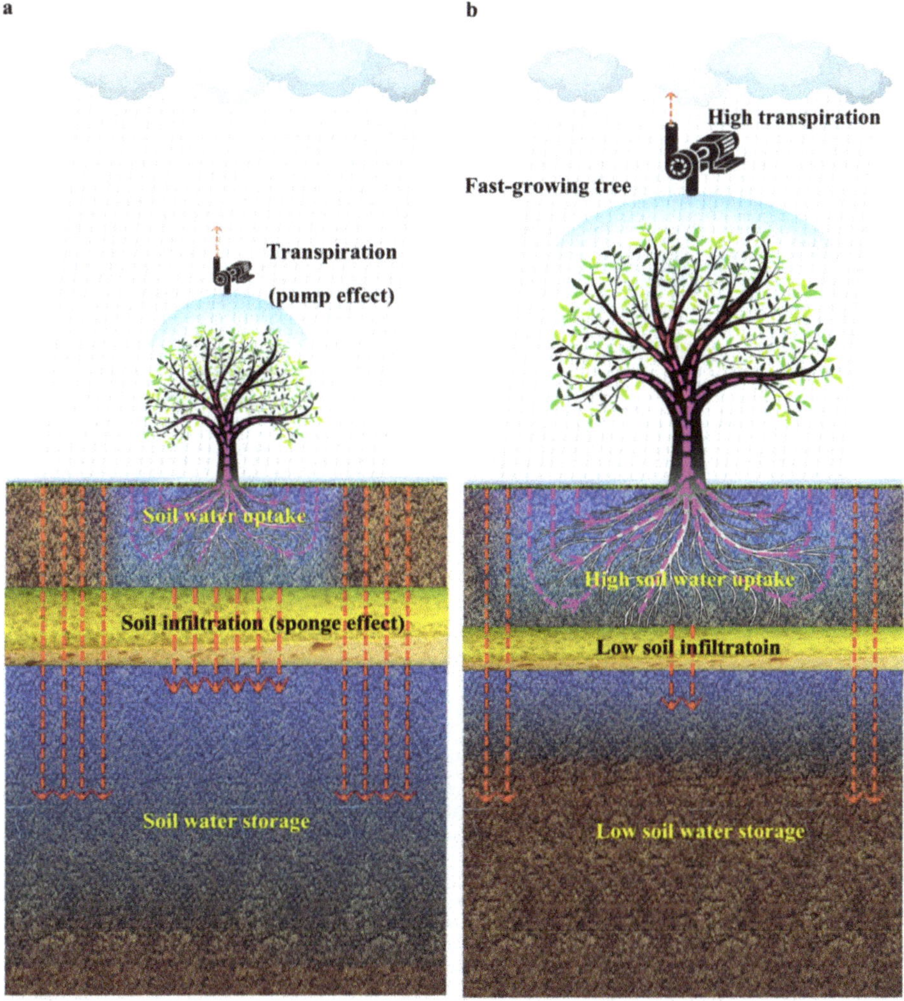

Figure 1. Hypothesizing (**a**) the roles of tree transpiration and soil infiltration, and (**b**) the influences of fast-growing tree species on soil water content.

2. Materials and Methods

2.1. Study Sites

Our study site was located in the Baopoling mountain, which is a limestone mountain in Sanya City, Hainan China (BPL, 109°51′01″ E, 18°31′99″ N; Figure 1). It has a tropical monsoon oceanic climate with a mean annual temperature of 28 °C. The average annual precipitation on the island is 1500 mm, and most (91%) of the precipitation occurs in the wet season (June to October) [32]. The inhabitants of the village near BPL get their water supply from the nearby pond and the water works in Sanya city. Additionally, the occurrence of a major cement factory (Huasheng cement factory, China) near BPL exponentially increases the water demand so that the city sometimes gets very limited water supply in the dry season. The typical vegetation of the BPL is a species-rich tropical monsoon broad-leaf forest.

Due to 20 years of limestone mining associated with the cement industry, this 0.2 km^2 highly degraded tropical forest now consists merely of bare rocks that do not support plant life (Figure 2). Areas of the BPL outside of this 0.2 km^2 degraded area have been significantly disturbed and, therefore, have remained as a species-rich tropical rainforest (Figure 2). In May 2016, we used the adjacent undisturbed forest as a reference to perform a reforestation project in BPL with the aim to recover soil water content and vegetation cover of BPL. The slope and deep soil layers of the undisturbed forest area were used as a reference to reconstruct slope and soil layers for the reforested area. Then, refilling of the area was performed with the help of the soil from the undisturbed tropical rainforest areas to monoculture seedlings (3 m height and 2 cm diameter at breast height (DBH)) of eight fast-growing tree species: *Terminalia neotaliala, Bombax malabarica, Cleistanthussumatranus, Ficusmicrocarpa, Muntingiacolabura, Acacia mangium, Leucaena glauca* and *Bougainvillea spectabilis*. Seedlings of these eight fast-growing species were purchased commercially. These species are known to be fast-growing and have high survival rates within the study region. Therefore, we reasoned that these eight species should have high potential to prevent landslides during frequent typhoon and heavy rains. These eight species were separately monocultured from the top to the bottom of BPL (Figure 2), and planting density for each of the species was maintained at 100 stems per hectare. The restoration project was finished at the end of the year 2016. In 2019, thirty plots, each of 20 × 20 m^2 (an area of 400 m^2 for each plot) that were at least 100–300 m apart from one another, were randomly sampled across the adjacent undisturbed old-growth forest. Within each plot, all freestanding trees with diameter of ≥1 cm at breast height (DBH) were measured and identified to species. *We finally found 80 tree species in the undisturbed old-growth forest, and we selected the 8 tree species (200–300 stems per hectare) Brideliatomentosa, Radermacherafrondosa, Lepisanthesrubiginosa, Rhaphiolepisindica, Pterospermumheterophyllum, Fissistigmaoldhamii, Psychotria rubra, and Cudraniacochinchinensis as our candidate dominant slow-growing tree species.*

2.2. Sampling

We selected two sites (A and B) in the reforested and the undisturbed areas, respectively, in BPL (Figure 2). In the peak of the wet season (August) in 2019, we sampled 20 fully expanded, healthy leaves from the same five independent individuals for each of the eight fast-growing species and the eight dominant slow-growing species found in the surrounding undisturbed region. Resampling was performed in the dry season (February) in 2020. Leaf samples were used to measure five hydraulic traits: transpiration rate (TR; $\mu mol\ m^{-2}\ s^{-1}$), maximum photosynthesis rate (A_{mass}; $\mu mol\ m^{-2}\ s^{-1}$), stomatal conductance (SC; $mmol\ m^{-2}\ s^{-1}$), leaf hydraulic conductivity (LHC; $mmol\ m^{-2}\ s^{-1}\ MPa^{-1}$), and leaf turgor loss point (TLP; Mpa). Detailed descriptions of the trait measurements are provided in the Supplementary Materials. We also collected 30 soil samples at a depth of 0–100 cm at site A and site B, respectively, to measure soil water content ($mg\ kg^{-1}$) gravimetrically. Every soil sample was homogenized for the whole depth (0–100 cm). Resampling was performed again in the dry season (February) in 2020.

2.3. Statistics Methods

First, we used a nonparametric test (generalized linear mixed effect model with Poisson error family) ($p < 0.05$) to test whether there were differences in transpiration-related functional traits (transpiration rate, photosynthesis rate, stomatal conductance, leaf hydraulic conductivity, and leaf turgor loss point) between the eight nonnative and the eight native tropical tree species. We also compared the differences in soil water content between site A (reforestation area) and site B (undisturbed area). A generalized linear mixed effect model with Poisson error family was carried out by function *"glmer"* in R package *"lme4"*.

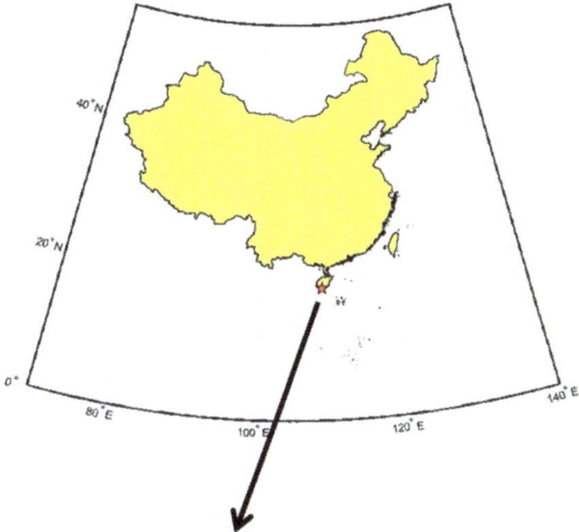

Figure 2. The map of the study site (Baopoling mountain) and the landscape of the 0.2 km^2 highly degraded area before and after reforestation: leaf samples for the eight nonnative and the native tree species, and soil samples were collected from both site A (reforestation area) and site B (undisturbed area).

3. Results

The entire project was completed in 2017, and in the past two years, both typhoon and heavy rains during the wet seasons have never caused landslide and tree lodgings on the reforested site. The mean precipitation per month in the wet and dry seasons were 1380 mm and 182 mm, respectively (Figure S1 in the Supplementary Materials). The native dominant tree species in the dry season had significantly much lower (from 1/5th to 1/2th) transpiration rate, photosynthesis rate, stomatal conductance, lead hydraulic conductivity, and leaf turgor loss point than those in the wet season (Figure 3; $p < 0.05$, generalized linear mixed effect model with Poisson error family). In contrast, compared to the wet season, nonnative fast-growing species had significantly higher (5 times) transpiration rate and leaf hydraulic conductivity, but photosynthesis rate, stomatal conductance and leaf turgor loss point were not significantly different (Figure 3). In the wet season, nonnative species have significantly higher (from 2 to 4 times) transpiration rate, photosynthesis rate, stomatal conductance, and leaf hydraulic conductivity but comparable leaf turgor loss point, compared to those of the native species (Figure 4). During the wet season, all five traits for nonnative fast-growing tree species were much higher (6 times) than those for native dominant tree species (Figure 4).

Soil water content slightly decreased from the wet to the dry seasons, in the undisturbed ecosystem, whereas soil water content in the wet season was 2.1 times the dry season in the reforested area (Figure 5). Moreover, soil water content for the undisturbed ecosystem is much higher (1.5 to 2.5 times) than those for the reforestation project in both the seasons (Figure 5).

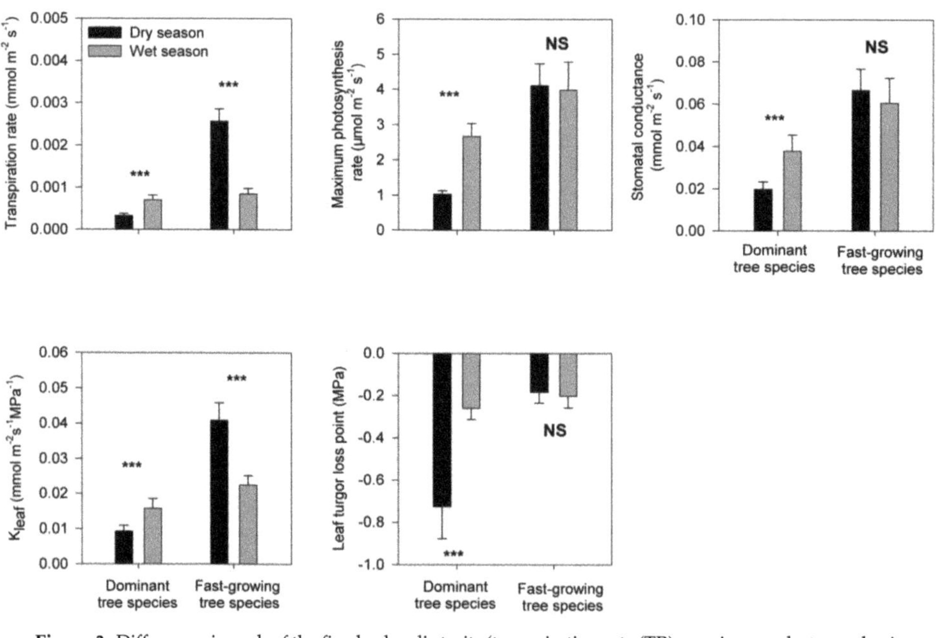

Figure 3. Differences in each of the five hydraulic traits (transpiration rate (TR), maximum photosynthesis rate (Amass), stomatal conductance (SC), leaf hydraulic conductivity (LHC), and leaf turgor loss point (TLP)) between the wet and the dry seasons for the fast-growing species used for reforestation as well as the dominant slow-growing species in the adjacent undisturbed tropical rain forest. *** indicates $p < 0.05$ and NS (nonsignificant) indicates $p > 0.05$ based on a generalized linear mixed effect model with Poisson error family. Bars indicate the mean values, and error bars denote standard errors.

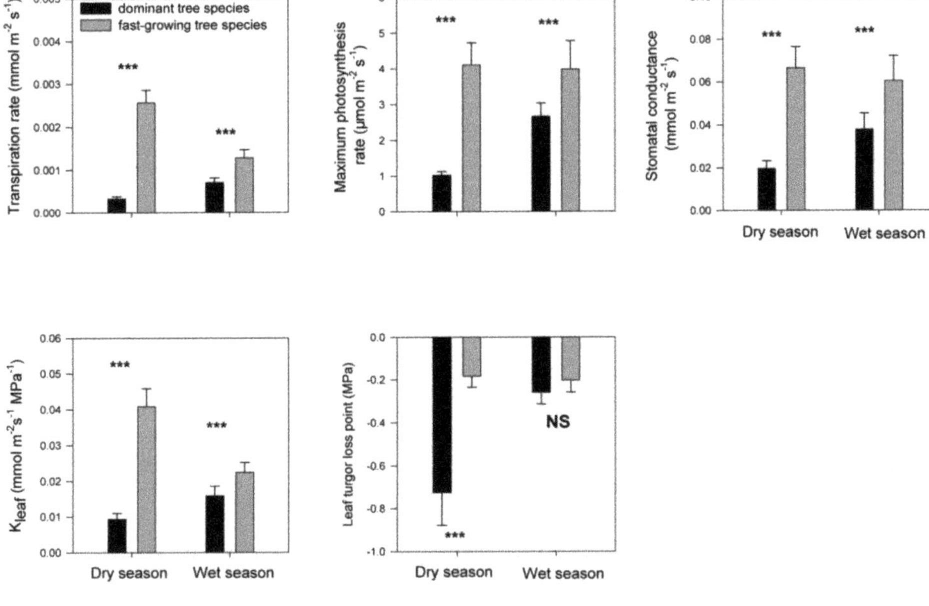

Figure 4. Differences in each of the five hydraulic traits (transpiration rate (TR), maximum photosynthesis rate (Amass), stomatal conductance (SC), leaf hydraulic conductivity (LHC), and leaf turgor loss point (TLP)) between the fast-growing species used for reforestation and the dominant slow-growing species in the adjacent undisturbed tropical rain forest in the dry and the wet seasons, respectively. *** indicates $p < 0.05$ and NS (nonsignificant) indicates $p > 0.05$ based on a generalized linear mixed effect model with Poisson error family. Bars indicate the mean values, and error bars denote standard errors.

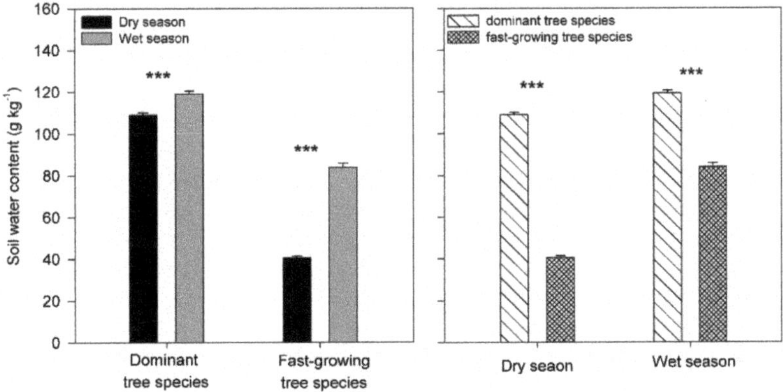

Figure 5. Differences in the soil water contents of reforested area and its adjacent undisturbed tropical rainforest ecosystem in the wet and the dry seasons. *** indicates $p < 0.05$ and NS (nonsignificant) indicates $p > 0.05$ based on generalized linear mixed effect model with Poisson error family. Bars indicate the mean values, and error bars denote standard errors.

4. Discussion

Using a large area of reforestation project, we evaluated whether reforestation could recover soil water content in humid tropic. The reforestation in this project was conducted by mono-planting eight nonnative fast-growing tree species to recover a highly degraded tropical rainforest ecosystem; We hypothesized that the very high photosynthesis water demand across wet and dry seasons by the mono-planted nonnative species might result in equally high soil water uptake thereby leading to a very low soil water content. Therefore, in the current stage of this reforestation project, it cannot help recover soil water content.

We found that the mean precipitation in the dry season is merely 1/3 of those in the wet season. Moreover, we found that the eight native species had much lower leaf turgor loss point in the dry season than that in the wet season. Lower leaf turgor loss point is usually observed when a leaf cannot get enough water supply [33]. These results indicated that all the native trees have very limited water supply in the dry season. Limited water supply usually would constrain the photosynthesis rate [34], which in turn could decrease the transpiration rate, stomatal conductance, and leaf hydraulic conductivity to decrease the photosynthesis water demand [25,27,35]. Indeed, in the dry season, the eight native tree species had lower transpiration rate, photosynthesis rate, stomatal conductance, leaf hydraulic conductivity, and leaf turgor loss point than those in the wet season. In addition, lower TLP also indicated higher drought stress tolerance [36]. As a result, for adapting to the limited water supply in the dry season, native species were inclined towards having low photosynthesis water need and high drought stress tolerance, which in turn resulted in lower soil water uptake but higher soil infiltration, thereby maintaining a very high soil water content. Consistent with this hypothesis, we observed a slightly decreased soil water content from the wet to the dry seasons. Thus, native tree species' appropriated hydraulic responses to seasonal drought could result in a slight decrease in the soil water content from the wet to the dry seasons.

Fast growth usually requires a high photosynthesis rate [30,31,37], which in turn leads to higher transpiration rate, stomatal conductance, and leaf hydraulic conductivity [13,25–28]. Indeed, we found that the eight nonnative species had much higher transpiration, photosynthesis rate, stomatal conductance, and leaf hydraulic conductivity than those for native species in both seasons. This demonstrated that these nonnative species require high photosynthesis water demand to maintain a very high maximum photosynthesis rate, which in turn may lead to lower soil water content [34]. Indeed, soil water content, as indicated in the reforestation project, is much lower than those for the undisturbed tropical rainforest ecosystem. Thus, use of mono-plantation of fast-growing tree species would lead to very low soil water content. This has also been observed in rubber and *Eucalyptus* monocultures [29,38]. In contrast, native species could facilitate higher soil water content than the nonnative species, which has also been observed in other reforestation projects [39].

We did not find significant differences in leaf turgor loss point (TLP) between nonnative and native species in the wet season, whereas nonnative species had much higher TLP than that for native-species in the dry season. Moreover, TLP for nonnative species did not vary from the wet to dry seasons. These results indicated that, in the wet season, both the nonnative and the native species could have enough leaf water supply, which leads to comparable TLP between the nonnative species and the native species. In contrast, a very limited leaf water supply might have helped the native species in their higher drought stress tolerance but native species had enough leaf water supply. It is very surprising that the photosynthesis rate for nonnative species are much higher (5 times) than those for native species in the dry season and that the photosynthesis rate for nonnative species did not vary between the wet and the dry seasons. Thus, the nonnative species should require much higher leaf water supply than the native species in the dry season: this appears impossible due to the limited water supply in the dry season. One possibility is that nonnative species tend to have higher transpiration and leaf hydraulic conductivity that absorb a large amount of soil water. This could help the plants maintain leaf water supply, thereby having higher photosynthesis rates in the dry season. Indeed, fast-growing species had high transpiration rates and leaf hydraulic conductivities. Moreover, soil water content in the reforestation area in the dry season was considerably less than in the wet season and it was also significantly less than that for the undisturbed ecosystem in the dry season. As a result, very high water demands for the mono-planted non-native species might have resulted in very high soil water uptake, which in turn would lead to a very low soil water content.

5. Conclusions

Here, we found that the very high water demand by the nonnative species across the whole year in this reforestation project raises difficulties for retaining high soil water content. The adjacent native species' low water needs and the high drought stress tolerance could help restore high soil water content in the dry season. Although reforestation based on mono-planting fast-growing tree species does not seems to maintain high soil water content, we still suggest fast-growing tree species to be the first step in performing reforestation of degraded tropical forests, as this could help prevent landslides and tree lodgings that occur due to frequent typhoons and heavy rains in the tropics. Moreover, plantation of fast-growing tree species could also increase microbial diversity and abundances [40]. In addition, fast-growing tree species dominated tropical rainforest in early successions, and the slow-growing tree species will gradually replace the fast-growing tree species in late succession, when soil nutrient and water are not enough to sustain the growth and survival of fast-growing tree species [41]. Thus, plantation of fast-growing tree species could also facilitate the recovery of the original tropical rainforest in BPL. However, for gradually recovering soil water content in BPL, we suggest mixing adjacent native tree species with these mono-plantation. However, the following studies should be performed in the future when planning to mix the adjacent dominant tree species with fast-growing tree species: (1) after what time should the native tree species be replanted together with fast-growing species? (2) Once native trees have been planted with fast-growing trees in the reforestation project, will and by when should the soil water fully recover to the levels of undisturbed forests.

Supplementary Materials: The following are available online at http://www.mdpi.com/2073-4441/12/11/3077/s1, Figure S1: The variations of mean precipitation in wet (June to October) and dry seasons (November to May) based on precipitation record per month from the local weather bureau in Sanya City, China, Text S1: Methods for functional trait measurements.

Author Contributions: Conceptualization, W.H., J.L., H.Z., and J.Y.; methodology, W.H., J.L., H.Z., and J.Y.; software, J.X. and K.J.; formal analysis, H.Z. and J.Y.; validation: W.H., J.L., H.Z., and J.Y.; investigation, J.L.; data curation, W.H., J.L., H.Z., and J.Y.; writing—original draft preparation, W.H., J.L., H.Z., and J.Y.; writing—review and editing, W.H., J.L., H.Z., and J.Y.; visualization, H.Z.; supervision, H.Z.; project administration, H.Z. All authors have read and agreed to the published version of the manuscript.

Funding: This work was funded by scientific research project of ecological restoration of Baopoling mountain in Sanya, China and a startup fund from Hainan University (KYQD (ZR) 1876).

Acknowledgments: The authors thank Liang Cong's assistance with the field experiment.

Conflicts of Interest: The authors declare no conflict of interest.

References

1. Mekonnen, M.M.; Hoekstra, A.Y. Four billion people facing severe water scarcity. *Sci. Adv.* **2016**, *2*, e1500323. [CrossRef] [PubMed]
2. Famiglietti, J.S.; Rodell, M. Water in the Balance. *Science* **2013**, *340*, 1300. [CrossRef] [PubMed]
3. Zhou, G.; Wei, X.; Chen, X.; Zhou, P.; Liu, X.; Xiao, Y.; Sun, G.; Scott, D.F.; Zhou, S.; Han, L.; et al. Global pattern for the effect of climate and land cover on water yield. *Nat. Commun.* **2015**, *6*, 5918. [CrossRef] [PubMed]
4. Sahin, V.; Hall, M.J. The effects of afforestation and deforestation on water yields. *J. Hydrol.* **1996**, *178*, 293–309. [CrossRef]
5. Lambin, E.F.; Meyfroidt, P. Global land use change, economic globalization, and the looming land scarcity. *Proc. Natl. Acad. Sci. USA* **2011**, *108*, 3465–3472. [CrossRef] [PubMed]
6. Vörösmarty, C.J.; Hoekstra, A.Y.; Bunn, S.E.; Conway, D.; Gupta, J. Fresh water goes global. *Science* **2015**, *349*, 478. [CrossRef] [PubMed]
7. Castello, L.; Macedo, M.N. Large-scale degradation of Amazonian freshwater ecosystems. *Glob. Chang. Biol.* **2016**, *22*, 990–1007. [CrossRef]

8. Solange, F.; Ometto, B.M.; Weiss, K.C.B.; Palmer, M.A.; Silva, L.C.R. Impacts of forest restoration on water yield: A systematic review. *PLoS ONE* **2017**, *12*, e0183210.
9. Peña-Arancibia, J.L.; Bruijnzeel, L.A.; Mulligan, M.; van Dijk, A.I.J.M. Forests as 'sponges' and 'pumps': Assessing the impact of deforestation on dry-season flows across the tropics. *J. Hydrol.* **2019**, *574*, 946–963. [CrossRef]
10. Bruijnzeel, L.A. Hydrological functions of tropical forests: Not seeing the soil for the trees? *Agric. Ecosyst. Environ.* **2004**, *104*, 185–228. [CrossRef]
11. Tardieu, F.; Parent, B. Predictable 'meta-mechanisms' emerge from feedbacks between transpiration and plant growth and cannot be simply deduced from short-term mechanisms. *Plant Cell Environ.* **2017**, *40*, 846–857. [CrossRef] [PubMed]
12. Krishnaswamy, J.; Bonell, M.; Venkatesh, B.; Purandara, B.K.; Rakesh, K.N.; Lele, S.; Kiran, M.C.; Reddy, V.; Badiger, S. The groundwater recharge response and hydrologic services of tropical humid forest ecosystems to use and reforestation: Support for the "infiltration-evapotranspiration trade-off hypothesis". *J. Hydrol.* **2013**, *498*, 191–209. [CrossRef]
13. Tuzet, A.; Perrier, A.; Leuning, R. A coupled model of stomatal conductance, photosynthesis and transpiration. *Plant Cell Environ.* **2003**, *26*, 1097–1116. [CrossRef]
14. Gebrehiwot, S.G.; Ellison, D.; Bewket, W.; Seleshi, Y.; Inogwabini, B.-I.; Bishop, K. The Nile Basin waters and the West African rainforest: Rethinking the boundaries. *Wiley Interdiscip. Rev. Water* **2019**, *6*, e1317. [CrossRef]
15. Mapulanga, A.M.; Naito, H. Effect of deforestation on access to clean drinking water. *Proc. Natl. Acad. Sci. USA* **2019**, *116*, 8249. [CrossRef] [PubMed]
16. Guidicini, G.; Iwasa, O.Y. Tentative correlation between rainfall and landslides in a humid tropical environment. *Bull. Eng. Geol. Environ.* **1977**, *16*, 13–20. [CrossRef]
17. Chang, K.T.; Chiang, S.H.; Lei, F. Analysing the Relationship between Typhoon-Triggered Landslides and Critical Rainfall Conditions. *Earth Surf. Proc. Land.* **2008**, *33*, 1261–1271. [CrossRef]
18. Yumul, J.G.P.; Servando, N.T.; Suerte, L.O.; Magarzo, M.Y.; Juguan, L.V.V.; Dimalanta, C.B. Tropical cyclone–southwest monsoon interaction and the 2008 floods and landslides in Panay island, central Philippines: Meteorological and geological factors. *Nat. Hazards* **2012**, *62*, 827–840. [CrossRef]
19. Acosta, L.A.; Eugenio, E.A.; Macandog, P.B.M.; Macandog, D.B.M.; Lin, E.K.H.; Abucay, E.R.; Cura, A.L.; Primavera, M.G. Loss and damage from typhoon-induced floods and landslides in the Philippines: Community Perceptions on climate impacts and adaptation options. *Int. J. Glob. Warm.* **2016**, *9*, 33–65. [CrossRef]
20. Gao, Q.; Yu, M. Reforestation-induced changes of landscape composition and configuration modulate freshwater supply and flooding risk of tropical watersheds. *PLoS ONE* **2017**, *12*, e0181315. [CrossRef] [PubMed]
21. Villamayor, B.M.R.; Rollon, R.N.; Samson, M.S.; Albano, G.M.G.; Primavera, J.H. Impact of Haiyan on Philippine mangroves: Implications to the fate of the widespread monospecific Rhizophora plantations against strong typhoons. *Ocean Coast. Manag.* **2016**, *132*, 1–14. [CrossRef]
22. Stokes, A.; Atger, C.; Bengough, A.G.; Fourcaud, T.; Sidle, R.C. Desirable plant root traits for protecting natural and engineered slopes against landslides. *Plant Soil* **2009**, *324*, 1–30. [CrossRef]
23. Walker, L.R.; Velázquez, E.; Shiels, A.B. Applying lessons from ecological succession to the restoration of landslides. *Plant. Soil.* **2009**, *324*, 157–168. [CrossRef]
24. Pang, C.C.; Ma, X.K.-K.; Lo, J.P.-L.; Hung, T.T.-H.; Hau, B.C.-H. Vegetation succession on landslides in Hong Kong: Plant regeneration, survivorship and constraints to restoration. *Glob. Ecol. Conserv.* **2018**, *15*, e00428. [CrossRef]
25. Miyashita, K.; Tanakamaru, S.; Maitani, T.; Kimura, K. Recovery responses of photosynthesis, transpiration, and stomatal conductance in kidney bean following drought stress. *Environ. Exp. Bot.* **2005**, *53*, 205–214. [CrossRef]
26. Fisher, R.A.; Williams, M.; Da Costa, A.L.; Malhi, Y.; Da Costa, R.F.; Almeida, S.; Metr, P. The response of an Eastern Amazonian rain forest to drought stress: Results and modelling analyses from a throughfall exclusion experiment. *Global Change Biol.* **2007**, *13*, 2361–2378. [CrossRef]
27. Maherali, H.; Sherrard, M.E.; Clifford, M.H.; Latta, R.G. Leaf hydraulic conductivity and photosynthesis are genetically correlated in an annual grass. *New Phytol.* **2008**, *180*, 240–247. [CrossRef]

28. Santos, V.A.H.F.d.; Ferreira, M.J.; Rodrigues, J.V.F.C.; Garcia, M.N.; Ceron, J.V.B.; Nelson, B.W.; Saleska, S.R. Causes of reduced leaf-level photosynthesis during strong El Niño drought in a Central Amazon forest. *Global Change Biol.* **2018**, *24*, 4266–4279. [CrossRef]
29. White, D.A.; McGrath, J.F.; Ryan, M.G.; Battaglia, M.; Mendham, D.S.; Kinal, J.; Downes, G.M.; Crombie, D.S.; Hunt, M.E. Managing for water-use efficient wood production in Eucalyptus globulus plantations. *Forest Ecol. Manag.* **2014**, *331*, 272–280. [CrossRef]
30. Zhang, H.; Chen, H.Y.H.; Lian, J.; John, R.; Li, R.; Liu, H.; Ye, W.; Berninger, F.; Ye, Q. Using functional trait diversity patterns to disentangle the scale-dependent ecological processes in a subtropical forest. *Func. Ecol.* **2018**, *32*, 1379–1389. [CrossRef]
31. Zhang, H.; John, R.; Zhu, S.; Liu, H.; Xu, Q.; Qi, W.; Liu, K.; Chen, H.Y.H.; Ye, Q. Shifts in functional trait–species abundance relationships over secondary subalpine meadow succession in the Qinghai-Tibetan Plateau. *Oecologia* **2018**, *188*, 547–557. [CrossRef]
32. Luo, J.H.; Cui, J.; Shree, P.P.; Jiang, K.; Tan, Z.Y.; He, Q.F.; Zhang, H.; Long, W.X. Seasonally distinctive growth and drought stress functional traits enable Leucaena Leucocephala to successfully invade a Chinese tropical forest. *Trop. Conserv. Sci.* **2020**, *9*, 1–7.
33. Bartlett, M.K.; Scoffoni, C.; Sack, L. The determinants of leaf turgor loss point and prediction of drought tolerance of species and biomes: A global meta-analysis. *Ecology* **2016**, *97*, 503–504. [CrossRef] [PubMed]
34. Guan, K.; Pan, M.; Li, H.; Wolf, A.; Wu, J.; Medvigy, D.; Caylor, K.K.; Sheffield, J.; Wood, E.F.; Liang, M.; et al. Photosynthetic seasonality of global tropical forests constrained by hydroclimate. *Nature Geosci.* **2015**, *8*, 284–289. [CrossRef]
35. Wu, J.; Serbin, S.P.; Ely, K.; Wolfe, B. The response of stomatal conductance to seasonal drought in tropical forests. *Global Change Biol.* **2020**, *26*, 823–839. [CrossRef]
36. Bartlett, M.K.; Zhang, Y.; Yang, J.; Kreidler, N.; Sun, S.-W.; Lin, L.; Hu, Y.-H.; Cao, K.-F.; Sack, L. Drought tolerance as a driver of tropical forest assembly: Resolving spatial signatures for multiple processes. *Ecology* **2016**, *97*, 503–514. [CrossRef]
37. Kirschbaum, M.U. Does enhanced photosynthesis enhance growth? Lessons learned from CO^2 enrichment studies. *Plant Physiol.* **2011**, *155*, 117–124. [CrossRef] [PubMed]
38. Tan, Z.H.; Zhang, Y.P.; Song, Q.H.; Liu, W.J.; Deng, X.B.; Tang, J.W.; Yun, D.; Zhou, W.J.; Yang, L.Y.; Yu, G.R.; et al. Rubber plantations act as water pumps in tropical China. *Geophys. Res. Lett.* **2011**, *38*, 24406. [CrossRef]
39. Wang, Y.; Shao, M.A.; Zhu, Y.; Liu, Z. Impacts of land use and plant characteristics on dried soil layers in different climatic regions on the Loess Plateau of China. *Agric. Forest Meteorol.* **2011**, *151*, 437–448. [CrossRef]
40. Zhang, W.; Zhang, H.; Jian, S.; Liu, N. Tree plantations influence the abundance of ammonia-oxidizing bacteria in the soils of a coral island. *Appl. Soil Ecol.* **2019**, *138*, 220–222. [CrossRef]
41. Mason, N.W.; Richardson, S.J.; Peltzer, D.A.; de Bello, F.; Wardle, D.A.; Allen, R.B. Changes in coexistence mechanisms along a long-term soil chronosequence revealed by functional trait diversity. *J. Ecol.* **2012**, *100*, 678–689. [CrossRef]

Publisher's Note: MDPI stays neutral with regard to jurisdictional claims in published maps and institutional affiliations.

Article

Trend and Variance of Continental Fresh Water Discharge over the Last Six Decades

Chen Wang [1] and Hui Zhang [2,*]

1. Key Laboratory of Vegetation Restoration and Management of Degraded Ecosystems, South China Botanical Garden, Chinese Academy of Sciences, Guangzhou 510650, China; chen.wang@scbg.ac.cn
2. Key Laboratory of Genetics and Germplasm Innovation of Tropical Special Forest Trees and Ornamental Plants (Hainan University), Ministry of Education, College of Forestry, Hainan University, Haikou 570228, China

* Correspondence: 993781@hainu.edu.cn

Received: 5 November 2020; Accepted: 15 December 2020; Published: 18 December 2020

Abstract: Trend estimation of river discharge is an important but difficult task because discharge time series are nonlinear and nonstationary. Previous studies estimated the trend of discharge using a linear method, which is not applicable to nonstationary time series with a nonlinear trend. To overcome this problem, we used a recently developed wavelet-based method, ensemble empirical mode decomposition (EEMD), which can separate nonstationary variations from the long-term nonlinear trend. Applying EEMD to annual discharge data of the 925 world's largest rivers from 1948–2004, we found that the global discharge decreased before 1978 and increased after 1978, which contrasts the nonsignificant trend as estimated by the linear method over the same period. Further analyses show that precipitation had a consistent and dominant influence on the interannual variation of discharge of all six continents and globally, but the influences of precipitation and surface air temperature on the trend of discharge varied regionally. We also found that the estimated trend using EEMD was very sensitive to the discharge data length. Our results demonstrated some useful applications of the EEMD method in studying regional or global discharge, and it should be adopted for studying all nonstationary hydrological time series.

Keywords: discharge trend; discharge variation; nonlinear and nonstationary; ensemble empirical mode decomposition

1. Introduction

River discharge to the world oceans is an important component of the global hydrological cycle. In a steady state, the amount of global river discharge is roughly balanced by land precipitation that originates from ocean evaporation. As a result of climate change, increasing atmospheric CO_2, land-use change, and so on, both the regional and the global water cycle deviated from the steady state [1–4]. Accurately quantifying the trend of global river discharge is important for understanding how various external factors, climate change, land-use change, atmospheric CO_2, and water availability on land influenced the hydrological cycle. However, the estimated trends of global river discharge have not been consistent among different studies, ranging from a significant increase [5] to no significant trend [6,7]. The reason for this discrepancy can be explained by the differences in the number of gauging stations used, in the period of investigation, or in the method used to estimate trends [8].

Using the constructed runoff time series of the 221 largest rivers over the 20th century, Labat et al. [5] found that global river discharge increased significantly at a rate of 4% increase per 1° global warming. However, the result was questioned by other studies [6,9,10]. On the other hand, Dai et al. [6]

collected observations from multiple sources and used Community Land Model (CLM) Version 3 in offline mode to fill the gaps in the observed discharge time series. According to the dataset of historical monthly streamflow at the farthest downstream stations for the world's 925 largest rivers from 1948–2004, Dai et al. [6] found that only about one-third of rivers showed statistically significant trends during 1948–2004, with the rivers having downward trends outnumbering those with upward trends. Their conclusion is also consistent with some other studies [8,11,12].

Trend identification is an important and difficult task in hydrological time-series analysis [13]. The trend is often taken as the tendency over the whole data span, and it is often estimated by fitting a preselected function to the data. However, mathematically, the trend is an intrinsically monotonic function only when there can be at most one extremum within a given data span [14]. Even using the same dataset, an inaccurate trend computing method can lead to erroneous results. Following Wu et al. [14], here, we define trend as the residue of data after removing the components of the data with frequency higher than a threshold frequency. If the functional form of the trend is not preselected, the processes of determining the trend must be adaptive to accommodate for nonstationary and nonlinear properties in the time series.

To quantify the trend of discharge, the rank-based nonparametric Mann–Kendall (MK) [15,16] method is often used [17,18]. The MK method assumes that the time series is stationary and that the trend is linear over the data span, which may not be true for most discharge time series. As defined by Sun et al. [19], a time series is considered to be stationary only if (1) the mean is constant, and (2) autocorrelation depends only on the relative position in the time series. Using this definition, we found that nearly all time series of river discharge we studied here are nonstationary. However, there are some modification versions of the MK method [17,18,20,21], for example, the MK test uses a pre-whitening process to account for significant autocorrelation or to take long-term persistence into account using Hurst coefficient [21,22]. These modifications still cannot overcome the difficulties of the MK method with a nonlinear trend of nonstationary time series. To overcome this issue of nonstationarity of global river discharge, we used a recently developed ensemble empirical mode decomposition (EEMD) [23,24] method in this study. The main benefits of EEMD are that it can separate nonstationary variations from the long-term trend, and the trend is estimated empirically without any prior assumption about the shape of the trend [25].

Recently, the EEMD method was applied to hydro-meteorological studies. Ji et al. [26] used the EEMD method to compute the global land surface air temperature trend from 1901–2009. They demonstrated that warming accelerated over most of the 20th century and the acceleration was much greater after 1980 than that calculated using a linear method. Chen et al. [25] used the EEMD method to calculate the increasing rate of global mean sea-level rise (GMSL) from 1993–2014. The results show that GMSL has been rising at a greater rate than previous decades and is expected to accelerate further over the coming century. More recently, Pan et al. [27] used multidimensional EEMD to compute vegetation trends and found that they were spatially and temporally nonuniform during 1982–2013. Their results showed that most vegetated areas exhibited greening trends in the 1980s. Both these studies had some new findings based on the EEMD method beyond the linear trend estimation methods used before. To our best knowledge, the EEMD method has not been used to compute a global discharge trend.

In this study, we used a recently developed wavelet-based EEMD method to compute the global and regional discharge trend of the 925 largest rivers in the world from 1948–2004. We then decomposed the river discharge time series into its trend and variance and quantified how much the trend and variance of discharge of six major continents or globally were correlated with precipitation or surface air temperature. The objectives of this study were to (1) compare the differences in the estimated trends of regional and global river discharges using EEMD and the linear MK method, (2) quantify the influences of precipitation and surface air temperature on the trend or variance of regional and global river discharges, and (3) analyze the sensitivity of the estimated trend using EEMD to the length of river discharge data. This paper is organized as follows: Section 2 lays out methods and datasets used in this study; Section 3 provides the results; discussions and conclusions are summarized in Sections 4 and 5.

2. Method and Dataset

2.1. Ensemble Empirical Mode Decomposition

Empirical mode decomposition (EMD) is an adaptive method that decomposes a time series of data $x(t)$ locally into a few oscillatory components (so-called intrinsic mode functions, IMFs, $c_j(t)$) and a residual component [28] as follows:

$$x(t) = \sum_{j=1}^{n} c_j(t) + r_n(t) + \varepsilon.$$

An IMF, c_i, must satisfy two conditions: (1) in the whole dataset, the number of extrema and the number of zero-crossings must either be equal or differ at most by one; (2) at any point, the mean value of the upper and lower envelopes is zero. Each IMF is associated with different oscillatory modes embedded in the original series [29]. The residual component, r_n, can be a constant, a monotonic function, or a function that contains only one extremum [14]. By definition, the residual in the EMD can be considered as an estimator of the trend of the original time series [14]. ε is the error term.

The ensemble EMD or EEMD is an extension of EMD by adding multiple white noise realizations to the original time series $x(t)$ before decomposition to reduce likely discrepancy caused by random errors. In this study, we perturbed the original time series by adding 10 white-noise random realizations with zero mean and a standard deviation of ε, and we used EMD to obtain the estimates of different IMFs for each of the 10 perturbed time series. We then obtained the ensemble average of respective IMFs using 10 different estimates. As compared to EMD, EEMD reduces the effects of random data errors on the estimated IMFs and estimates low-frequency IMFs more accurately [25].

2.2. Global Discharge Dataset

In this study, we used the monthly streamflow data at the farthest downstream stations for the world's 925 largest ocean-reaching rivers from 1948–2004 as compiled and gap-filled by Dai et al. [6] (http://www.cgd.ucar.edu/cas/catalog/surface/dai-runoff/). Figure S1 (Supplementary Materials) shows the distribution of the gauge stations included in this study. The river runoff data in the dataset cover ~80% of global ocean-draining land areas and amount to about 73% of global total land runoff on average. The dataset consists of harmonized observations from several data centers, including the Global Runoff Data Center (GRDC; http://grdc.bafg.de), the National Center for Atmospheric Research (NCAR; http://dss.ucar.edu/catalogs/ranges/range550.html), the University of New Hampshire (UNH; http://www.r-arcticnet.sr.unh.edu/v3.0/index.html), the United States (US) Geological Survey (http://nwis.waterdata.usgs.gov.nwis/), the Water Survey of Canada (http://www.wsc.ec.gc.ca/hydat/H2O/), the Hydrologic Cycle Observation System for West and Central Africa (AOSHYCOS; http://aochycos.ird.ne/INDEX/INDEX/HTM), and the Brazilian Hydro Web (http://hidroweb.ana.gov.br/). For most rivers, the records for 1948–2004 are fairly complete. The remaining gaps were filled using the Community Land Model Version 3 (CLM3) forced with observed precipitation and other atmospheric forcing.

The dataset also adjusted the original observed data to account for the effect of gauge location when the gauges were located some distance away from the river outlets to the oceans. For example, the farthest downstream stations for many rivers are often hundreds of kilometers away from the river mouths; thus, the observed discharge was corrected by multiplying the observed station flow by a ratio of the flow rates at the river mouth and the station simulated by a river routing model forced by the observation-based estimates. To our best knowledge, the data of Dai et al. [6] are the most complete dataset of global river discharge at present and were, therefore, chosen for this study. We found that the constructed time series faithfully reproduced the observed discharge rates for all

925 rivers. Figure S2 (Supplementary Materials) shows the constructed discharge time series of the 10 largest rivers (sorted by annual mean discharge volume) in the world, along with the gauge observed discharge values.

2.3. Global Precipitation and Surface Air Temperature Datasets

As global datasets of precipitation (P) and surface air temperature (T) are very important for understanding global discharge changes, we used two P products (Global Precipitation Climatology Centre, GPCC and Climatic Research Unit Time series 4.0, CRU TS4.0) and two T products (Climate Research Unit Temperature, CRUTEM4 and Goddard Institute for Space Studies Temperature, GISTEMP) in this study to quantify their relationships with the discharge variation and trend in global and regional scale. The reason that two sets of P and T datasets were used was to understand the effect of data uncertainty on the subsequent analysis results. The monthly GPCC rainfall data were derived from a large number of quality-controlled station data (https://www.esrl.noaa.gov/psd/data/gridded/data.gpcc.html) from 1901 to present. The monthly CRU TS4.0 rainfall dataset is based on the analysis of over 4000 individual weather station records (https://crudata.uea.ac.uk/cru/data/hrg/) from 1902 to 2016. Both data products we used were gridded to 0.5° by 0.5° spatial resolution.

The monthly CRUTEM4 data product developed by the Climate Research Unit in conjunction with the Hadley Center includes land air temperature anomalies on a 5° by 5° spatial resolution [30] from 1850 to present. Raw station data used to produce CRUTEM4 are available from the Met Office website (https://www.metoffice.gov.uk/hadobs/crutem4/). The monthly GISTEMP data product produced by Goddard Institute for Space Studies includes estimates of global land surface temperature change [31] (https://www.esrl.noaa.gov/psd/data/gridded/data.gistemp.html) with a spatial resolution of 2° by 2° from 1880 to present.

3. Results

3.1. Global Continental Discharge Trend

A time series is considered stationary if (1) the mean is constant and (2) the autocorrelation only depends on the relative position in the time series [19]. Using this definition of stationarity, we evaluated all 925 river discharge data from 1948 to 2004 and found that none of the discharge time series were found to be strictly stationary.

Using the total discharge from 925 river basins to approximate the global river discharge, we decomposed the global discharge time series into four IMFs and one trend component using EEMD (see Figure 1). The first IMF mode had a wide spectrum and represented synoptic oscillations with periods ranging from 2 years (1/0.47) to 8 years (1/0.12) (see Figure 1g). The second IMF represented variance with periods from 5 years (1/0.2) to 20 years (1/0.05) and centered at around 9 years (see Figure 1h). The third IMF contained mostly interdecadal variations with periods between 9 years (1/0.11) and 20 years (1/0.05) (see Figure 1i). The 20 year to 50 year interdecadal variations were mostly captured by the fourth IMF (see Figure 1j). The differences between original time series and the sums of all four IMFs represented the trend (see Figure 1f).

Results from Figure 1f show that global river discharge decreased from 1948 to 1978, and then increased after 1978. This contrasts the result of no significant trend as estimated by the MK method that imposes a linear fit to a nonlinear trend (see Figure 1a). The estimated trend using the EEMD method in this study is consistent with previous studies that found significant shifts in global discharge trend around the 1970s from 1901 to 2002 [1,32,33].

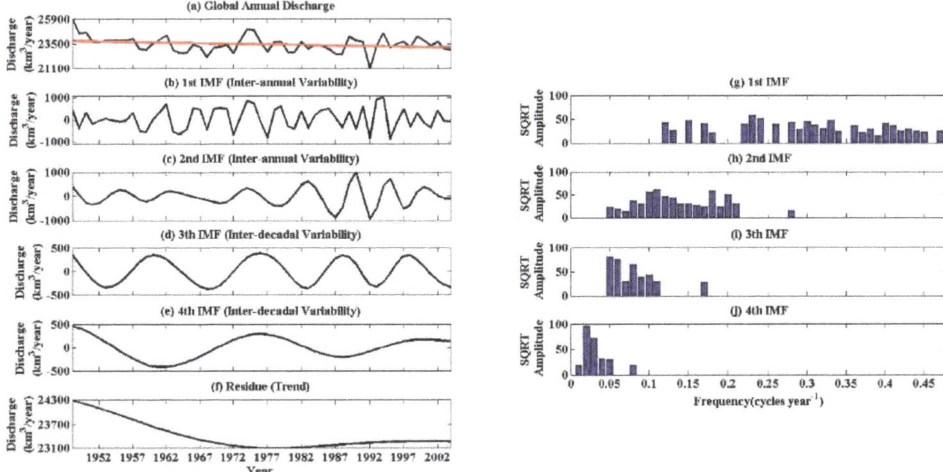

Figure 1. Annual global freshwater discharge from 1948 to 2004 (**a**), and its decomposition into four intrinsic mode functions (**b**–**e**) and trend (**f**) using ensemble empirical mode decomposition (EEMD). The red line in (**a**) represents the trend estimate using the Mann–Kendall (MK) method, and the power spectrum of each intrinsic mode function (IMF) is also shown in (**g**–**j**).

Another advantage of EEMD is its ability to quantify the contribution of variance at different timescales to total variance of the observed time series. Using the results in Figure 1, we estimated that variance contributions were 41%, 25%, 10%, 9%, and 15% from the first, second, third, and fourth IMFs and the trend. The greater contribution by the trend than the third and fourth IMFs also suggests that the trend was nonlinear.

3.2. Spatial and Temporal Variation of River Discharge Trend from 1948 to 2004

We compared the estimated trends by EEMD with those by MK for all 925 rivers. Consistent with the estimates by Dai et al. [6], our results using the MK approach showed that the discharge of about 80% of 925 rivers had no significant trends from 1948 to 2004, and that the number of rivers with significantly downward trends (116) was about twice as many as the number of rivers with significantly upward trends (48) (see Figure 2a). Contrary to the estimated trends using the MK method, estimates using EEMD showed that more than 70% of 925 rivers had significant trends from 1948–2004, including 292 of them with significantly upward trends in the regions of northern high latitudes, southeast North America, and southeast South America. Among those 70% rivers with significant trends, 442 of them had significantly downward trends, mainly in low latitudes. Only 191 rivers did not have significant trends from 1948–2004 (see Figure 2b). Because all discharge time series were not stationary, the trends estimated using MK method were not reliable. Our results are consistent with Nohara et al. [34] who found significant increases in the discharge of rivers in the northern high-latitude regions because of the increased precipitation. The downward trends for rivers in the subtropics found in this study are also consistent with Sun et al. [35] and Solomon et al. [36].

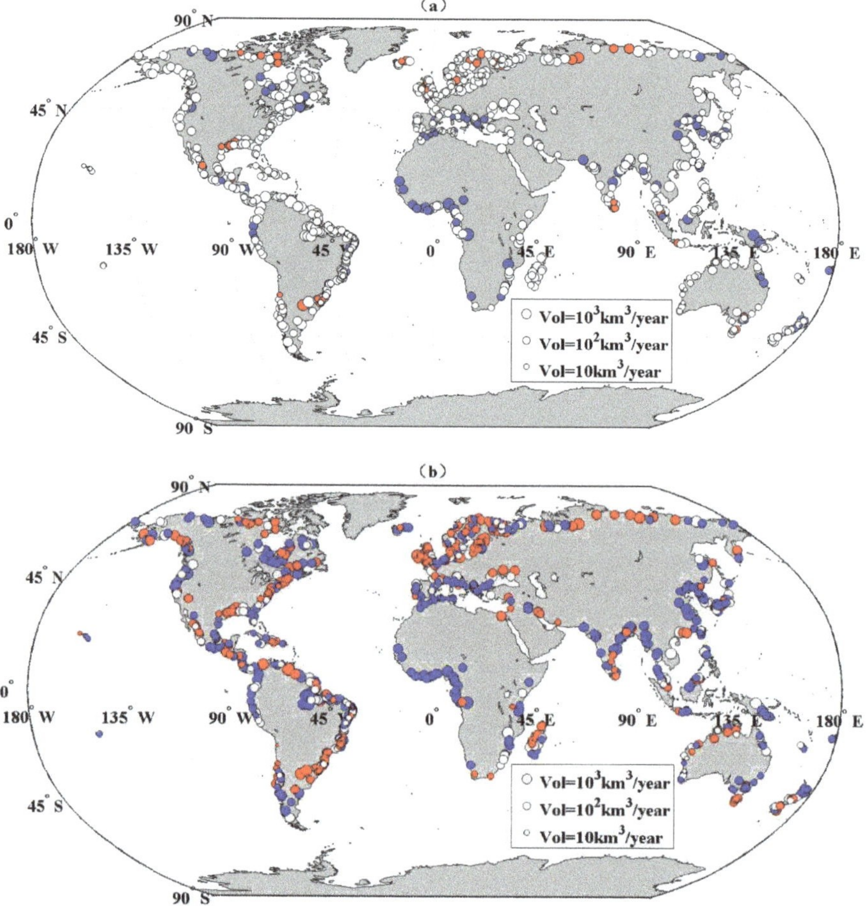

Figure 2. Trends of 925 river discharges to the ocean from 1948–2004 as estimated using MK (**a**) or EEMD (**b**) methods. Stations with significant upward (downward) trends are shown in red (blue), while white circles represent no significant trend. The log transform of annual volume discharge is proportional to the area of each circle, and each circle is located at the mouth of the watershed.

We also analyzed the temporal variations of the trends of all 925 river discharges and found that the trends of all river discharges were nonlinear over time. Figure 3 shows the annual mean discharge and its long-term trend using the EEMD method for the 10 largest rivers (sorted by annual mean volume) in the world. Results show that, for each river, that EEMD estimated a nonlinear trend, while the MK test gave nonsignificant trends for all 10 rivers except for river Congo with a significant downward trend and rivers Yenisey and Parana with significant upward trends (see Table S1, Supplementary Materials). Because of their trends being nonlinear over time, particularly for those rivers with a variable direction of change in the trend, the linear method is likely to give erroneous estimates.

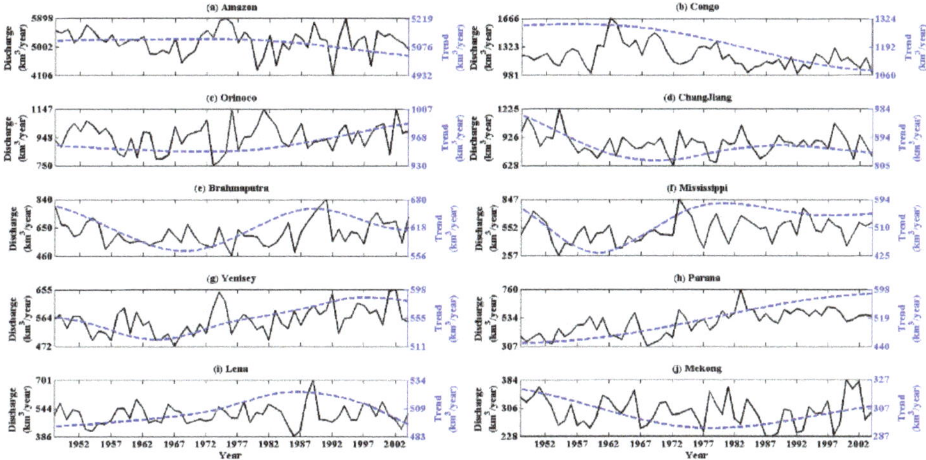

Figure 3. The constructed discharge time series of the 10 largest rivers (black curve) and their trends (blue curve) as estimated using the EEMD method from 1948–2004.

3.3. Influences of Changing Trend and Variance of P and T on the Trend and Variance of River Discharge

Both theoretical studies and observations show that P and T likely influence variation and trend of river discharge [5], in addition to surface properties, such as vegetation and soil types, leaf area index, and so on. Here, we separately analyzed the correlations of detrended P or T with the detrended discharge or correlations of their respective trends. Figures S3 and S4 (Supplementary Materials) show the two global P products and two global T products we used in this study and their long-term trend computed by the EEMD method.

Because of the likely regional variations in discharge sensitivities to P or T variations, we grouped the discharge data, P and T, into six major continents or globally for correlation analysis. Across six continents or globally, variations in the detrended P and the detrended discharge time series were significantly and positively correlated, with the highest correlation valued about 0.7 for the Asian continent (see Figure 4). These correlations varied very little between the two different precipitation data products. However, correlations between the trends of P and discharge varied across different continents, being significantly positive for the African and Asian continents, significantly negative for the European and Oceanian continents, and not significant for the North American and South American continents. Correlations in the trends between discharge and P also differed between the two precipitation data products, being significantly positive for GPCC rainfall on a global scale, but insignificantly different from zero for the CRU TS4.0 data product on a global scale, suggesting relatively large uncertainties in the correlations of trends.

We also correlated the detrended and trend of discharge time series with two different products of global T. Correlations between the two detrended time series of discharge and T varied from region to region, being positive for African and Asian continents and negative for the other four continents. The correlation was only significant for Asia and South America for both products and Oceania for the CRUTEMP4 product. Globally, the detrended discharge was not significantly correlated with the detrended T for both temperature products (see Figure 5a). For the trends, T was negatively correlated with discharge across all six continents, and the correlation was only significant for Africa, Europe, and Oceania for both T products and South America for the CRUTEMP4 product (see Figure 5b). Globally, the trends of T and discharge were also significantly negatively correlated (see Figure 5b).

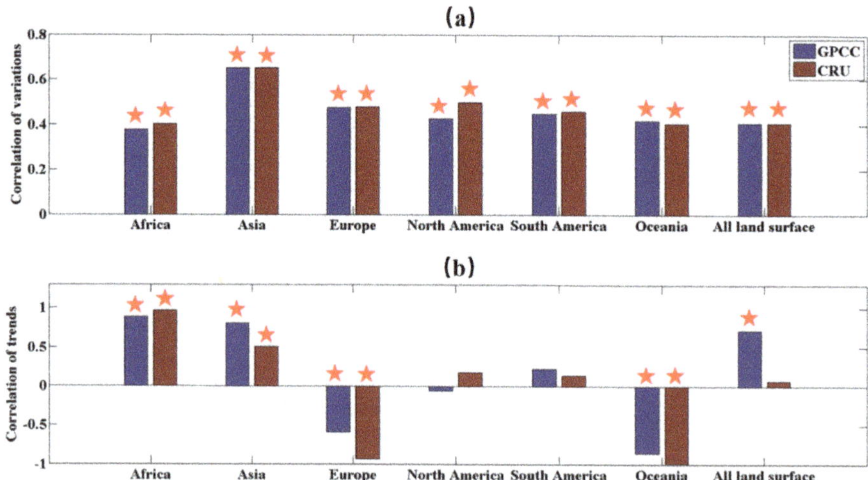

Figure 4. Correlations of (**a**) the detrended time series of annual precipitation and discharge from 1948 to 2004; (**b**) the trends of precipitation and discharge from 1948 to 2004; the red star represents significant correlation at the 95% level.

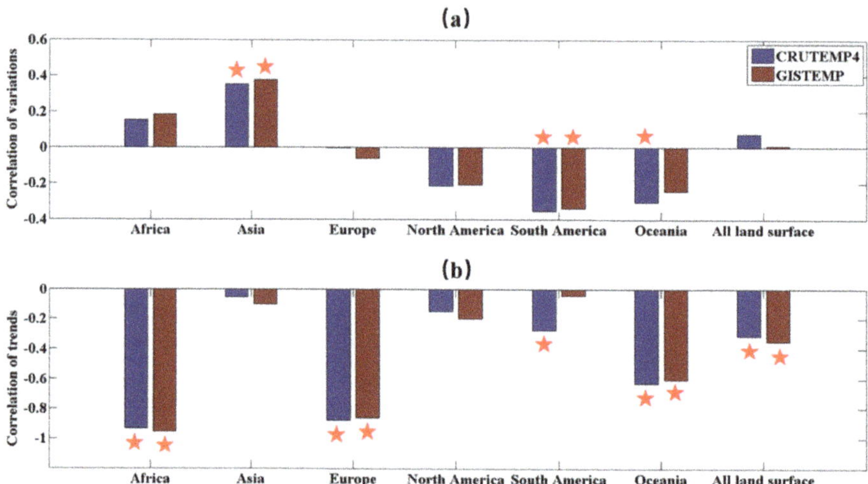

Figure 5. Same as Figure 4 but for the correlations between (**a**) the detrended time series of annual surface air temperature and discharge from 1948 to 2004; (**b**) the trends of surface air temperature and discharge from 1948 to 2004; the red star represents significant correlation at the 95% level.

3.4. Sensitivity of the Estimated Global Continental Discharge to the Length of Data Record

The estimated trend by EEMD can be sensitive to the data length. We compared the estimated trends using MK or EEMD methods for Dai's global discharge data series with different ending and starting times (see Figures 6 and 7). The discharge data included three periods with very high or low discharge rates, e.g., very high rates from 1948 to 1951 and from 1974 to 1975, and very low rates in 1992. The results in Figure 6 show that the estimated trends using the MK method were rather insensitive to the ending year. Depending on whether the two periods of very high (1974–1975) or low discharge (1992) were included or not, the trend estimated using EEMD could go up (see the red curve in Figure 6b) or down (see the blue curve in Figure 6b). Different sensitivities for the starting year of the estimated trends using the two methods are further illustrated in Figure 7. Overall, the linear trends as estimated using the MK method were not significantly different from zero. However, the estimated trend using the EEMD method could vary significantly, depending on whether the very high discharge period (1948–1951) was included or not. Only when the observed discharges of 1948–1951 were included, the trend in global discharge as estimated using the EEMD method decreased from 1948 to 1978 and then increased. The abnormally high discharge from 1948 to 1952 resulted in the global discharge time series being nonstationary with a nonlinear trend. This nonlinear trend was more accurately quantified using EEMD than using the MK method.

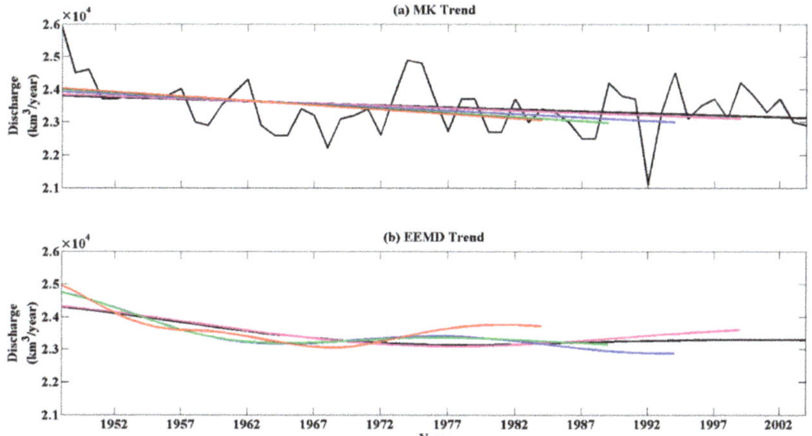

Figure 6. Sensitivity of the estimated trends to the same starting year (1948) but different ending year of global runoff data. Different colors represent the estimated trends for ending times of 1979, 1984, 1989, 1994, 1999, and 2004. The estimated trends are shown in (**a**) for the MK method and (**b**) for the EEMD method.

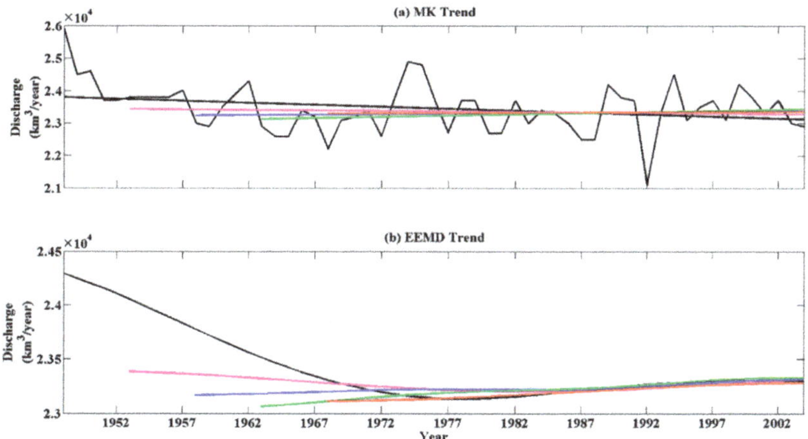

Figure 7. Same as Figure 6, but the trends calculated based on global discharge data with the same ending time (2004), but different starting years of 1948, 1953, 1958, 1963, 1968, and 1973. The estimated trends are shown in (**a**) for the MK method and (**b**) for the EEMD method.

4. Discussion

The EEMD method presented here showed its advantage in dealing with nonstationary time series with a nonlinear trend. As the widely used MK method for trend quantification assumes stationarity, it is not strictly applicable to nonstationary time series such as river discharge. This study found that the MK method significantly underestimated the number of world's major rivers with significant trends and the trends themselves, as compared with the EEMD method.

At a regional scale, contrary to the estimated trend of the MK method, estimates using EEMD showed that more than 70% of 925 rivers had significant trends from 1948–2004. The rivers with significantly upward trends were mainly located in the regions of northern high latitudes, southeast North America, and southeast South America, while those with significantly downward trends were mainly located in low latitudes. The decrease in precipitation over low- and mid-latitude areas and the increase in precipitation at high latitudes may have been the major cause for the change in river discharge over those different regions from 1948–2004 [6,37].

The observed great warming at high latitudes may have also contributed to the observed increase in discharge in those regions. Nijssen et al. [38] found that the predicted warming in high-latitude basins was greatest during the winter months, which may lead to the stored snow melting as stream flow. The rising temperature also results in increases in oceanic evaporation and, consequently, increased precipitation and runoff over land [6]. Furthermore, surface warming at high latitudes has caused a degradation of the permafrost, which may also contribute to increases in discharge trend at these latitudes [12]. As a result, this leads to the hypothesis that there will be an intensification of the global water cycle, especially at high latitudes [5]. However, our study showed that the process may be more complex because the climate system is highly nonlinear. The effect of climate change on global water cycle also exhibits some nonlinearities.

More importantly our analysis showed that variations in precipitation had dominant influences on the variances in river discharge both regionally and globally, while the trends of discharge might be influenced by many other factors, including atmospheric CO_2, land-use change, and climate change [33,39,40]. This means that the variation in continental discharge is mainly caused by climate variance, while the causes of the change in discharge trend are more difficult to identify. However, correctly quantifying the observed trend of regional or global discharge is a prerequisite. Because of large regional and temporal variations in the trends of the observed river discharges, attribution studies of discharge trend should focus on different regions using global land models, such as those by Gedney et al. [39] and Piao et al. [40].

The estimated change in the trend of global river discharge around 1978 had significant implications for our interpretation of how various external factors influenced the global hydrological cycle over time. Globally, the river discharge decreased before 1978 and increased after 1978, while global surface temperature increased steadily over the study period, which is different from the estimated sensitivity by Labat et al. [5] showing that the river runoff is increasing with global warming. According to the increasing trend of global discharge as estimated by Labat et al. [5] using the MK method, Gedney et al. [39] attributed the decrease in evapotranspiration to increasing atmospheric CO_2 as the dominant cause, while land-use changes, climate change, or increased precipitation were considered as the major causes by other studies [33,40]. If the trend differs in direction before and after 1978, the claim highlighting increasing CO_2 alone as the likely cause is unlikely to hold [41]. Future studies should focus on why the trend changed from a downward trend before 1978 to a small upward trend after 1978.

Because the trend of global river discharge is nonlinear, the estimated trend can be affected by the uncertainties and lengths of the observed datasets [10]. This study also showed that the estimated trend of global river discharge from 1948 to 2004 using EEMD was very sensitive to the unusually high discharge rate in the first 4 years. The difference in the estimated trend of global river discharge between the previous two studies by Labat et al. [5] and Dai et al. [6] may have resulted from the different amount of data (221 vs. 925 rivers) and different period (the whole 20th century vs. 1948 to 2004). Therefore, it is important to compare the estimated trend for the same period.

Water demands (especially for irrigation systems) are also a factor that influence the continental discharge trend. To demonstrate this kind of effect, we used the Global Map of Irrigation Areas (GMIA) version 5 (http://www.fao.org/nr/water/aquastat/irrigationmap/index10.stm) from the Food and Agriculture Organization (FAO) of the United Nations to show whether irrigation had a significant effect on continental discharge trend (Figure S5, Supplementary Materials). At the global scale, northern India and the North China Plain are the most intensive irrigation areas, while others have very low percentages of irrigation practices (Figure S5a, Supplementary Materials). However, these two places mainly used groundwater to irrigate (Figure S5c, Supplementary Materials). We concluded that irrigation may have little effect on the global continental discharge trend. In this study, we did not conclude irrigation as one of the impact factors on the global continental discharge trend. Our result is consistent with Haddeland et al. [42], who found that the impact of manmade reservoirs and water withdrawals on the long-term global terrestrial water balance was small.

Our results have other implications, e.g., they provided a new idea to better quantify model simulation uncertainties. Land surface and hydrological models always simulate gridded runoff [43]. The model-simulated runoff may contain significant bias due to errors in the meteorological forcing data and model structure. However, runoff is difficult to measure directly, while discharge over most of the world's major rivers has been monitored by stream gauges for many decades [6]. Thus, historical records of discharge provide a measure of basin-integrated runoff, and they have been used to calibrate model-simulated runoff field. Nevertheless, it is difficult for models to fully simulate the long-term time series of runoff/discharge. If the models can exactly simulate the long-term trend of runoff/discharge, they will also be of great reference value. Accurately identifying the discharge trend is the basis of the abovementioned model evaluations. In the future, it is necessary for model developers to not only reduce the uncertainties of time series modeled discharge, but also to accurately simulate the nonlinear discharge trend.

Our study also has implications for guiding future water use in people's production and life. The global water cycle is expected to change over the 21st century due to the combined effects of climate change and increasing human activities. In a warm world, the water holding capacity of the atmosphere will increase, resulting in a change in the frequency of extreme events, e.g., precipitation extremes and dry periods. Under the moderate representative concentration pathway (RCP) 4.5 emissions scenario, annual precipitation could increase by 10–25% from 1970–1999 to 2070–2099 over most of Eurasia, North America, and central to northern Africa, but decrease by 3–15% over most of Australia, southern Africa, and Central America [6]. The changes in future precipitation will greatly affect the trends of continent discharge. On the basis of the accurate calculation of global and regional discharge trends, we can apply quantitative attribution analysis to the trend results. The attribution results can demonstrate which are the key determinants of nonlinear discharge trends [39,40] and help us understand the physical mechanism of water cycle under the influence of climate change and human activities. These results can be used to construct a new model or improve the existing models to predict the shortage of water resources under the future climate change scenarios, which in turn can provide valuable references for water resource management policies. For example, if we predict significant downward discharge trends at some point in the future in one or more watersheds, long-term planning for water resources can be prepared in advance in order to guarantee domestic and agricultural water. Thus, an accurate calculation of discharge trend has momentous application values to the guidance of water use in people's production and life.

Our study had some limitations. For example, the calculation results deeply depended on whether there were sufficient discharge data. Our calculation of discharge trend required complete time-series data. However, in actual observations, especially for long-time-scale data, there is often a lack of measurements. In this paper, we used Dai's discharge data which were compiled and gap-filled by many technical means, including CLM3 model simulation and others. The final dataset met our requirements to compute the long-term discharge trend. However, Dai's dataset was populated at the farthest downstream stations for the major rivers. According to this dataset, we could not test if the discharge trends changed in one river's upstream and downstream gauge sections. If more discharge observations can be acquired in the future, we should conduct more analysis to show what happens for a river's different gauge stations.

5. Conclusions

In this study, we performed discharge trend analysis of 925 global rivers in the period from 1948–2004 using the EEMD method. Results showed that the EEMD method can better estimate the nonlinear and nonstationary nature of discharge trends at a global and regional scale, compared with the commonly used MK method. At a global scale, the EEMD method showed that the continental discharge trend decreased before 1978 and increased after 1978 in the period from 1948 to 2004, whereas the commonly used MK method gave no significant trend along the data span. At a regional scale, the EEMD method estimated that more than 70% of the 925 rivers had significant trends, including 292 of them with significantly upward trends and the others with significantly downward trends. On the other hand, the MK method consistently underestimated the discharge trends. We strongly recommend that future studies should adopt EEMD for analyzing the nonlinear trends of all hydrological nonstationary time series and perhaps all geophysical nonstationary time series.

We also found that precipitation had consistently dominant influences on the variations in discharge across six continents and globally, but the influencing factors of discharge trend were more complex. Future studies should particularly focus on explaining why the global runoff trend changed direction around 1978 using process-based global land or hydrological models.

Supplementary Materials: The following are available online at http://www.mdpi.com/2073-4441/12/12/3556/s1: Figure S1. Distribution of the farthest downstream gauge stations (circle) of the 925 rivers included in this study. Size of a circle is proportional to the log transformed mean annual discharge of each river; Figure S2. The constructed river discharge (black line) from Dai et al. (2009) with observed data (red star) for 10 largest rivers (sorted by discharge) in the world from 1948–2004; Figure S3 (a) Global land precipitation from 1948 to 2004 (CRU data product in blue and GPCC data product in red); and (b) their respective trend as estimated using EEMD; Figure S4 (a) Global land surface temperature anomalies from 1948 to 2004 (CRUTEM4 data in blue and GISTEMP data product in red); and (b) their respective trend as estimated using EEMD; Figure S5 (a) The amount of area equipped for irrigation in percentage of the total area on a raster with a resolution of 5 minutes; (b) Area irrigated with groundwater expressed as percentage of total area equipped for irrigation; and (c) Area irrigated with surface water expressed as percentage of total area equipped for irrigation; Table S1. Trend estimated using MK method for the ten largest rivers (by discharge volume) in the world from 1948–2004.

Author Contributions: Conceptualization, C.W. and H.Z.; methodology, C.W.; formal analysis, C.W. and H.Z.; validation: C.W. and H.Z.; investigation, C.W. and H.Z.; data curation, C.W. and H.Z.; writing—original draft preparation, C.W.; writing—review and editing, C.W. and H.Z.; visualization, C.W. All authors read and agreed to the published version of the manuscript.

Funding: This study was supported by the National Natural Science Foundation of China (41905094, 31770469 and 41991285), the Scientific Research Project of Ecological Restoration of Baopoling Mountain in Sanya, China, and a startup fund from Hainan University (KYQD (ZR) 1876).

Acknowledgments: The authors thank Christine Verhille at the University of British Columbia for assistance with English language and grammatical editing of the manuscript.

Conflicts of Interest: The authors declare no conflict of interest.

References

1. Milly, P.C.; Dunne, K.A.; Vecchia, A.V. Global pattern of trends in streamflow and water availability in a changing climate. *Nature* **2005**, *438*, 347–350. [CrossRef] [PubMed]
2. Oki, T.; Kanae, S. Global hydrological cycles and world water resources. *Science* **2006**, *313*, 1068. [CrossRef] [PubMed]
3. Betts, R.A.; Boucher, O.; Collins, M.; Cox, P.M.; Falloon, P.D.; Gedney, N.; Hemming, D.L.; Huntingford, C.; Jones, C.D.; Sexton, D.M.H.; et al. Projected increase in continental runoff due to plant responses to increasing carbon dioxide. *Nature* **2007**, *448*, 1037–1041. [CrossRef]

4. Milly, P.C.; Betancourt, J.; Falkenmark, M.; Hirsch, R.M.; Kundzewicz, Z.W.; Lettenmaier, D.P.; Stouffer, R.J. Stationarity is dead: Whither water management? *Science* **2008**, *319*, 573–574. [CrossRef] [PubMed]
5. Labat, D.; Goddéris, Y.; Probst, J.L.; Guyot, J.L. Evidence for global runoff increase related to climate warming. *Adv. Water Resour.* **2004**, *27*, 631–642. [CrossRef]
6. Dai, A.; Qian, T.; Trenberth, K.E.; Milliman, J.D. Changes in continental freshwater discharge from 1948–2004. *J. Clim.* **2009**, *22*, 2773. [CrossRef]
7. Alkama, R.; Kageyama, M.; Ramstein, G. Relative contributions of climate change, stomatal closure, and leaf area index changes to 20th and 21st century runoff change: A modelling approach using the organizing carbon and hydrology in dynamic ecosystems (ORCHIDEE) land surface model. *J. Geophys. Res. Atmos.* **2010**, *115*, D17112. [CrossRef]
8. Alkama, R.; Marchand, L.; Ribes, A.; Decharme, B. Detection of global runoff changes: Results from observations and CMIP5 experiments. *Hydrol. Earth Syst. Sci.* **2013**, *17*, 2967–2979. [CrossRef]
9. Legates, D.R.; Lins, H.F.; McCabe, G.J. Comments on "Evidence for global runoff increase related to climate warming" by Labat et al. *Adv. Water Resour.* **2005**, *28*, 1310–1315. [CrossRef]
10. Peel, M.C.; McMahon, T.A. Recent frequency component changes in interannual climate variability. *Geophys. Res. Lett.* **2006**, *33*, 373–386. [CrossRef]
11. Milliman, J.D.; Farnsworth, K.L.; Jones, P.D.; Xu, K.H.; Smith, L.C. Climatic and anthropogenic factors affecting river discharge to the global ocean, 1951–2000. *Glob. Planet. Chang.* **2008**, *62*, 187–194. [CrossRef]
12. Alkama, R.; Decharme, B.; Douville, H.; Ribes, A. Trends in global and basin-scale runoff over the late twentieth century: Methodological issues and sources of uncertainty. *J. Clim.* **2011**, *24*, 3000–3014. [CrossRef]
13. Sang, Y.F.; Wang, Z.; Liu, C. Comparison of the MK test and EMD method for trend identification in hydrological time series. *J. Hydrol.* **2014**, *510*, 293–298. [CrossRef]
14. Wu, Z.; Huang, N.E.; Long, S.R.; Peng, C.K. On the trend, detrending, and variability of nonlinear and nonstationary time series. *Proc. Natl. Acad. Sci. USA* **2007**, *104*, 14889. [CrossRef] [PubMed]
15. Mann, H.B. Nonparametric tests against trend. *Econometrica* **1945**, *13*, 245–259. [CrossRef]
16. Kendall, M.G. *Rank Correlation Methods*; Charles Grifin: London, UK, 1975.
17. Yue, S.; Pilon, P.; Cavadias, G. Power of the Mann–Kendall and Spearman's rho tests for detecting monotonic trends in hydrological series. *J. Hydrol.* **2002**, *259*, 254–271. [CrossRef]
18. Hamed, K.H. Trend detection in hydrologic data: The Mann–Kendall trend test under the scaling hypothesis. *J. Hydrol.* **2008**, *349*, 350–363. [CrossRef]
19. Sun, F.; Roderick, M.L.; Farquhar, G.D. Rainfall statistics, stationarity, and climate change. *Proc. Natl. Acad. Sci. USA* **2018**, *115*, 2305–2310. [CrossRef]
20. Hamed, K.H.; Rao, A.R. A modified Mann-Kendall trend test for autocorrelated data. *J. Hydrol.* **1998**, *204*, 192–196. [CrossRef]
21. Su, L.; Miao, C.; Kong, D.; Duan, Q.; Lei, X.; Hou, Q.; Li, H. Long-term trends in global river flow and the causal relationships between river flow and ocean signals. *J. Hydrol.* **2018**, *563*, 818–833. [CrossRef]
22. Hurst, H.E. Long-term storage capacity of reservoirs. *Trans. Am. Soc. Civ. Eng.* **1951**, *116*, 770–799. [CrossRef]
23. Huang, N.E.; Shen, Z.; Long, S.R.; Wu, M.C.; Shih, H.H.; Zheng, Q.; Yen, N.; Tung, C.C.; Liu, H.H. The Empirical Mode Decomposition Method and the Hilbert Spectrum for Non-stationary Time Series Analysis. *Proc. R. Soc. Lond. A* **1998**, *454*, 903–995. [CrossRef]
24. Huang, N.E.; Shen, Z.; Long, S.R. A new view of nonlinear water waves: The Hilbert spectrum. *Annu. Rev. Fluid Mech.* **1999**, *31*, 417–457. [CrossRef]
25. Chen, X.; Zhang, X.; Church, J.A.; Watson, C.S.; King, M.A.; Monselesan, D.; Legresy, B.; Harig, C. The increasing rate of global mean sea-level rise during 1993–2014. *Nat. Clim. Chang.* **2017**, *7*, 492–497. [CrossRef]
26. Ji, F.; Wu, Z.; Huang, J.; Chassignet, E.P. Evolution of land surface air temperature trend. *Nat. Clim. Chang.* **2014**, *4*, 462–466. [CrossRef]
27. Pan, N.; Feng, X.; Fu, B.; Wang, S.; Ji, F.; Pan, S. Increasing global vegetation browning hidden in overall vegetation greening: Insights from time-varying trends. *Remote Sens. Environ.* **2018**, *214*, 59–72. [CrossRef]
28. Mhamdi, F.; Poggi, J.M.; Jaïdane, M. Trend extraction for seasonal time series using ensemble empirical mode decomposition. *Adv. Data Anal.* **2011**, *3*, 363–383. [CrossRef]
29. Carmona, A.M.; Poveda, G. Detection of long-term trends in monthly hydro-climatic series of Colombia through Empirical Mode Decomposition. *Clim. Chang.* **2014**, *123*, 301–313. [CrossRef]

30. Jones, P.D.; New, M.G.; Parker, D.E.; Marun, S.; Rigor, I.G. Surface air temperature and its changes over the past 150 years. *Rev. Geophys.* **1999**, *37*, 173–199. [CrossRef]
31. Hansen, J.; Ruedy, R.; Glascoe, J.; Sato, M. GISS analysis of surface temperature change. *J. Geophys. Res.* **1999**, *104*, 30997–31022. [CrossRef]
32. Krakauer, N.Y.; Fung, I. Mapping and attribution of change in streamflow in the coterminous united states. *Hydrol. Earth Syst. Sci.* **2008**, *12*, 1111–1120. [CrossRef]
33. Gerten, D.; Rost, S.; von Bloh, W.; Lucht, W. Causes of change in 20th century global river discharge. *Geophys. Res. Lett.* **2008**, *35*, L20405. [CrossRef]
34. Nohara, D.; Kitoh, A.; Hosaka, M.; Oki, T. Impact of climate change on river discharge projected by multimodel ensemble. *J. Hydrometeorol.* **2006**, *7*, 1076. [CrossRef]
35. Sun, Y.; Solomon, S.; Dai, A.; Portmann, R.W. How often will it rain? *J. Clim.* **2007**, *20*, 4801–4818. [CrossRef]
36. Solomon, S.; Qin, D.; Manning, M.; Marquis, M.; Averyt, K.; Tignor, M.M.B.; Miller, H.L., Jr.; Chen, Z. (Eds.) *Climate Change 2007: The Physical Science Basis*; Cambridge University Press: Cambridge, UK, 2007; p. 996. [CrossRef]
37. Mccabe, G.J.; Wolock, D.M. Century-scale variability in global annual runoff examined using a water balance model. *Int. J. Clim.* **2011**, *31*, 1739–1748. [CrossRef]
38. Nijssen, B.; O'Donnell, G.M.; Hamlet, A.F.; Lettenmaier, D.P. Hydrologic sensitivity of global rivers to climate change. *Clim. Chang.* **2001**, *50*, 143–175. [CrossRef]
39. Gedney, N.; Cox, P.M.; Betts, R.A.; Boucher, O.; Huntingford, C.; Stott, P.A. Detection of a direct carbon dioxide effect in continental river runoff records. *Nature* **2006**, *439*, 835–838. [CrossRef]
40. Piao, S.; Friedlingstein, P.; Ciais, P.; Nobletducoudré, N.D.; Labat, D.; Zaehle, S. Changes in climate and land use have a larger direct impact than rising CO_2 on global river runoff trends. *Proc. Natl. Acad. Sci. USA* **2007**, *104*, 15242. [CrossRef]
41. Franzke, C.L.E.; O'Kane, T.J.; Monselesan, D.P.; Risbey, J.S.; Horenko, I. Systematic attribution of observed Southern Hemisphere circulation trends to external forcing and internal variability. *Nonlinear Process. Geophys.* **2015**, *2*, 675–707. [CrossRef]
42. Haddeland, I.; Heinke, J.; Biemans, H.; Eisner, S.; Flörke, M.; Hanasaki, N.; Konzmann, M.; Ludwig, F.; Masaki, Y.; Schewe, J.; et al. Global water resources affected by human interventions and climate change. *Proc. Natl. Acad. Sci. USA* **2014**, *111*, 3251–3256. [CrossRef]
43. Lawrence, D.M.; Oleson, K.W.; Flanner, M.G.; Thornton, P.E.; Swenson, S.C.; Lawrence, P.J.; Zeng, X.; Yang, Z.-L.; Levis, S.; Sakaguchi, K.; et al. Parameterization improvements and functional and structural advances in version 4 of the Community Land Model. *J. Adv. Model. Earth Syst.* **2011**, *3*, 365–375. [CrossRef]

MDPI

Article

How Do Ground Litter and Canopy Regulate Surface Runoff?—A Paired-Plot Investigation after 80 Years of Broadleaf Forest Regeneration

Anand Nainar [1,*], Koju Kishimoto [2], Koichi Takahashi [2,3], Mie Gomyo [2] and Koichiro Kuraji [4,*]

1 Faculty of Science and Natural Resources, Universiti Malaysia Sabah, Jalan UMS, Kota Kinabalu 88400, Sabah, Malaysia
2 Ecohydrology Research Institute, Graduate School of Agricultural and Life Sciences, The University of Tokyo, 11-44, Goizuka, Seto 489-0031, Japan; kishimoto@uf.a.u-tokyo.ac.jp (K.K.); takakou@uf.a.u-tokyo.ac.jp (K.T.); gomyo_mie@yahoo.co.jp (M.G.)
3 Graduate School of Agricultural and Life Sciences, The University of Tokyo Forest, 9-61 Yamabe Higashimachi, Furano 079-1563, Japan
4 Executive Office, Graduate School of Agricultural and Life Sciences, The University of Tokyo Forests, 1-1-1 Yayoi, Bunkyo-ku, Tokyo 113-8657, Japan
* Correspondence: nainar.sci@gmail.com (A.N.); kurajikoichiro@g.ecc.u-tokyo.ac.jp (K.K.)

Abstract: Relatively minimal attention has been given to the hydrology of natural broadleaf forests compared to conifer plantations in Japan. We investigated the impacts of ground litter removal and forest clearing on surface runoff using the paired runoff plot approach. Plot A (7.4 m^2) was maintained as a control while plot B (8.1 m^2) was manipulated. Surface runoff was measured by a tipping-bucket recorder, and rainfall by a tipping-bucket rain gauge. From May 2016 to July 2019, 20, 54, and 42 runoff events were recorded in the no-treatment (NT), litter removed before clearcutting (LRBC), and after clearcutting (AC) phases, respectively. Surface runoff increased 4× when moving from the NT to LRBC phase, and 4.4× when moving from the LRBC to AC phase. Antecedent precipitation index (API_{11}) had a significant influence on surface runoff in the LRBC phase but not in the NT and AC phases. Surface runoff in the AC phase was high regardless of API_{11}. The rainfall required for initiating surface runoff is 38% and 56% less when moving from the NT to LRBC, and LRBC to AC phases, respectively. Ground litter and canopy function to reduce surface runoff in regenerated broadleaf forests.

Keywords: litter; canopy; logging; overland flow; surface runoff; interception; broadleaf; forest

Citation: Nainar, A.; Kishimoto, K.; Takahashi, K.; Gomyo, M.; Kuraji, K. How Do Ground Litter and Canopy Regulate Surface Runoff?—A Paired-Plot Investigation after 80 Years of Broadleaf Forest Regeneration. *Water* **2021**, *13*, 1205. https://doi.org/10.3390/w13091205

Academic Editor: David Dunkerley

Received: 19 March 2021
Accepted: 23 April 2021
Published: 27 April 2021

1. Introduction

As of 2017, approximately 25 million ha (67%) of land area in Japan is forested. Plantation forests make up approximately 40% of the forested area while the remaining 60% is natural forests [1,2]. From 1966 to 2017, plantation forests have increased by 29% to 10.2 million ha while natural forests shrunk by 13% to 13.5 million ha [1]. While forested areas remain high (twice the world's average of 29%), the proportion of forest type (broadleaf, conifer, mixed, natural, and plantation forests) fluctuates according to market forces, situational demands, and governmental policies [1,3]. Conifer plantation forests are expected to expand, but instead, some conifer forests are being reverted to broadleaf forests [4]. At the same time, some broadleaf and mixed-forests are also being cleared to make way for development [5].

Since hydrology is highly influenced by forests, there is a need to understand how these forests (and their modification) affect runoff processes and water resources. In the past, there have been extensive studies carried out on hydrological processes in cypress plantations, including those investigating the hydrological impacts of forest conversion [6], forest litter [7,8], tree physiology [9,10], and various management practices [11,12]. The focus on cypress plantation was a result of its rapid expansion between 1970 and 2000 that was driven by the government's economic policy. Up to date, temperate broadleaf forests and mixed-forests especially those which had undergone restoration have received relatively little attention, although there are some exemplary studies [13,14]. More specifically, the hydrological role of ground litter and the canopy are not yet well-understood. Such knowledge is needed because more and more forests in suburban areas are being cleared for development. The resulting hydrological changes need to be quantified to better formulate preventive and corrective measures. At the same time, due to the lack of economic incentives, vast forest areas in the interior remain unharvested and untended, causing ground litter to thicken [13,15]. Therefore, knowledge on how ground litter affect runoff is equally important. These two knowledge gaps, (i) hydrological role of ground litter and (ii) hydrological role of canopy in temperate broadleaf forests make up the motivation for the present study.

A number of studies that focuses on the hydrological role of ground litter have been conducted in the past: Li et al. [16] investigated the relationship between ground litter and surface runoff in northern China. Prosdocimi et al. [17] performed a similar experiment in Mediterranean vineyards. Zhou et al. [18] carried out a microplot-scale observation of surface runoff generation in differing ground litter and topsoil depths in Guizhou province, China, and have highlighted challenges in upscaling these findings to the catchment scale. In Japan, Miyata et al. [19] studied surface runoff generation and soil erosion in mature Japanese cypress plantations under varying ground litter coverage conditions. To date, there is little investigation on the role of ground litter on surface runoff dynamics in naturally regenerated broadleaf forest in Japan. One existing example is a study by Gomyo and Kuraji [13], in which they demonstrated the effects of litter removal at the catchment scale. The study found that 3-year runoff increased by 2.7% (80.3 mm) post treatment. In addition, peak runoff was up to 1.5 times greater. This suggests that the effects of litter interception was greater than that of runoff-on-litter [20], and that the increased ground evaporation from litter removal did not offset the increased effective rainfall [11]. Such catchment study is rare due to the required time and labour.

In Japan, studies focussing on the hydrological role of the canopy or on the impacts of clearcutting broadleaf forests are scarcer because most efforts had been focussed on cypress plantation forests. Most studies that were conducted in Japan are related to the effects of thinning of conifer plantations, whereby thinning reduces canopy interception and may increase the likelihood of floods [21]. At the time of writing, the authors found only one study in Tochigi prefecture, eastern Japan, that reported canopy interception in a regenerated multi-layered broadleaf forest to be 21% [14]. The impacts on surface runoff from clearcutting broadleaf forests and mixed-forest remain unexplored.

Results from catchment-scale studies may better represent that of real-world conditions. However, due to the presence of various environmental factors, sources, and sinks, it is difficult to understand the individual underlying mechanisms. In particular, does the presence of litter buffer (due to interception) or enhance (due to runoff-on-litter) surface runoff? With regards clearcutting, will surface runoff increase (from increased throughfall) or decrease (from decreased stemflow)? These uncertainties have been highlighted by past studies [13,20] and can be better elucidated via a plot-scale experiment. Furthermore, Liu et al. [22], in an extensive review, emphasised that it is necessary to isolate the different ways in which vertical vegetation component control runoff. Understanding the role of each of these vertical vegetation structures and their inter-dependency will be useful in determining the efficient architectures and morphologies of vegetation for ecological restoration. Liu et al. [22] also emphasized that more effort should be made in quantifying the roles of these vertical structures of vegetation to be used in improving models.

This study took place in western Aichi prefecture—a region that was historically subjected to heavy forest exploitation up to 1910s. As a result, high sediment yields and landslides were a common phenomenon [23]. Since then, over a century of forest restoration was carried out, resulting in the secondary mixed-forest cover (predominantly broadleaf) today [24,25]. However, due to their proximity to urban areas, more and more of such areas are being cleared to make way for development (i.e., residential, commercial, and industrial buildings) in recent years [4,5]. For this reason, it is also important to understand the impacts of clearcutting secondary broadleaf forests on surface runoff. In this study, we extend previous works [13,26,27] by observing at the plot scale, changes in surface runoff after litter removal and clearcutting. We hypothesise that surface runoff will increase with litter removal and clearcutting in a regenerated temperate broadleaf forest. At a larger scale, this may have implication on floods during storm events.

2. Materials and Methods

2.1. Study Area

Experimental plots were set up in the 13.9-ha Ananomiya Experimental Forest of the Ecohydrology Research Institute, the University of Tokyo Forests, in Seto city, Aichi prefecture (Figure 1). The general area, traditionally known as the Eastern Owari Hills, is known for a long history of ceramic and pottery production. As a result of intensive timber harvesting to be used as kiln fuel, the area was characterised by denuded hills with bare, exposed subsoil in the past (Figures 2 and 3) [24,25]. Another round of intensive timber exploitation took place during the Meiji era (1868–1912), when industrialisation in Japan was at its peak [25]. High sediment yields and landslides had been a common phenomenon. With auspices from the government, the Ananomiya Experimental Forest (AEF) area was established in 1922 by the Tokyo Imperial University (University of Tokyo, today for the purpose of restoring forests and preventing erosion-related hillslope disasters. Construction of a weir dam was completed in 1924, and hydrological and meteorological observations commenced in 1925 [24]. Although this site was severely denuded in the past, forest cover has now recovered due to restoration-planting efforts (known as "sabo planting"), where Japanese red pine (*Pinus densiflora*) and Japanese black pine (*Pinus thunbergii*) were planted as the pioneer species [24,25]. Figure 2 shows photographs of one of the hillslopes in the experimental forest in its denuded state in 1924 and recovered state in 2021. Aerial photographs (Figure 3) show the gradual recovery of forest through the years 1948, 1976, and 2010.

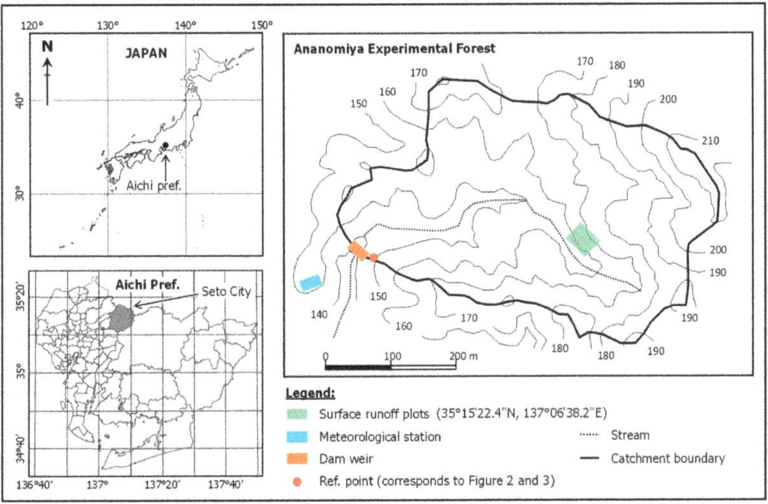

Figure 1. Study area. Map of Japan and Aichi prefecture were drawn via the software R v.4.0.3 using packages "maps", "mapdata", and "mapproj". Map of AEF and contours were obtained from the archives of the Ecohydrology Research Institute [28].

Figure 2. (**a**) Denuded state in 1924, (**b**) regenerated forest cover in 2021. Red dot is reference point.

Figure 3. Gradual forest recovery in the Ananomiya experimental forest area. (**a**) White areas are bare and exposed weathered granite subsoil. (**b**,**c**) show gradual forest recovery—corresponding to the reducing bare (white) areas. The reference point (red dot) is the same as in Figure 2.

The area has a warm temperate climate characterised by hot summers, moderate winters, and high annual rainfall that occurs mainly in two distinct wet periods—the *Baiu* season (May–June) with frequent and prolonged rain events; and the typhoon season (September–October) with high-intensity rainfall [27]. Annual rainfall is 1594 mm year^{-1} (averaged 1 May 2016–30 April 2019). The annual mean temperature is 12–15 °C, but the extremes may reach −12 °C (dawn, in winter) and 39 °C (midday, in summer). Daily averages of relative humidity range between 55 and 99.7% (data from on-site meteorological station). The dominant geology is a deep cretaceous layer overlain by weathered granite. Through soil core sampling around the plots followed by the jar test, percentage sand (15.03–20.61%), silt (26.22–39.05%), and clay (43.21–53.18%), show that the soil texture is of the clay type [29]. Bulk density range is 808.3–1069.2 kg m^{-3} (mean = 928.7 kg m^{-3}).

At present, the area has a temperate mixed-forest composition with dominant canopy species comprising *Quercus serrata* and *Ilex pedunculosa*. The sub-canopy layer is dominated by *Eurya japonica*, *Lyonia ovalifolia* and *Gamblea innovans*; while the ground cover, by *Sasa nipponica* and *Dicranopteris dichotoma*. Although this is a secondary mixed-forest, the number of pioneer conifer trees (black pine and red pine) is small owing to species succession. Trees in and around the experimental plots are broadleaved; therefore, effectively making this study a comparison in a broadleaf forest. Stand density is 1925 stems/ha, which comprises canopy trees (>9 m in height; 725 stems/ha) and sub-canopy trees (3–9 m in height; 1200 stems/ha).

2.2. Treatment and Data Collection

Two sloping experimental plots, A (7.4 m^2) and B (8.1 m^2), were established with a 4.2-m buffer in-between (Figure 4). Boundaries of each plot were created by inserting hard plastic (roofing) sheets into the ground. The boundary was made contiguous to prevent leakage of surface runoff. The mean slope for plots A and B were 35° and 28°, respectively. At the bottom boundary of each plot, a gutter was installed to collect overland flow which was then channelled via a PVC pipe to an 'Uizin UIZ-TB200' tipping-bucket gauge. Throughout the study period, no treatment was applied to plot A (control plot) whereas litter removal followed by clearcutting were carried out in plot B.

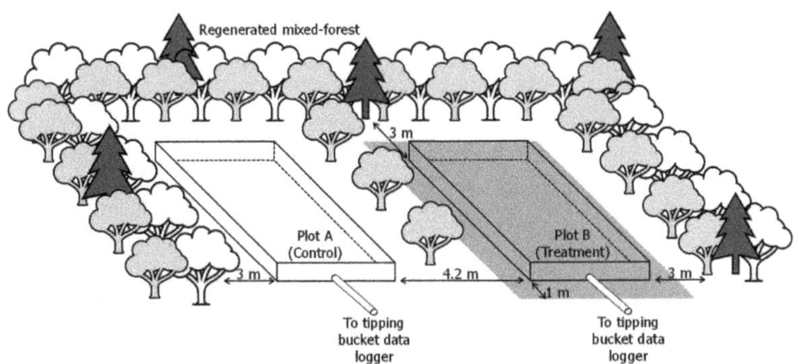

Figure 4. Experimental plots. Treatments were applied to the shaded area (larger than plot B).

To ensure minimal disturbance to the soil surface, litter removal was carried out from outside the plot boundary and precautionary measures were taken to not step into the plot area. The dry weight of litter removed (including leaf, twigs, fruits, and seeds) was 9.48 kg (oven dried at 60 °C for 72 h). In the clearcutting phase, a total of 13 trees (species: *Deutzia crenata*, *Eurya japonica*, *Gamblea innovans*, *Ilex pedunculosa*, and *Lyonia ovalifolia*) with DBH ranging 0.8–3.5 cm as well as 10 small shrubs were removed from plot B. In plot A (control plot), six trees (species: *E. japonica* and *L. ovalifolia*) and small shrubs were left as they were. Although having fewer trees than in plot B, one of the trees (*E. japonica*) in plot A was multi-stemmed (nine stems), resulting in a canopy coverage that was more extensive than that of a normal tree. To minimise external environmental influence, areas adjacent to the plots were made to be as similar as possible to that inside the plots. Areas surrounding plot A (3 metres away from the top, left, and bottom boundary) were left forested. For plot B, a 1-metre buffer from the perimeter (plot boundary) was given the same treatment (litter removal and clearcutting) as inside the plot. Photographs of plot B in the NT, LRBC, and AC phases are shown in Figure 5.

At a nearby weather station (approximately 470 m westwards of the runoff plots), rainfall data was recorded at 5-minute intervals by an 'Ota Keiki OW-34-BP' tipping-bucket rain gauge (0.5 mm/tip) connected to a 'Campbell Scientific CR10X' datalogger. Data collection, site observation, and maintenance were carried out in three phases: no treatment phase (NT) from May to October 2016, litter removed before clearcutting phase (LRBC) from March 2017 to June 2018, and after clearcutting phase (AC) from September 2018 to July 2019.

2.3. Data Processing and Analysis

The main mode of analysis is comparing storm surface runoff between the NT, LRBC, and AC phases. For this, we considered storm events to be separate if there were at least 6 hours of rain-free period in between events—a common rule used in the region. In all analyses, runoff in plot B was normalised by runoff in plot A. Erroneous data and periods of equipment malfunction were excluded from analysis. In total, 116 runoff events were available for analysis—20 in the NT, 54 in the LRBC, and 42 in the AC phases, respectively.

We explored the data in four different aspects: (i) overall runoff amount and duration, (ii) storm event surface runoff, (iii) the effects of antecedent precipitation on surface runoff, and (iv) rainfall threshold required for the initiation of surface runoff (amount of rain from the start of a rain event to the first recorded surface runoff). For comparison between the treatment phases, the variable of interest (surface runoff, rainfall threshold) and their various influencing factors (treatment, average rainfall intensity, antecedent precipitation index) were fitted via the generalised linear model (GLM). The best-fit GLM was chosen via the Akaike Information Criterion (AIC) and Bayes Factor in the software R v.4.0.3 (required packages: glm2, devtools, flexplot). Differences between phases were assessed via the Mann–Whitney U test.

Figure 5. State of plot B in (**a**) no treatment (NT) phase; (**b**) litter removed before clearcutting (LRBC) phase; and (**c**) after clearcutting (AC) phase.

2.3.1. Storm Event Surface Runoff

Before analysis, we ascertained the relationships between surface runoff (Q) and rainfall (P) as well as between surface runoff in plot B (Q_B) and surface runoff in plot A (Q_B). Strong positive relationships were found between Q_B and Q_A, especially in the NT phase ($r^2 = 0.98$), which signify suitability for a paired-plot study. After establishing a baseline relationship and ensuring that there were no anomalies, the latter regression was carried forward to be used in the paired-plot comparison.

Following this, mean rainfall intensity (P_i) was also tested for influence on runoff generation. However, it was not statistically significant ($p > 0.05$, $r^2 \leq 0.1$); thus, not included for further analysis.

Because the data were non-normally distributed (Shapiro–Wilk, $p < 0.05$) and have unequal population variance (Levene, $p < 0.05$), Q_B vs. Q_A was provisionally test-fitted using the 'glm2' command (package: glm2) in R using several 'family' (gaussian, poisson, gamma) and 'link' (identity, log, inverse) functions. Both original and log-transformed data were tried. The 'gamma-identity' fit on original data was found to produce the best fit in the NT, LRBC, and AC phases, and was thus used.

Differences in runoff between NT, LRBC, and AC were assessed via the Mann–Whitney U test using calculated treatment effects (Te). To calculate Te, the first step was to establish a calibration equation (using the earlier gamma-identity GLM) between plot B and plot A in the NT phase:

$$Q^{\mathrm{NT}}_{\mathrm{obs\ B}} = a \times Q^{\mathrm{NT}}_{\mathrm{obs\ A}} + b \quad (1)$$

where $Q^{\mathrm{NT}}_{\mathrm{obs\ B}}$ and $Q^{\mathrm{NT}}_{\mathrm{obs\ A}}$ are event surface runoff observed in the NT phase in plot B and A, respectively; and a and b are regression coefficients. After treatment, runoff in plot B in the LRBC and AC phases were estimated using Equations (2) and (3), respectively:

$$Q^{\mathrm{LRBC}}_{\mathrm{est\ B}} = a \times Q^{\mathrm{LRBC}}_{\mathrm{obs\ A}} + b \quad (2)$$

$$Q^{\mathrm{AC}}_{\mathrm{est\ B}} = a \times Q^{\mathrm{AC}}_{\mathrm{obs\ A}} + b \quad (3)$$

where $Q^{\mathrm{LRBC}}_{\mathrm{est\ B}}$ and $Q^{\mathrm{AC}}_{\mathrm{est\ B}}$ are 'estimated runoff' in plot B in the LRBC and AC phases, respectively, and $Q^{\mathrm{LRBC}}_{\mathrm{obs\ A}}$ and $Q^{\mathrm{AC}}_{\mathrm{obs\ A}}$ are 'observed runoff' in plot A in the LRBC and AC phases, respectively. Treatment effects (Te) in LRBC ($Te^{\mathrm{LRBC/NT}}$) and AC ($Te^{\mathrm{AC/NT}}$) with reference to the NT phase were calculated as follows:

$$Te^{\mathrm{LRBC/NT}} = Q^{\mathrm{LRBC}}_{\mathrm{obs\ B}} - Q^{\mathrm{LRBC}}_{\mathrm{est\ B}} \quad (4)$$

$$Te^{\mathrm{AC/NT}} = Q^{\mathrm{AC}}_{\mathrm{obs\ B}} - Q^{\mathrm{AC}}_{\mathrm{est\ B}} \quad (5)$$

In the same way, with reference to the LRBC phase, Te in AC ($Te^{\mathrm{AC/LRBC}}$) was calculated by first establishing a calibration equation between plot B and plot A in the LRBC phase (Equation (6)); then, using the equation coefficients (c and d) to estimate surface runoff in AC (Equation (7)), and subtracting the estimated values from observed values (Equation (8)):

$$Q^{\mathrm{LRBC}}_{\mathrm{obs\ B}} = c \times Q^{\mathrm{LRBC}}_{\mathrm{obs\ A}} + d \quad (6)$$

$$Q^{\mathrm{AC}}_{\mathrm{est\ B}} = c \times Q^{\mathrm{AC}}_{\mathrm{obs\ A}} + d \quad (7)$$

$$Te^{\mathrm{AC/LRBC}} = Q^{\mathrm{AC}}_{\mathrm{obs\ B}} - Q^{\mathrm{AC}}_{\mathrm{est\ B}} \quad (8)$$

2.3.2. Effects of Antecedent Precipitation on Surface Runoff

The first step was to assess the relationship between surface runoff and antecedent precipitation index (API), as well as to determine antecedent days with the highest influence. Several combinations of GLM family and link function were tried as with the earlier section. All equations were then inspected visually and compared based on AIC and Bayes Factor. Relationships were weak and monotonic instead of linear. API_{11} were found to produce the highest correlation and best fit (AIC: 450.05 via 'gamma-log').

After confirming that API_{11} has a negative relationship with Q, the GLM, Q_B/Q_A against API_{11}, were fitted for each treatment phase (NT, LRBC, and AC):

$$Q_B/Q_A = e \times API_{11} + f \tag{9}$$

where e and f are equation coefficients.

In the LRBC phase, relationships were statistically significant ($p < 0.01$) with the 'gamma-identity' GLM being parsimonious (AIC: 316.82).

2.3.3. Rainfall Threshold

The rainfall threshold required for the initiation of surface runoff (P_T)—defined as the total rainfall from the start of an event to the first tip of the tipping-bucket datalogger—were also compared between the different treatments. Before comparison, the influence of API_{11} and P_i on P_T were assessed. Relationships were tested using data from individual plots and both plots collectively, in individual phases and all phases collectively, as well as every combination of phases and plots. However, no clear relationships were found. Therefore, the comparison of P_T was conducted via the Mann–Whitney U Test.

3. Results

3.1. Overall Surface Runoff and Precipitation

At the monthly timescale, surface runoff in plot B was higher than in plot A after treatment (LRBC and AC phase) (Figure 6). This differed from the pre-treatment period where surface runoff in plot A was slightly higher. The normalised runoff ratio (C_N)—defined as the runoff ratio in plot B divided by runoff ratio in plot A—increased from 0.558 in NT to 2.647 and 11.875 in LRBC and AC, respectively (Table 1). The standard deviation (SD) also increased, signifying increased variability (Table 1).

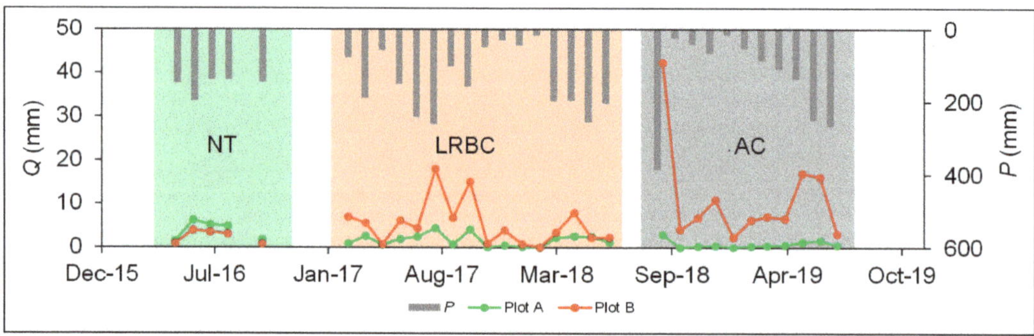

Figure 6. Monthly runoff in plot A, B and precipitation in the different phases.

Table 1. General statistics in the NT, LRBC, and AC phases.

Parameter	No Treatment		Litter Removal before Clearcut		After Clearcut	
	Plot A	Plot B	Plot A	Plot B	Plot A	Plot B
Rainfall, P (mm)	519.0		1993.5		1194.0	
Total runoff, Q (mm)	22.4	12.3	33.0	88.8	10.1	113.7
Runoff ratio, C	0.043	0.024	0.017	0.045	0.008	0.095
Normalised runoff ratio, C_N	0.558		2.647		11.875	
n	20		54		42	
min (mm)	0.135	0.074	0.054	0.049	0.054	0.123
max (mm)	4.027	2.420	3.568	14.840	1.162	11.235
mean (mm)	1.120	0.615	0.612	1.644	0.241	2.707
SD (mm)	1.102	0.708	0.635	2.264	0.243	2.430
Freq. Q_B: Freq. Q_A, $T_{B/A}$	0.977		2.292		7.250	

Besides runoff, the runoff frequency represented by the ratio of runoff frequency in plot B to runoff frequency in plot A ($T_{B/A}$) also increased with disturbance: 0.977, 2.292, and 7.250 in the NT, LRBC, and AC phases, respectively (Table 1). In the NT phase, surface runoff frequency was slightly higher in plot A as visualised in the flow-duration curve (FDC) in Figure 7. In the LRBC and AC phases, despite a decrease in surface runoff frequency in plot A, surface runoff frequency in plot B have increased. The FDC of plot B also became more convex, which is a sign of increased frequency of high-magnitude surface runoff.

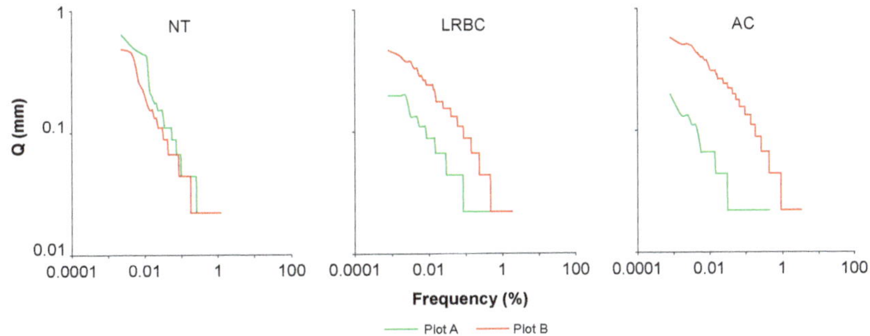

Figure 7. Flow-duration curve of plot A and B in the NT, LRBC, and AC phase.

3.2. Event-Scale Surface Runoff in Different Phases

In all phases and in both plots, surface runoff (Q) increased linearly with event rainfall (P) (Figure 8). In the NT phase, strong linear relationships were found (plot A, R^2 = 0.9019; plot B, R^2 = 0.8488).

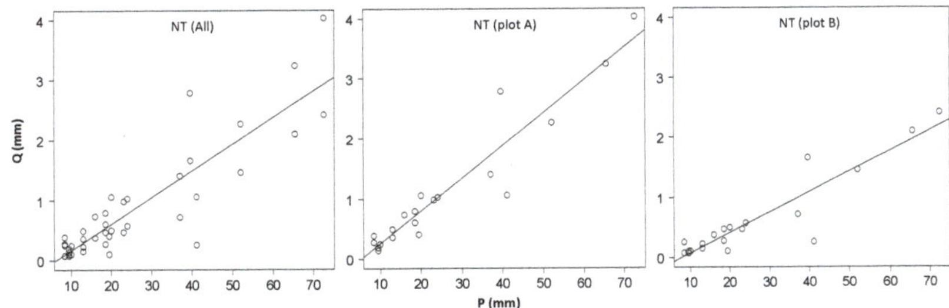

Figure 8. Rainfall-surface runoff plots.

Figure 9 shows the relationships between surface runoff in plot B (Q_B) and surface runoff in plot A (Q_A) in the NT, LRBC, and AC phase. The resulting GLM equations are:

$$Q_B = 0.5015 \cdot Q_A - 0.0038 \text{ (NT phase; } R^2 = 0.9223, p < 0.01) \quad (10)$$

$$Q_B = 2.1971 \cdot Q_A + 0.3336 \text{ (LRBC phase; } R^2 = 0.3630, p < 0.01) \quad (11)$$

$$Q_B = 9.3967 \cdot Q_A + 0.5503 \text{ (AC phase; } R^2 = 0.4388, p < 0.01) \quad (12)$$

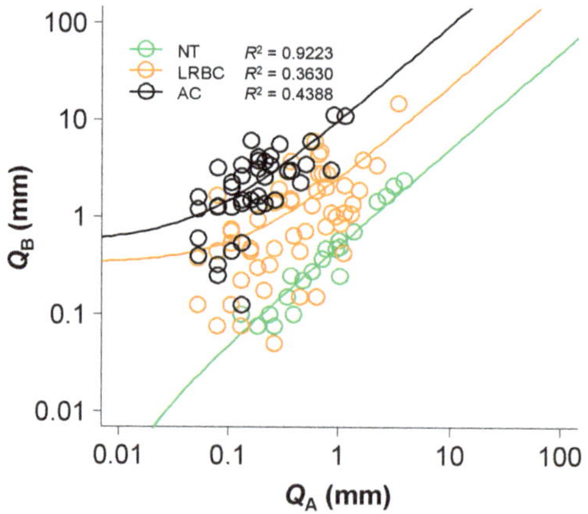

Figure 9. Surface runoff in plot B vs. surface runoff in plot A.

The strong correlation between plot A and B in the NT phase ($R^2 = 0.9223$) indicates suitability for a paired-plot comparison. Based on regression coefficients, surface runoff in the LRBC and AC phase is 4.4 and 18.7 times higher, respectively, than in the NT phase. From the LRBC to AC phase, surface runoff increased 4.3 times. These increases are in the range of that in the monthly timescale. When expressing in terms of treatment effects (*Te*), the LRBC and AC phases have *Te* values of 1.341 and 2.589, respectively. From LRBC to AC, *Te* is 1.843 (Table 1). *Te* (Figure 10) were statistically significantly different between all phases (Mann–Whitney, $p < 0.01$). In addition, the y-intercept increased by two orders when moving from NT to LRBC and AC, signifying earlier commencement of surface runoff.

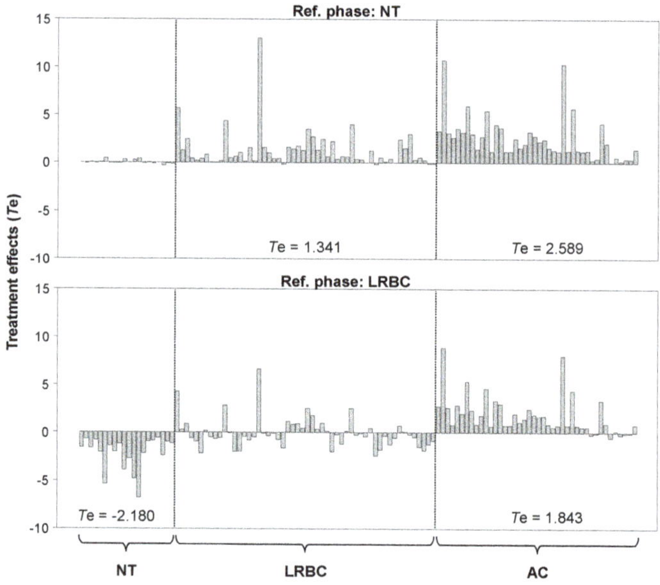

Figure 10. Treatment effects.

3.3. Surface Runoff and Antecedent Precipitation Index (API)

Figure 11 shows surface runoff against API_{11} in plots A and B in different treatment phases. In plot A (control), surface runoff varies little with API_{11} throughout all phases whereas in plot B (treatment), the effects of API_{11} (negative relationship) increased from NT to LRBC to AC.

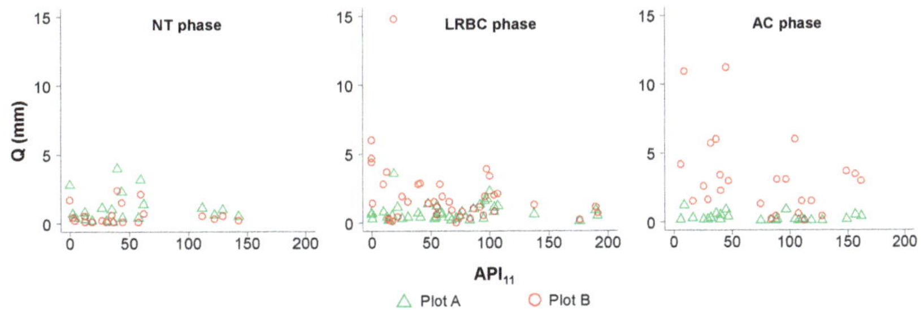

Figure 11. Surface runoff against the 11-day antecedent precipitation index (API_{11}).

Naturally, surface runoff is directly proportional to rainfall and inversely proportional to API. In our study, the relationship Q_B/Q_A vs. API_{11} is only significant in the LRBC phase ($Q_B/Q_A = -0.014861 \cdot API_{11} + 3.863$; $p > 0.01$).

3.4. Rainfall Threshold for Surface Runoff Generation

The amount of rainfall required for surface runoff generation (rainfall threshold, denoted by "θP_{SR}") was also evaluated (Figure 12). With decreasing interception (NT to LRBC to AC phase), the rainfall threshold decreased accordingly (Figure 12). *Pi* and API did not have significant influence on rainfall threshold; thus, rainfall threshold between the different phases was assessed via the Mann–Whitney U test. Rainfall threshold in plot B normalised by that in plot A ($\theta P_{SR\_B}/\theta P_{SR\_A}$) is 38% lower when moving from the NT to LRBC phase. When moving from NT to AC, and LRBC to AC, $\theta P_{SR\_B}/\theta P_{SR\_A}$ is 73% and 56% lower, respectively. Differences in rainfall threshold were statistically significantly different between the NT, LRBC, and AC phases (Mann–Whitney, $p < 0.01$).

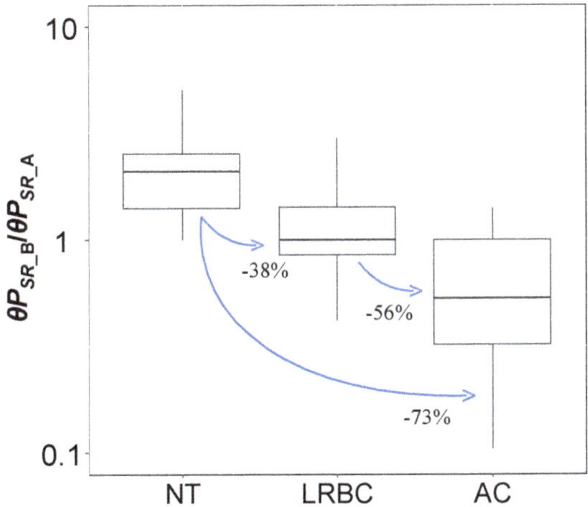

Figure 12. Reduction in rainfall threshold for the initiation of surface runoff. Rainfall threshold in plot B (θP_{SR_B}) was normalised by rainfall threshold in plot A (θP_{SR_A}).

4. Discussion

4.1. No-Treatment Phase

In the NT phase, surface runoff increased linearly with precipitation. This follows findings of existing studies that found similar relationships at the catchment scale in a secondary mixed-forest [13]. Antecedent moisture did not significantly affect surface runoff in the NT period, which may have been due to more important natural factors (rainfall, canopy, ground litter) that govern surface runoff in natural forests. Rainfall threshold for runoff generation was the highest in the NT phase, which reflects good interception properties (by ground litter and canopy) compared to in the treatment phases. These characteristics serve to establish a baseline for post-treatment comparisons.

4.2. Litter Removed, before Clearcutting Phase

In the LRBC phase, surface runoff was higher than in the NT phase (Figure 9), which demonstrated the importance of ground litter in regulating surface runoff [30]. Although the loss in ground litter interception is perceived to be the immediate cause, other mechanisms may have also contributed. In particular, the loss of protective cover may have caused soil compaction and reduced permeability when rainfall and concentrated throughfall directly impacts the soil surface; additionally, the increased floor evaporation may have resulted in drier and hydrophobic conditions. Both of these reduce infiltration and promote rapid direct runoff [20,31,32]. Compared to the straightforward increase in effective rainfall, soil compaction could have played a more important role because canopy in broadleaf forests are known to concentrate rainwater into larger droplets with higher momentum; thus, compacting the soil surface upon impact [19,31]. This increase in surface runoff is consistent with an experiment at the catchment scale [13]. This present study, however, differed in terms of spatial and temporal scales. Therefore, the results are not directly comparable. A more similar plot study was conducted using simulated rainfall in northwest Beijing, China, but with litter from different tree species (*Quercus variabilis* and *Pinus tabulaeformis*) [31]. Our results agree with theirs, but the degree of runoff increase is different. Surface runoff in our study increased over four times as opposed to only 43% in Li et al. [31]. This may be attributed to different species, litter morphology, soil type, and rainfall characteristics [22]. Miyata et al. [19] observed that overland flow increases with silt and clay content, and have hypothesised that this is due to silt and clay being key components in the formation of soil crust. This could be a possible mechanism in the present study considering the high clay content in our plots. As for species and litter morphology, broadleaf litter are known to have higher water storage capacity than conifer litter [33]. The increased flow frequency may have resulted from the detection of throughfall of small events that were previously effectively intercepted by ground litter (higher effective rainfall).

Besides an increase in the amount and duration of surface runoff, sensitivity towards API_{11} also increased (Figure 11). Although dry soils enhance hydrophobicity, resulting in increased surface runoff in forested environments, the effects of this mechanism may have been masked by the presence of ground litter in the NT phase [11,34–36]. In addition, the absence of ground litter in the LRBC phase may have accelerated soil drying due to enhanced floor evaporation; thus, creating hydrophobic conditions that promote direct runoff [33]. Therefore, for a given value of API_{11}, actual soil moisture could have been drier in LRBC compared to in the NT phase.

The rainfall threshold required for surface runoff generation was 37.55% lower in LRBC (Figure 12), which reflects the buffering properties of ground litter interception. This is also a sign of increased water repellency usually associated with dry and compacted soils. This indicated that the suggested runoff-on-litter mechanism [20,34], if ever occurred, did not play a significant hydrological role in the NT phase. Li et al. [31] also reported similar results, whereby shorter time was required to initiate surface runoff in bare plots as opposed to litter-covered plots. They also observed that broadleaf litter was marginally more effective in delaying the generation of surface runoff.

4.3. *After Clearcutting Phase*

Surface runoff was the highest in the AC phase. The increase in surface runoff (4× from LRBC to AC) demonstrated the role of forest canopy in intercepting and evaporating rainwater in secondary broadleaf forests (Figure 9) [14,37,38]. Compared to other hydrological losses, canopy interception (12–24% throughout Japan) may have a distinctively large influence as it directly controls effective rainfall [14,21,39]. Studies from other regions also recorded similar values: 21.3% of rainfall in a tropical broadleaf forest in Kalimantan [40]; 10% and 20% in a primary and regenerated tropical rainforest, respectively, in Sabah, Malaysia [41]; and 22.4% in a conifer forest in California [42]. Our findings differed from what Nanko et al. [43] have suggested, whereby throughfall in forests have larger droplets and higher momentum that compacts the soil and promotessurface runoff (corresponding to LRBC in this study). The increase in flow frequency compared to in the AC phase (Figure 7) could have been caused by similar mechanisms as when transitioning from the NT to LRBC phase, which are increased amount and duration of effective rainfall. Besides a shift in the position of the FDC, its shape has become more convexed; therefore, signifying an increase in the frequency of large-magnitude runoff events (despite plot B having a lower slope angle). In contrast to the LRBC phase, where soil compaction by throughfall was assumed to be the main cause of increased runoff, increases in the AC phase is attributed to increased effective rainfall due to the removal of canopy.

Surface runoff in AC was generally higher than that in the NT and LRBC phases regardless of API_{11} (Figure 11). This may be caused by high water repellency as a result of increased solar radiation, hence the drying-up of soils and formation of crust. Additionally, the effects of removing the canopy and increasing effective rainfall may have been disproportionately dominant to the extent of exempting influence from soil moisture. Although some studies have attributed this to the latter [42,44,45], at present, we suggest causality to the former based on findings in existing studies [46,47]. The findings of Daikoku et al. [48]—vapour pressure deficit and the below-canopy available energy strongly controls forest floor evaporation in Japan—partly support this. Although saturated soils may also override the influence of API, such conditions never occurred in our well-drained sloping plots.

Compared to the NT and LRBC phases, the decrease in rainfall threshold required for surface runoff generation demonstrated the dominant influence of canopy interception (Figure 12). Although stemflow is known to be a major mechanism that channels canopy-intercepted water to the ground [49–52], our data has shown that its influence was overridden by the effects of loss in canopy interception. Instead of ground factors, we attribute changes in the AC phase to the loss in canopy interception because of two reasons: (i) following common logging practices, clearcutting has left the lower portion of trunks and roots untouched, hence ground roughness remained unchanged at least in the medium term before root decay takes place; (ii) prior to clearcutting, litter removal in the LRBC phase have discounted the possibility of changes in ground litter being a factor. Therefore, we are able to deduce that it is the increased effective rainfall from the removal of canopy that has caused enhanced surface runoff (amount, duration, and initiation threshold) in the AC phase.

4.4. *Additional Consideration and Suggestions*

The heterogenous rainfall regime in the region resulted in different rainfall input and moisture conditions throughout the year, which may in turn affect surface properties (such as hydrophobicity) that govern surface runoff. Although surface runoff data of different treatments were collected in different phenological periods, differences in rainfall and moisture conditions were accounted for by the control plot and should not affect the results [13,53,54]. Although effort was made to establish the plots to be as similar as possible, there were small differences in tree species, slope, and plot size. Being situated in a natural forest, it was impossible to establish both plots with the exact tree species and size. The uneven microtopography and undulations on the ground resulted in a 0.7 m^2 difference in plot size. This should not affect the results and conclusion because comparisons were performed between different treatment (phases) instead of between different plots. Plot A merely served as a control to account for environmental conditions.

Rainfall intensity is known to significantly affect surface runoff generation in various land cover and regions [55–58]. The reason for the lack of influence from rainfall intensity in our plot study requires further investigation, but this may be provisionally attributed to surface runoff being naturally high in the area as well as the lack of complex hydrological pathways and sinks found in catchment scale studies [8,59].

This study has quantified the role of ground litter and canopy on surface runoff generation at the plot scale. Results from this study are not directly extrapolatable to the hillslope, catchment, and basin scale due to various hydrological and landscape factors. At the hillslope scale, species composition in a patch area may govern canopy and litter interception. At the catchment scale, various factors such as subsurface flow, groundwater, sinks, and hydrological connectivity need to be accounted for. At the basin scale, land-use and regional weather may be the governing factors instead [22]. In agreement with Liu et al. [22], future studies should cover more species, regions, and spatial scales—larger scales to understand environmental impacts; smaller scales to understand ecohydrological mechanisms.

Although we have investigated changes in surface runoff from litter removal and clearcutting, other hydrological components (infiltration, preferential flow, soil moisture) are still less understood. Investigating these (via moisture sensors, dyes, etc.) will give insights on the below-ground hydrology as well as verify the possible mechanisms that have been discussed in this article.

5. Conclusions

Due to the scarcity of information on the role of ground litter in regenerated broadleaf forests and the effects of clearcutting on surface runoff, we conducted a paired-plot experiment and discussed the possible hydrological mechanisms.

The absence of ground litter increased surface runoff by up to four times. Antecedent moisture had a significant influence on surface runoff generation after ground litter was removed. Without ground litter, 38% less rainfall was required to initiate surface runoff.

Clearcutting increased surface runoff by another four times when compared to the litter-removed period. Without both ground litter and the canopy, antecedent moisture no longer affect surface runoff. Canopy loss resulted in 56% less rainfall required to initiate surface runoff when compared to the ground litter-removed phase.

Author Contributions: Conceptualization, K.K. (Koichiro Kuraji) and M.G.; methodology, K.K. (Koichiro Kuraji); software, A.N., K.K. (Koju Kishimoto), K.T., M.G., and K.K. (Koichiro Kuraji); validation, A.N., K.K. (Koju Kishimoto), K.T., and K.K. (Koichiro Kuraji); formal analysis, A.N., K.K. (Koju Kishimoto), K.T., and K.K. (Koichiro Kuraji); investigation, A.N., K.K. (Koju Kishimoto), K.T., M.G., and K.K. (Koichiro Kuraji); resources, K.K. (Koichiro Kuraji); data curation, A.N., K.K. (Koju Kishimoto), K.T., M.G., and K.K. (Koichiro Kuraji); writing—original draft preparation, K.K. (Koju Kishimoto) and A.N.; writing—review and editing, A.N.; visualization, A.N.; supervision, K.K. (Koichiro Kuraji); project administration, K.K. (Koichiro Kuraji); funding acquisition, K.K. (Koichiro Kuraji). All authors have read and agreed to the published version of the manuscript.

Funding: This work was supported by JSPS KAKENHI (Grant-in-Aid for Young Scientists (B), grant number 26870834); and by the JSPS Core-to-core program (grant number: JPJSCCB20190007).

Institutional Review Board Statement: Not applicable.

Informed Consent Statement: Not applicable.

Acknowledgments: The authors acknowledge all staff of the Ecohydrology Research Institute for providing support in establishing and maintaining the experimental plots. Special mentions: Takanori Sato and Shigenari Satomi.

Conflicts of Interest: The authors declare no conflict of interest. The funders had no role in the design of the study; in the collection, analyses, or interpretation of data; in the writing of the manuscript, or in the decision to publish the results.

References

1. Forestry Agency Annual Report on Forest and Forestry in Japan Fiscal Year 2018 (Summary). Available online: http://www.rinya.maff.go.jp/j/kikaku/hakusyo/30hakusyo/index.html (accessed on 13 April 2020).
2. Forestry Agency. Annual Report on Forest and Forestry in Japan—Fiscal Year 2017. Available online: http://www.rinya.maff.go.jp/j/kikaku/hakusyo/29hakusyo/attach/pdf/index-1.pdf (accessed on 18 April 2021).
3. Food and Agricultural Organization of the United Nations. World Deforestation Slows Down as More Forests Are Better Managed. 2015. Available online: http://www.fao.org/news/story/en/item/326911/icode/ (accessed on 18 April 2021).
4. Forestry Agency. Annual Report on Forest and Forestry in Japan Fiscal Year 2015. Available online: http://www.rinya.maff.go.jp/j/kikaku/hakusyo/27hakusyo/attach/pdf/index-1.pdf (accessed on 18 April 2021).
5. The Japanese Forest Society. *Forest Science; Special Issue—Possibilities of Conversion to Broadleaf Forests*. 2010, Volume 59. Available online: https://www.forestry.jp/publish/ForSci/BackNo/sk59/59.pdf (accessed on 26 April 2021).
6. Tanaka, N.; Kuraji, K.; Shiraki, K.; Suzuki, M.; Suzuki, M.; Ohta, T.; Suzuki, M. Throughfall, stemflow and rainfall interception at mature Cryptomeria japonica and Chamaecyparis obtusa stands in Fukuroyamasawa watershed. *Bull. Tokyo Univ. For.* **2005**, *113*, 197–240. (In Japanese)
7. Miyata, S.; Kosugi, K.; Gomi, T.; Onda, Y.; Mizuyama, T. Surface runoff as affected by soil water repellency in a Japanese cypress forest. *Hydrol. Process.* **2007**, *21*, 2365–2376. [CrossRef]
8. Sidle, R.C.; Hirano, T.; Gomi, T.; Terajima, T. Hortonian overland flow from Japanese forest plantations—an aberration, the real thing, or something in between? *Hydrol. Process.* **2007**, *21*, 3237–3247. [CrossRef]
9. Murakami, S.; Tsuboyama, Y.; Shimizu, T.; Fujieda, M.; Noguchi, S. Variation of evapotranspiration with stand age and climate in a small Japanese forested catchment. *J. Hydrol.* **2000**, *227*, 114–127. [CrossRef]
10. Saito, T.; Kumagai, T.; Tateishi, M.; Kobayashi, N.; Otsuki, K.; Giambelluca, T.W. Differences in seasonality and temperature dependency of stand transpiration and canopy conductance between Japanese cypress (Hinoki) and Japanese cedar (Sugi) in a plantation. *Hydrol. Process.* **2017**, *31*, 1952–1965. [CrossRef]
11. Sun, X.; Onda, Y.; Otsuki, K.; Kato, H.; Gomi, T. The effect of strip thinning on forest floor evaporation in a Japanese cypress plantation. *Agric. For. Meteorol.* **2016**, *216*, 48–57. [CrossRef]
12. Kuraji, K.; Gomyo, M.; Nainar, A. Thinning of cypress forest increases subsurface runoff but reduces peak storm-runoff: A lysimeter observation. *Hydrol. Res. Lett.* **2019**, *13*. [CrossRef]
13. Gomyo, M.; Kuraji, K. Effect of the litter layer on runoff and evapotranspiration using the paired watershed method. *J. For. Res.* **2016**, *21*, 306–313. [CrossRef]
14. Abe, Y.; Gomi, T.; Nakamura, N.; Kagawa, N. Field estimation of interception in a broadleaf forest under multi-layered structure conditions. *Hydrol. Res. Lett.* **2017**, *11*, 181–186. [CrossRef]
15. Onda, Y.; Gomi, T.; Mizugaki, S.; Nonoda, T.; Sidle, R.C. An overview of the field and modelling studies on the effects of forest devastation on flooding and environmental issues. *Hydrol. Process.* **2010**, *24*, 527–534. [CrossRef]
16. Li, X.; Niu, J.; Xie, B. Study on Hydrological functions of litter layers in North China. *PLoS ONE* **2013**, *8*, e70328. [CrossRef]
17. Prosdocimi, M.; Jordán, A.; Tarolli, P.; Keesstra, S.; Novara, A.; Cerdà, A. The immediate effectiveness of barley straw mulch in reducing soil erodibility and surface runoff generation in Mediterranean vineyards. *Sci. Total Environ.* **2016**, *547*, 323–330. [CrossRef]
18. Zhou, Q.; Zhou, X.; Luo, Y.; Cai, M. The effects of litter layer and topsoil on surface runoff during simulated rainfall in Guizhou Province, China: A plot scale case study. *Water* **2018**, *10*, 915. [CrossRef]
19. Miyata, S.; Kosugi, K.; Gomi, T.; Mizuyama, T. Effects of forest floor coverage on overland flow and soil erosion on hillslopes in Japanese cypress plantation forests. *Water Resour. Res.* **2009**, *45*. [CrossRef]
20. Walsh, R.P.D.; Voigt, P.J. Vegetation litter: An underestimated variable in hydrology and geomorphology. *J. Biogeogr.* **1977**, *4*, 253. [CrossRef]
21. Sun, X.; Onda, Y.; Kato, H.; Gomi, T.; Komatsu, H. Effect of strip thinning on rainfall interception in a Japanese cypress plantation. *J. Hydrol.* **2015**, *525*, 607–618. [CrossRef]
22. Liu, J.; Gao, G.; Wang, S.; Jiao, L.; Wu, X.; Fu, B. The effects of vegetation on runoff and soil loss: Multidimensional structure analysis and scale characteristics. *J. Geogr. Sci.* **2018**, *28*, 59–78. [CrossRef]
23. Nainar, A.; Tanaka, N.; Sato, T.; Kuraji, K. Comparing runoff characteristics between an evergreen cypress forest and a mixed-broadleaf forest during different phenological periods in central Japan. *Geophys. Res. Abstr.* **2019**, *21*, 2019–4671. Available online: https://meetingorganizer.copernicus.org/EGU2019/EGU2019-4671.pdf (accessed on 26 April 2021).
24. Ecohydrology Research Institute. Available online: http://www.uf.a.u-tokyo.ac.jp/eri/ (accessed on 7 April 2021).
25. Nainar, A.; Tanaka, N.; Sato, T.; Kishimoto, K.; Kuraji, K. A comparison of the baseflow recession constant (K) between a Japanese cypress and mixed-broadleaf forest via six estimation methods. *Sustain. Water Resour. Manag.* **2021**, *7*, 1–13. [CrossRef]
26. Gomyo, M.; Kuraji, K. Long-term variation of annual loss and annual evapotranspiration in small watershed with the forest restoration and succession on denuded hills. *J. Jpn. For. Soc.* **2013**, *95*, 109–116. [CrossRef]
27. Gomyo, M.; Kuraji, K. Effect of litter layer on water balance and runoff at the watershed scale. *Water Sci.* **2016**, *60*, 31–45. [CrossRef]
28. Ecohydrology Research Institute. *Map of Ananomiya Experimental Forest*; Ecohydrology Research Institute (Archive): Seto, Japan. Available online: https://www.a.u-tokyo.ac.jp/english/institute_e/ai-uforests.html (accessed on 19 April 2021).

29. United States Department of Agriculture Soil Texture Calculator | NRCS Soils. Available online: https://www.nrcs.usda.gov/wps/portal/nrcs/detail/soils/survey/?cid=nrcs142p2_054167 (accessed on 22 April 2020).
30. Putuhena, W.M.; Cordery, I. Estimation of interception capacity of the forest floor. *J. Hydrol.* **1996**, *180*, 283–299. [CrossRef]
31. Li, X.; Niu, J.; Xie, B. The effect of leaf litter cover on surface runoff and soil erosion in Northern China. *PLoS ONE* **2014**, *9*, e107789. [CrossRef]
32. Sayer, E.J. Using experimental manipulation to assess the roles of leaf litter in the functioning of forest ecosystems. *Biol. Rev.* **2005**, *81*. [CrossRef] [PubMed]
33. Kim, J.K.; Onda, Y.; Kim, M.S.; Yang, D.Y. Plot-scale study of surface runoff on well-covered forest floors under different canopy species. *Quat. Int.* **2014**, *344*, 75–85. [CrossRef]
34. Butzen, V.; Seeger, M.; Marruedo, A.; de Jonge, L.; Wengel, R.; Ries, J.B.; Casper, M.C. Water repellency under coniferous and deciduous forest—Experimental assessment and impact on overland flow. *Catena* **2015**, *133*, 255–265. [CrossRef]
35. Deguchi, A.; Hattori, S.; Daikoku, K.; Park, H.-T. Measurement of evaporation from the forest floor in a deciduous forest throughout the year using microlysimeter and closed-chamber systems. *Hydrol. Process.* **2008**, *22*, 3712–3723. [CrossRef]
36. Benyon, R.G.; Doody, T.M. Comparison of interception, forest floor evaporation and transpiration in *Pinus radiata* and *Eucalyptus globulus* plantations. *Hydrol. Process.* **2015**, *29*, 1173–1187. [CrossRef]
37. Crockford, R.H.; Richardson, D.P. Partitioning of rainfall into throughfall, stemflow and interception: Effect of forest type, ground cover and climate. *Hydrol. Process.* **2000**, *14*, 2903–2920. [CrossRef]
38. Levia, D.F.; Nanko, K.; Amasaki, H.; Giambelluca, T.W.; Hotta, N.; Iida, S.; Mudd, R.G.; Nullet, M.A.; Sakai, N.; Shinohara, Y.; et al. Throughfall partitioning by trees. *Hydrol. Process.* **2019**, *33*, 1698–1708. [CrossRef]
39. Komatsu, H.; Shinohara, Y.; Kume, T.; Otsuki, K. Relationship between annual rainfall and interception ratio for forests across Japan. *For. Ecol. Manag.* **2008**, *256*, 1189–1197. [CrossRef]
40. Vernimmen, R.R.E.; Bruijnzeel, L.A.; Romdoni, A.; Proctor, J. Rainfall interception in three contrasting lowland rain forest types in Central Kalimantan, Indonesia. *J. Hydrol.* **2007**, *340*, 217–232. [CrossRef]
41. Chappell, N.A.; Bidin, K.; Tych, W. Modelling rainfall and canopy controls on net-precipitation beneath selectively-logged tropical forest. *Plant. Ecol.* **2001**, *153*, 215–229. [CrossRef]
42. Reid, L.M.; Lewis, J. Rates, timing, and mechanisms of rainfall interception loss in a coastal redwood forest. *J. Hydrol.* **2009**, *375*, 459–470. [CrossRef]
43. Nanko, K.; Onda, Y.; Ito, A.; Ito, S.; Mizugaki, S.; Moriwaki, H. Variability of surface runoff generation and infiltration rate under a tree canopy: Indoor rainfall experiment using Japanese cypress (*Chamaecyparis obtusa*). *Hydrol. Process.* **2010**, *24*, 567–575. [CrossRef]
44. Oishi, A.C.; Oren, R.; Stoy, P.C. Estimating components of forest evapotranspiration: A footprint approach for scaling sap flux measurements. *Agric. For. Meteorol.* **2008**, *148*, 1719–1732. [CrossRef]
45. Zheng, C.; Jia, L. Global canopy rainfall interception loss derived from satellite earth observations. *Ecohydrology* **2020**, *13*, e2186. [CrossRef]
46. Vesala, T.; Suni, T.; Rannik, Ü.; Keronen, P.; Markkanen, T.; Sevanto, S.; Grönholm, T.; Smolander, S.; Kulmala, M.; Ilvesniemi, H.; et al. Effect of thinning on surface fluxes in a boreal forest. *Global Biogeochem. Cycles* **2005**, *19*, GB2001. [CrossRef]
47. Ishii, H.T.; Maleque, M.A.; Taniguchi, S. Line thinning promotes stand growth and understory diversity in Japanese cedar (Cryptomeria japonica D. Don) plantations. *J. For. Res.* **2008**, *13*, 73–78. [CrossRef]
48. Daikoku, K.; Hattori, S.; Deguchi, A.; Aoki, Y.; Miyashita, M.; Matsumoto, K.; Akiyama, J.; Iida, S.; Toba, T.; Fujita, Y.; et al. Influence of evaporation from the forest floor on evapotranspiration from the dry canopy. *Hydrol. Process.* **2008**, *22*, 4083–4096. [CrossRef]
49. Germer, S.; Werther, L.; Elsenbeer, H. Have we underestimated stemflow? Lessons from an open tropical rainforest. *J. Hydrol.* **2010**, *395*, 169–179. [CrossRef]
50. Su, L.; Xu, W.; Zhao, C.; Xie, Z.; Ju, H. Inter- and intra-specific variation in stemflow for evergreen species and deciduous tree species in a subtropical forest. *J. Hydrol.* **2016**, *537*, 1–9. [CrossRef]
51. Kuraji, K.; Tanaka, Y.; Tanaka, N.; Karakama, I. Generation of stemflow volume and chemistry in a mature Japanese cypress forest. *Hydrol. Process.* **2001**, *15*, 1967–1978. [CrossRef]
52. Tanaka, N.; Levia, D.; Igarashi, Y.; Yoshifuji, N.; Tanaka, K.; Tantasirin, C.; Nanko, K.; Suzuki, M.; Kumagai, T. What factors are most influential in governing stemflow production from plantation-grown teak trees? *J. Hydrol.* **2017**, *544*, 10–20. [CrossRef]
53. Dung, B.X.; Gomi, T.; Miyata, S.; Sidle, R.C.; Kosugi, K.; Onda, Y. Runoff responses to forest thinning at plot and catchment scales in a headwater catchment draining Japanese cypress forest. *J. Hydrol.* **2012**, *444–445*, 51–62. [CrossRef]
54. Dung, B.X.; Miyata, S.; Gomi, T. Effect of forest thinning on overland flow generation on hillslopes covered by Japanese cypress. *Ecohydrology* **2011**, *4*, 367–378. [CrossRef]
55. Mu, W.; Yu, F.; Li, C.; Xie, Y.; Tian, J.; Liu, J.; Zhao, N. Effects of rainfall intensity and slope gradient on runoff and soil moisture content on different growing stages of spring maize. *Water* **2015**, *7*, 2990–3008. [CrossRef]
56. Liang, Z.; Liu, H.; Zhao, Y.; Wang, Q.; Wu, Z.; Deng, L.; Gao, H. Effects of rainfall intensity, slope angle, and vegetation coverage on the erosion characteristics of Pisha sandstone slopes under simulated rainfall conditions. *Environ. Sci. Pollut. Res.* **2020**, *27*, 17458–17467. [CrossRef]

57. Dos Santos, J.C.N.; de Andrade, E.M.; Medeiros, P.H.A.; Guerreiro, M.J.S.; de Queiroz Palácio, H.A. Effect of rainfall characteristics on runoff and water erosion for different land uses in a tropical semiarid region. *Water Resour. Manag.* **2017**, *31*, 173–185. [CrossRef]
58. Zhao, B.; Zhang, L.; Xia, Z.; Xu, W.; Xia, L.; Liang, Y.; Xia, D. Effects of rainfall intensity and vegetation cover on erosion characteristics of a soil containing rock fragments slope. *Adv. Civ. Eng.* **2019**, *2019*, 7043428. [CrossRef]
59. Sidle, R.C.; Ziegler, A.D.; Negishi, J.N.; Nik, A.R.; Siew, R.; Turkelboom, F. Erosion processes in steep terrain—Truths, myths, and uncertainties related to forest management in Southeast Asia. *For. Ecol. Manag.* **2006**, *224*, 199–225. [CrossRef]

Article

Long-Term Changes in Relationship between Water Level and Precipitation in Lake Yamanaka

Koichiro Kuraji [1,*] **and Haruo Saito** [2]

[1] Executive Office, The University of Tokyo Forests, Graduate School of Agricultural and Life Sciences, The University of Tokyo, Bunkyo-ku, Tokyo 113-8657, Japan
[2] Fuji Iyashinomori Woodland Study Center, The University of Tokyo Forests, Graduate School of Agricultural and Life Sciences, The University of Tokyo, Yamanakako, Yamanashi 401-0501, Japan; haruo_s@uf.a.u-tokyo.ac.jp
* Correspondence: kuraji_koichiro@uf.a.u-tokyo.ac.jp

Abstract: Lake water levels fluctuate due to both natural and anthropogenic influences. Climate change can alter precipitation, driving fluctuations in lake water levels. Extreme fluctuations can cause flooding, water shortages, and changes in lake water quality and ecosystems, as well as affecting fisheries and tourism. Despite the need to predict future water-level rises, especially in the context of climate change, long-term hydrological studies are scarce. Here, we analyzed 93 years of data from 1928 to 2020 to identify changes in the relationship between water level and precipitation in Lake Yamanaka, Japan. We found that the six-day maximum rise in water level for the same six-day maximum precipitation was significantly greater in the later period than in the earlier period; the difference increased with increasing precipitation. Particularly large increases in precipitation were sometimes caused by a single event or by multiple events occurring in succession.

Keywords: lake water level rise; Lake Yamanaka; maximum six-day precipitation

Citation: Kuraji, K.; Saito, H. Long-Term Changes in Relationship between Water Level and Precipitation in Lake Yamanaka. *Water* **2022**, *14*, 2232. https://doi.org/10.3390/w14142232

Academic Editor: Steve W. Lyon

Received: 6 December 2021
Accepted: 13 July 2022
Published: 15 July 2022

1. Introduction

Fluctuations in the water levels of lakes can be driven by both natural and anthropogenic influences. For example, climate change can drive changes in precipitation levels, which may cause lake water levels to fluctuate. Issues such as flooding, water shortages, and changes in lake water quality can arise as a result of extreme lake water level fluctuations, with potential effects on ecosystems, fisheries, and tourism [1]. Therefore, it is important to predict future changes in lake water levels, for which the effects of climate change and anthropogenic impacts need to be assessed separately. To this end, it is beneficial to examine past fluctuations in lake water levels over a long period; identifying their causes may provide useful information for the future prediction of lake level fluctuations [2].

The Fuji Five Lakes are a group of five small lakes in Yamanashi prefecture on the north side of Mount Fuji in Japan. From east to west, they are Lake Yamanaka, Lake Kawaguchi, Lake Sai, Lake Shoji, and Lake Motosu. In years and months with heavy precipitation, the water levels rise in the Fuji Five Lakes, and lakeside areas have suffered severe inundation damage under these conditions [3]. On the other hand, surface water flowing from the Fuji Five Lakes has long been used for agriculture, daily life, and hydropower generation, as has groundwater in lakeside areas. Thus, drops in the lake level have been a serious problem for water users because they lead directly to water shortages [4,5].

A number of hydrological studies have been conducted on the Fuji Five Lakes, focusing on lake water level, water balance, and groundwater flow [3,6–16]. Yamamoto and Uchiyama [3] examined the relationship between lake water level and precipitation using 100 years of lake water-level data (1911–2017) at Lake Kawaguchi. They identified three major increases in lake water levels since 1930, all of which corresponded to rainfall events (in which monthly precipitation generally exceeded 750 mm). Few of these studies have focused on Lake Yamanaka, although the area of the lake is the largest among the Fuji Five Lakes. In addition, Yamamoto and Uchiyama [3] only used monthly precipitation and water-level data; they did not analyze sub-monthly variations in precipitation.

In addition to lake water-level observation data, long-term precipitation data are essential when studying long-term fluctuations in lake water levels (e.g., over 100 years) because precipitation is the most important factor affecting lake water-level fluctuations. However, compared to Lake Kawaguchi, where precipitation observations have been recorded by the Yamanashi Prefecture and the Japan Meteorological Agency (JMA) since 1933, long-term precipitation observation data at Lake Yamanaka have not been systematically organized. Recently, Kuraji et al. [17] systematically organized long-term precipitation data at the Fuji Iyashinomori Woodland Study Center (FIWSC; 35°24′27.4″ N 138°51′51.6″ E) of the University of Tokyo, which is located in the vicinity of Lake Yamanaka. These data make it possible to study the relationship between long-term precipitation trends and lake water-level fluctuations in Lake Yamanaka.

Therefore, in this study, we focused on the relationship between increases in water levels in Lake Yamanaka and heavy rainfall events. We collected data on water levels over a long period (93 years) to analyze long-term fluctuations in relation to precipitation, with the aim of characterizing long-term fluctuations in lake water levels in Lake Yamanaka.

2. Materials and Methods

2.1. Study Site

Lake Yamanaka, one of the Fuji Five Lakes, is located in Yamanakako Village, Minamitsuru-gun, Yamanashi Prefecture, Japan (Figure 1). It is a natural lake with an area of 6.6 km^2, an elevation of 981 m, a circumference of 14 km, a maximum depth of 13.3 m, and an average depth of 9.4 m. The land cover of the Yamanaka Lake catchment area has changed drastically over the past 100 years. Before humans settled in the area, the catchment area had been a cool-temperate upper broadleaf forest. The land was unsuitable for cultivation due to a combination of adverse conditions, including a cold climate above 1000 m elevation and a nutrient-poor soil accumulated from volcanic debris and lava from Mt. Fuji. In order to harvest grasses for use as a natural fertilizer for cultivation, vast forests have been converted to grasslands. In addition, extensive pastures were created to raise horses for the logistics industry. Around 1900, grasslands occupied the largest area of land in the Yamanakako Village area (Figure 2). Later, when fossil fuels and chemical fertilizers became widespread, and the main industry of the area shifted from logistics to tourism, these grasslands became inaccessible and were abandoned and gradually transitioned to forests. Furthermore, since 1960, coniferous trees, mainly larch, have been planted for timber production; half of the forests are coniferous planted forests, and the other half are broadleaf natural forests (Figure 2).

2.2. Materials

Precipitation affects the water levels of Lake Yamanaka only when it falls within the lake's catchment area (including surface water and groundwater). The lake's surface water catchment area includes the northern foot of Mt. Fuji. Various estimates of the size of the catchment areas have been proposed in previous studies (e.g., 66.07 km^2 by Kambara [6] and 64.87 km^2 by Tsutsumi [9]). It is difficult to identify the recharge area of groundwater flowing into Lake Yamanaka based purely on the region's surface topography. Here, the water budget of Lake Yamanaka was not the main subject of our analysis, so it was not necessary for the precipitation data corresponding to increases in water level to be in the form of basin-averaged precipitation. Instead, we considered it sufficient to use precipitation data recorded at representative points. Therefore, we used daily precipitation data from the FIWSC, prepared by Kuraji et al. [17], for a period of 86 years from 1927 to 2013 (data were missing for 8–31 December 1940; 1–31 January, 1 March–31 May and 1–31 July 1943; and 1 January 1944–1 March 1952).

Figure 1. Map showing the Lake Yamanaka and the surrounding areas.

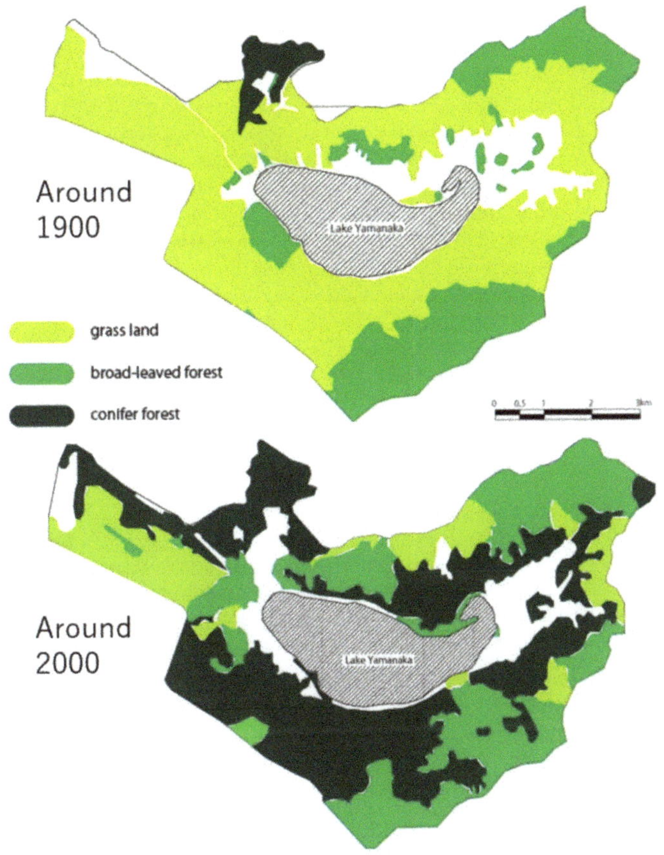

Figure 2. Land cover change in the Yamanaka Village. Source: Upper map is based on a 1/50,000 scale topographic map of the Imperial Japanese Land Survey Department [18]. The lower map is based on Yamanakako Village [19].

Daily FIWSC precipitation data were not available for 2014–2020. Therefore, we estimated daily precipitation data for this period using precipitation observations recorded at the Yamanaka site (observation point: Fujiyoshida Fire Station East Sub-branch) of the Automated Meteorological Data Acquisition System (AMeDAS) of the Japan Meteorological Agency (JMA). We accessed the JMA website on 11 September 2021, and downloaded hourly precipitation data. We compiled daily precipitation data from 25 March 2008, when the observation unit of AMeDAS Yamanaka changed from 1 mm to 0.5 mm, to 31 December 2013, at 09:00 local time (on the daily boundary). We determined the relationship between the daily precipitation data and the FIWSC daily precipitation data of Kuraji et al. [17] as follows:

$$Rf = 1.1588 \times Ra \quad (R^2 = 0.9742, \quad n = 276) \tag{1}$$

where Rf is the FIWSC daily precipitation (mm) and Ra is the AMeDAS Yamanaka daily precipitation (mm).

We applied Equation (1) to the daily precipitation data from the AMeDAS mountains from 1 January 2014, to 31 March 2021, to estimate the daily precipitation of the FIWSC. Data were absent for the periods of 14–15 February, 15 April, 23 May, and 16 September 2014; 23 February, 25 March, and 9, 12 and 18 May 2015; and 11, 13–14 and 31 March, 12 May and 6 September 2016. However, when examining the water-level data for these dates, we found no significant precipitation-driven water-level rises.

We used documents #1–#6 as data for the Yamanaka Lake water level and outflow. Document #1 [20] comprises daily lake water-level and outflow recordings (observation time unknown) from 1927 to 1962, as described in "Water utilization survey at the northern foot of Mt. Fuji and its data analysis (Appendix): Table of water level, outflow, and rainfall of the five lakes of Fuji". The reference point is listed as having an altitude of 978.49 m. Document #2 [21] comprises daily lake water-level and outflow recordings (observation time unknown) from 1959 to 1968, as listed in "Fuji Five Lakes: Water Level, Outflow, Rainfall Observation Table". The altitude reference point was listed to be 878.485 m, but we assume that this was a misprint and that it should have read "978.485 m". Comparing Documents #1 and #2 revealed that the overlapping data for 1959–1962 were consistent.

Document #3 [22–26] contains records of the lake water level for the years 1955–1958 and 1963–1968, as described in the "Yamanashi Prefecture Management Water Level Rainfall Observation Annual Report". The observation times in this document were recorded once per morning in June 1955–1956 and 1963–1964, and twice per morning and evening in July–December 1956 and 1965–1968. The altitude of the reference point was listed as 878.485 m, but as in Document #2, we assume this was a misprint and should have read 978.485 m. Comparing the lake levels in Document #1 and #2 with those in Document #3 revealed that for the years 1955–1958 and 1963–1964, adding 3.00 and 0.10 m, respectively, to the latter gave the same results as those in Document #1. In the 1965–1966 period, the data were not exactly the same, but there was no significant difference in the fluctuation patterns, so we used the values from the morning observations in Document #3.

Document #4 [27] contains lake water-level recordings from April 1981 to June 1996, as described in the "Water Level Monthly Report" of the Yamanashi Prefecture River Division. These observations were recorded twice daily at 06:00 and 18:00 from April 1981 to March 1982 and once daily at 10:00 after April 1982. These measurements were not taken in May or June 1996 because of the lake's low water level. In this study, therefore, we adopted the value of the morning observations during the period of twice-daily observations.

Document #5 [28] contains lake water levels from 16 April 1998 to 31 December 2020, as published by Yamanashi Prefecture on its website. These recordings were taken daily, with the reference point being 978.485 m. Data were missing from 2 to 17 December 2009; 30 March and 5 December 2010; 25 January and 12–24 March 2011; 7 September 2012; 7 February and 6 October 2013; 8–14 January 2015; from 5 April to 10 May 2020; and on 23 August 2020.

Document #6 [29] contains daily lake water levels and outflow from 1 January 1972 to 31 December 2020, observed by the Yamanashi Branch Office, Tokyo Electric Power

Company (TEPCO), and disclosed by Yamanashi Prefecture. Data were missing from 1 January 2011 to 31 March 2013; 18 (outflow only) and 21–24 December 2014; 18–19 and 22–26 January (outflow only), 14 (outflow only) and 15–16 March, and 8–9 September 2015.

2.3. Methods

We compared long-term variations in the relationship between the amount of precipitation in the FIWSC dataset and increases in the water level of Lake Yamanaka during periods of heavy rainfall. To this end, we divided the analysis period into two: the first period, 1927–1968, and the second period, 1972–2020.

2.3.1. Long-Term Changes in Relationship between Precipitation and Water-Level Rise

We defined a flood event as an event in which the rise in water level stopped within seven days of the day it began and in which the difference in water levels between these days was 0.2 m or greater. We limited our analysis to floods in which the rise in water level stopped after seven days. This was because, in floods where the water-level rise exceeded this period, several precipitation events could be responsible, rendering it impossible to identify one event. Our analysis of this type of long-term rise in water level is presented in Section 2.3.2.

Among the identified flood events, we selected those that exceeded 200 mm of precipitation over six days. To determine this period, we examined the correlation between precipitation and the difference in water level for n days from the day when the water level stopped rising to the day n days before, for all of the rising water events. The correlation coefficient for $n = 6$ was the largest ($R^2 = 0.751$), so we used a six-day period. We set the threshold value for the six-day precipitation at 200 mm because the number of events would decrease for values larger than 200 mm, while the influence of the base water level (prior to the increase in question) could not be ignored for values smaller than 200 mm.

We defined the maximum difference between the highest and lowest water levels for any six-day period during which the six-day precipitation exceeded 200 mm as the maximum six-day increase in water level. Furthermore, we defined the amount of precipitation during these six days for which the maximum six-day increase was obtained as the maximum six-day precipitation.

Analysis of covariance (ANCOVA) was used as a statistical method for comparison between the first and second periods. The comparison of time periods was made by removing the effect of the difference in six-day maximum precipitation, which is a covariate between the first and second periods.

2.3.2. Examination of Particularly Large Water Rises

The highest water level to occur in Lake Yamanaka in recent years was 4.48 m, which occurred on 22 September 2011. In the period for which data on daily precipitation and lake level are available, we found that this level was twice exceeded: on 6 September 1938 (maximum level = 5.02 m) and on 13 October 1991 (maximum level = 4.56 m). In addition, the years 1935, 1938 (see above), and 1983, which were listed as the three years having floods over the past 100 years in a previous study into Lake Kawaguchi [3], were also recorded to have had floods at Lake Yamanaka (the highest water levels were 4.40 m on 27 October 1935, and 4.34 m on 19 August 1983). These five occurrences of high-water levels coincided with the top five highest water-level increases observed in the period covered by this study. Therefore, we conducted a comparative analysis of these five water-level increases.

Unlike the events defined in Section 2.3.1, these substantial, relatively long-lasting increases in water level are likely to have been caused not only by precipitation during the preceding six days but also by precipitation over a longer period. Thus, we examined the correlation coefficients between the accumulated precipitation from the day before the occurrence of the highest water level to the day n days before, for the years 1935, 1938, 1983, 1991, and 2011; the maximum correlation coefficient ($R^2 = 0.983$) was obtained when $n = 71$.

Thus, we used a period of 71 days when describing the progress of precipitation and water levels for these events.

3. Results

3.1. Lake Water Level and Outflow Data Overview

Figure 3 shows all lake water level and outflow data over the 93 years used in this study. The maximum water level was 0.64 m on 28–29 August 1934 and 5.02 m on 6 September 1938, with a range of 4.38 m. No trend or periodicity was observed throughout the 93 years. The runoff is used for irrigation and power generation, and there are seasonal fluctuations in the amount of water used for irrigation. The TEPCO has held the water rights for power generation from 23 March 1925 to the present and can withdraw water for power generation when the lake level is between 2.12 m and 4.24 m (3.33 m from July to September) [30]. When the lake level becomes high, the river administrator, Yamanashi Prefecture, releases water for flood-control purposes.

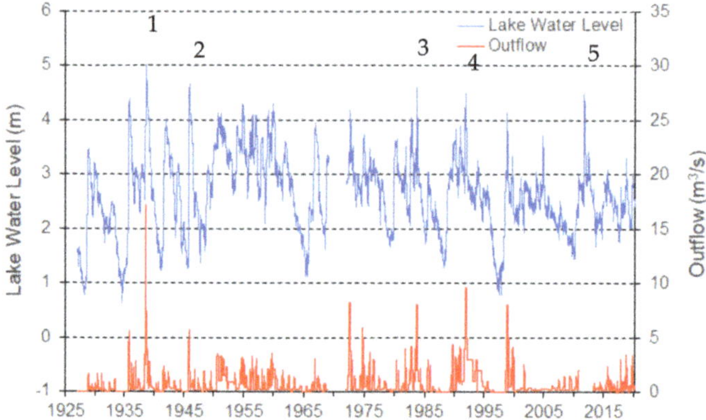

Figure 3. Lake water level and outflow of the Lake Yamanaka over 93 years. The number on the figure indicates the events with particularly large increments in the water level (see Section 2.3.2).

3.2. Long-Term Changes in the Relationship between Precipitation and Water-Level Rise

Figure 4 shows the relationship between the maximum six-day precipitation and the maximum six-day increase in water level. Although this relationship was linear in both the first and second halves of the study, a significant difference was observed between the first and second periods (ANCOVA, $p < 0.0001$). When the precipitation was generally 250 mm or more, the maximum six-day increase in water level for the same maximum six-day precipitation was significantly larger in the second period than in the first; moreover, this difference increased with increasing precipitation. When the maximum six-day precipitation values were 200, 400, and 600 mm, the maximum six-day increments predicted by the regression line were 0.20, 0.39, and 0.57 m, respectively, in the first period, and 0.18, 0.50, and 0.82 m in the second period, respectively.

3.3. Examination of Particularly Large Water-Level Increases

Figures 5 and 6 show the time series of variations in water levels, starting from 71 days before the highest water-level occurrences in 1935, 1938, 1983, 1991, and 2011.

3.3.1. Event of 27 October 1935 (Highest Water Level = 4.40 m)

On 17 August, 72 days before the event, the water level was 1.99 m. On 23 August, the water level then dropped to 1.94 m. However, on 3 September, it rose to 2.56 m due to 625.5 mm of precipitation falling over 10 days from 24 August to 2 September (following the approach of a typhoon, as included in the analysis in Section 3.1). Precipitation during the 11 days from 3 to 13 September only amounted to 10.4 mm, but the water level rose to 2.62 m on 14 September. There was 907.5 mm of precipitation during the 12-day period from 14 to 25 September, which caused the water level to rise to 3.68 m on 26 September (included in the analysis in Section 3.1). The 14 days from 26 September to 9 October saw only 26.8 mm of precipitation, but the water level rose to 4.22 m on 9 October. There was 61.2 mm of precipitation during the two days from 10 to 11 October, which further increased the water level to 4.32 m. Only 13.5 mm of precipitation occurred during the 14 days from 12 to 25 October, with the water level dropping to 4.18 m on 25 October. The total precipitation during the 71-day period was 1729.5 mm.

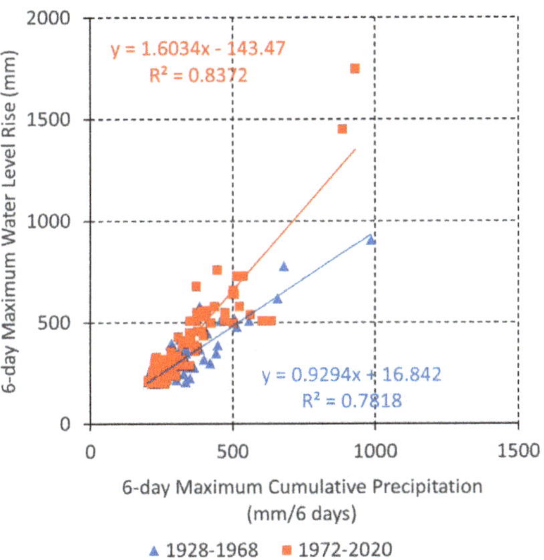

Figure 4. Relationship between six-day maximum cumulative precipitation (mm/6 days) and six-day maximum water-level rise for first (1928–1968) and second (1972–2020) periods.

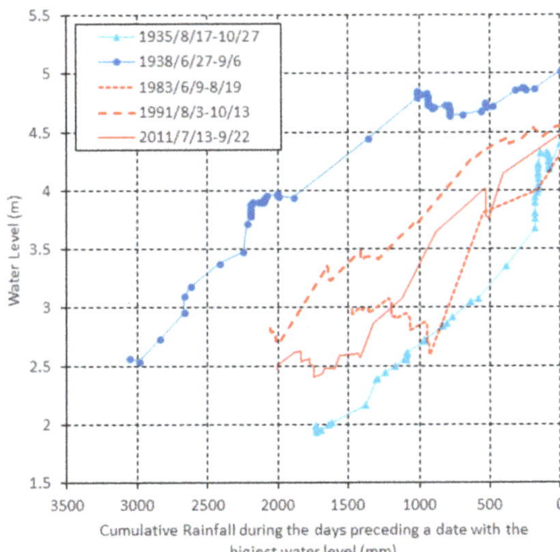

Figure 5. Relationship between cumulative rainfall during the days preceding a date with the highest water level. Note x-axis is reversed.

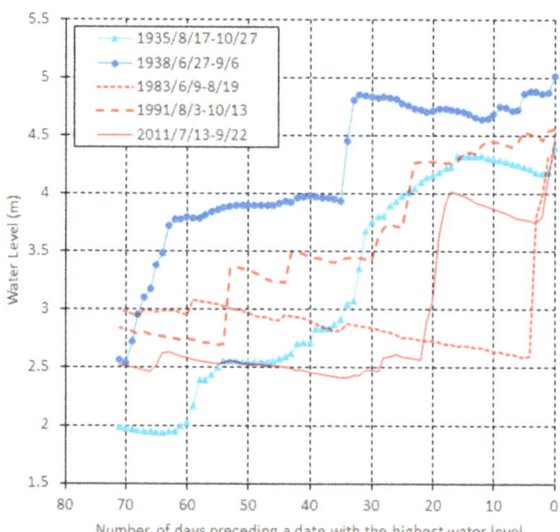

Figure 6. Relationship between the number of days preceding a date with the highest water level and water level. Note x-axis is reversed.

3.3.2. 6 September 1938 (Highest Water Level = 5.02 m)

The water level was 2.57 m on 27 June, 72 days before this event. Then, 860.8 mm of precipitation fell from 27 June to 5 July, with the water level rising to 3.78 m on 6 July (included in the analysis in Section 3.1). In total, 193.8 mm of precipitation fell from 6 to 31 July, but the water level only rose to 3.96 m on 31 July. From 4 to 24 August, precipitation amounted to 227.9 mm, with the water level decreasing to 4.64 m on 25 August, but it then increased to 3.96 m on 31 July. The total precipitation over 71 days was 3053.4 mm.

3.3.3. Event of 19 August 1983 (Highest Water Level = 4.34 m)

The water level 72 days prior to this event, on 9 June, was 2.98 m. The water level then gradually decreased, reaching 2.59 m on 14 August. The total precipitation for 71 days was 1476.6 mm.

3.3.4. Event of 13 October 1991 (Highest Water Level = 4.56 m)

On 3 August, 72 days prior to this event, the highest water level was 2.84 m. The water level then dropped to 2.69 m on 19 August. On 20 August, 342.8 mm of daily precipitation fell due to the approach of Typhoon No. 12, with the water level rising to 3.37 m on 22 August (included in the analysis in Section 3.1). The water level dropped to 3.23 m on 30 August but then rose to 3.49 m on 31 August, following 211 mm of daily precipitation; this precipitation fell due to the approach of Typhoon No. 14 on 30 August (included in the analysis in Section 3.1). The water level then fluctuated slightly before rising to 4.27 m on 21 September (included in the analysis in Section 3.1). This increase occurred due to 664.9 mm of precipitation falling over 10 days from 12 to 21 September, associated with the approach of Typhoon No. 18. It then increased further to 4.46 m on 2 October (included in the analysis in Section 3.1) due to 267.2 mm of precipitation falling over six days from 26 September to 1 October. The total precipitation over 71 days was 2053.1 mm.

3.3.5. Event of 22 September 2011 (Highest Water Level = 4.48 m)

On 13 July, 72 days before this event, the water level was 2.52 m. Initially, the water level did not change significantly and reached 2.57 m on 31 August. However, it rose to 4.02 m on 5 September due to 885.6 mm of precipitation falling during the six-day period from 31 August to 5 September; this was caused by Typhoon No. 12 (included in the analysis in Section 3.1). The water level dropped to 3.75 m on 19 September, but the highest water level (4.48 m) was then recorded on 22 September, following 498.3 mm of precipitation falling during 19 and 20 September, caused by Typhoon No. 15 (included in the analysis in Section 3.1) The total precipitation over 71 days was 2001.4 mm.

4. Discussion

Figure 4 demonstrates that the relationship between the maximum six-day rainfall and the maximum six-day increase in water level was linear for both the first and second periods. This meant that it was possible to predict the maximum water level after the floods, based on the six-day rainfall data and the water level on the day of the flood. The following Equation could, therefore, be used to estimate the maximum water level after the flood, based on the rainfall data in the mountains recorded at AMeDAS (which are published in real-time) (also see Equation (1)), and the regression line shown in Figure 4):

$$\Delta H = 1.6034 \times Rf6 - 143.47 = 1.8580 \times Ra6 - 143.47 \quad (R^2 = 0.8372) \tag{2}$$

where ΔH is the estimated lake level rise (mm), *Rf6* is the six-day precipitation at FIWSC (mm), and *Ra6* is the six-day precipitation at AMeDAS Yamanaka (mm).

Figure 4 clearly shows that the six-day maximum rise in water volume for the same maximum six-day rainfall was significantly greater in the second period than in the first. There are two possible reasons for this: first, lakeside road construction and shoreline reclamation work progressed from the first to the second period. Okai et al. [31] reported that the lakeside road construction work was conducted from 1959 until 1962. Takahashi et al. [32] reported that the Yamanaka Village began reclaiming the lake's shoreline and building parking lots in 1965 (Figure 7). At the same time, piers were built by private companies along the shore of the lake. The number of piers increased from 13 to 42 from 1962 to 1970. The reclamation projects likely changed the relationship between the water level and the water volume when the water level rose; the magnitude of the water-level rise may have been larger for the same increase in water volume during the second period. The second possibility is that river and erosion control efforts progressed in a channel where surface water only flowed during heavy rains, allowing surface water to enter the lake rapidly during heavy rainfall. In Yamanakako Village, Typhoon No. 9, which occurred in September 2010, caused sediment to flow out of the Yoshimasawa River, partially destroying some accommodation lodges and flooding a condominium and parking lot [33]. An erosion control weir was constructed as a disaster-related emergency erosion control project, disaster recovery projects were adopted for a village road downstream of the Yoshimasawa River, and other forms of construction were also carried out. These structures allowed precipitation to flow quickly into Lake Yamanaka, increasing the six-day maximum flood volume for the same amount of six-day maximum rainfall.

In contrast, the conversion of the former grassland watershed to forest (Figure 2) and the continued growth of trees in the lake's watershed would have reduced the amount of rainfall contributing to lake level rise due to increased canopy interception. The increased water-holding capacity of the forest soils may also have enhanced the ability of the watershed's soils to retain precipitation beyond six days. These "decreasing factors" may have lowered the maximum six-day increase for the same maximum six-day precipitation. However, our results suggest that these effects were overridden by the aforementioned "increasing factors", which increased the six-day maximum water level rise for the same maximum six-day rainfall.

We focused on five events with extremely high water levels, of which two occurred in the first period and three in the second period. Of these, the rise that led to the water level of 4.34 m on 19 August 1983, was caused by only one precipitation event, while the other four were caused by multiple intermittent precipitation events. In some cases, particularly large increases in precipitation were caused by a single event that exceeded 200 mm of precipitation for six days. In other cases, high water levels were caused by multiple such events occurring in succession.

Figure 7. Reclamation of the lake's shoreline and parking lot construction. Photo taken on 19 June 2022, when the lake water level was 2.26 m.

Figures 5 and 6 show that the two flood events occurring in the first period featured large precipitation events (625.5 mm over 10 days and 860.8 mm over 9 days). Both were followed by a period of relatively light rainfall during which the water level continued to rise. In contrast, in two of the three floods in the second period (except for the 1983 flood mentioned above), a large amount of precipitation fell in the early stages, followed by a period of relatively little rainfall, during which the water level continued to decline. This difference may correspond to the fact that the maximum six-day increase for the corresponding maximum six-day rainfall was smaller in the earlier period than in the later period, as shown in Figure 4.

In two of the three later events (apart from the one in 1983), both a reduction in inflow and an increase in outflow may have contributed to the rapid decrease observed in the water level during the low rainfall period immediately after major precipitation events. As groundwater is the primary source of inflow to Lake Yamanaka during periods of low rainfall, this reduced inflow can be attributed to an increase in the allocation rate of precipitation to surface water, with an accompanying decrease in its corresponding allocation rate to groundwater. Alternatively, the observed trend could correspond to an increase in groundwater flow rate. Assuming that there was no long-term change in the characteristics of groundwater discharge from Lake Yamanaka toward Oshino [16], the observed increase in discharge could have arisen from one or both of two possible causes: an increase in flood flow from the Katsura River (which is the only outlet river) and an increase in anthropogenic discharge, i.e., an increase in water withdrawal by water users, such as power generation and agricultural water use.

5. Conclusions

In this study, we collected data on water level observations for 93 years, from 1928 to 2020, and analyzed long-term fluctuations in water-level rises and precipitation in Lake

Yamanaka during periods of heavy rainfall. We found that the six-day maximum rise in water level for the same six-day maximum rainfall was significantly greater in the later period than in the earlier period. Moreover, this difference increased with increasing precipitation. In some cases, we found that particularly large increases in precipitation were caused by a single event, whereas in other cases, these increases were caused by multiple events occurring in succession. We posit two possible reasons for the observed difference between the two periods: first, lakeside road-building and shoreline reclamation projects both progressed from the earlier to the later period. Second, river and erosion-control structures were introduced into a channel where surface water flowed only during heavy rainfall. This allowed surface water to enter the lake rapidly during heavy rainfall.

Author Contributions: Conceptualization, K.K.; methodology, K.K.; software, K.K.; validation, K.K.; formal analysis, K.K.; investigation, K.K. and H.S.; resources, K.K. and H.S.; data curation, K.K.; writing—original draft preparation, K.K.; writing—review and editing, K.K. and H.S.; visualization, K.K.; supervision, K.K. and H.S.; project administration, K.K. and H.S.; and funding acquisition, K.K. and H.S. All authors have read and agreed to the published version of the manuscript.

Funding: This research received no external funding. It was supported by internal funding provided by the Meteorological Data Analysis Research Group at the University of Tokyo Forests.

Data Availability Statement: No new data were created or analyzed in this study. Data sharing is not applicable to this article.

Acknowledgments: This paper is a part of the results of the Research Promotion Committee Hydrological Water Quality Data Analysis Study Group established in the University of Tokyo Experimental Forest. I would like to express my profound gratitude to the Yamanashi Prefectural Library, the Yamanashi Prefectural River Improvement and Management Division, the Yamanashi Prefectural Citizens Information Center, and the University of Tokyo Library of Agricultural and Life Sciences for their generous assistance in collecting and viewing past observation data recorded in paper materials. At least 44 years of the Yamanaka Lake water-level data used in this study, from 1957 to 2000, were observed by the late Tadamichi Sugiura. I would like to express my heartfelt gratitude to the late Tadamichi Sugiura, who served as principal of Yamanaka Elementary School, Yamanaka Lake Middle School, and chairman of the Yamanaka Lake Village Cultural Properties Council, and who was awarded the Minister of Construction Prize for his many years of service as a Lake Yamanaka hydrological observer.

Conflicts of Interest: The authors declare no conflict of interest.

References

1. Leira, M.; Cantonati, M. Effects of water-level fluctuations on lakes: An annotated bibliography. *Hydrobiologia* **2008**, *613*, 171–184. (In Japanese) [CrossRef]
2. Calder, I.R. *Blue Revolution: Integrated Land and Water Resources Management (Revised Edition)*; Routledge: London, UK, 2005; p. 376.
3. Yamamoto, M.; Uchiyama, T. Precipitation and Water Level Fluctuations in Lake Kawaguchi over the Past 100 Years. *Tsuru Univ. Rev.* **2018**, *88*, 131–141. (In Japanese)
4. Fujiyoshida Museum of Local History. *MARUBI*; Fujiyoshida Museum of Local History: Fujiyoshida, Japan, 1997; p. 8. (In Japanese)
5. Yoneyama, F. Formation and changes of irrigation systems in Oshino village, Yamanashi Prefecture, Japan. *Ecumene Res.* **2010**, *1*, 3–21. (In Japanese)
6. Kanbara, S. *Geology and Hydrology of Mount Fuji*; Hakushinkan: Tokyo, Japan, 1929; p. 481. (In Japanese)
7. Tanaka, M.; Kasai, K.; Tsutsumi, M.; Sasamoto, J.; Nagata, T. Results of water quality survey and water balance at the time of abnormal rise of water level in the Fuji Five Lakes in 1982. *Annu. Rep. Yamanashi Prefect. Inst. Public Health Pollut. Res.* **1982**, *26*, 22–28. (In Japanese)
8. Kanno, T.; Ishii, T.; Kuroda, K. Study on groundwater flow in the northern foot area of Mt. Fuji and water level changes of Lake Kawaguchi, based on the hydrogeological structure. *J. Jpn. Groundw. Assoc.* **1986**, *28*, 25–32, (In Japanese, with English abstract).
9. Tsutsumi, M. An essay on the runoff analysis of the Sagami River basin in Yamanashi Prefecture. *Annu. Rep. Yamanashi Prefect. Inst. Public Health Pollut. Res.* **1987**, *31*, 34–38. (In Japanese)
10. Horiuchi, S.; Lee, Y.K.; Watanabe, M.; Fujita, E. Some limnological characteristics of Fuji. *Bull. Inst. Nat. Sci. Nihon Univ.* **1992**, *27*, 45–56. (In Japanese)
11. Susuki, E.; Taba, J. A study on the mechanism of exchange between lake waters of Fuji-Goko (Five Lakes) and the groundwater around the lakes. *Bull. Inst. Nat. Sci. Nihon Univ.* **1994**, *29*, 43–60. (In Japanese)

12. Takeuchi, K.; Kiriishi, F.; Imamura, H. The water balance analysis of the Five Lakes of Mt. Fuji. *J. Hydraul. Eng.* **1995**, *39*, 31–36. [CrossRef]
13. Yoshida, M.; Hirabayashi, K.; Yoshizawa, K. A study on the changes of water level and quantity of rainfall in Lake Kawaguchi. *Bull. Yamanashi Women's Coll.* **1998**, *31*, 79–84, (In Japanese, with English abstract).
14. Hamada, H.; Kitagawa, Y. Investigation of seasonal change of water temperature and water quality and water balance of Lake Yamanaka. *Bull. Fac. Educ. Chiba Univ.* **2010**, *58*, 371–380, (In Japanese, with English abstract).
15. Ogata, M.; Kobayashi, H. Chronological change of water level of Lake Kawaguchi and that of ground water level around southeast area of Lake Kawaguchi. *Compr. Res. Organ. Sci. Technol. Yamanashi Prefect. Gov.* **2015**, *10*, 91–94, (In Japanese, with English abstract).
16. Nakamura, T.; Hasegawa, T.; Yamamoto, M.; Uchida, T.; Seko, Y.; Shimizu, G.; Yoneyama, Y.; Kazama, F. Source of water and nitrate in springs at the northern foot of Mt. Fuji and nitrate loading in the Katsuragawa River. *J. Geogr.* **2017**, *126*, 73–88, (In Japanese, with English abstract). [CrossRef]
17. Kuraji, K.; Saito, H.; Nishiyama, N.; Tsuji, R. Long-term trend of precipitation in the Fuji Iyashinomori Woodland Study Center. *Bull. Univ. Tokyo For.* **2021**, *145*, 1–18, (In Japanese, with English abstract).
18. The Imperial Japanese Land Survey Department. *1/50,000 Scale Topographic Map (Yamanakako)*; The Imperial Japanese Land Survey Department: Tokyo, Japan, 1989. (In Japanese)
19. Yamanakako Village. *Natural History of Yamanakako Village*; Yamanakako Village: Yamanakako, Japan, 2006. (In Japanese)
20. Yanai, K. *Water Utilization Survey at the Northern Foot of Mount Fuji and Its Data Analysis (Appendix) Water Level, Discharge and Rainfall Table of Fuji Five Lakes from 1927 to 1961*; Yamanashi Prefecture: Kofu, Japan, 1962; p. 136. (In Japanese)
21. Yamanashi Prefecture. *Fuji Five Lakes Water Level, Discharge, Rainfall and Observation Table*; Yamanashi Prefecture: Kofu, Japan, 1970; p. 55. (In Japanese)
22. Yamanashi Prefecture. *Annual Report of Rainfall and Water Level Observation of Yamanashi Prefecture (1955–1956)*; Yamanashi Prefecture: Kofu, Japan, 1958; p. 164. (In Japanese)
23. Yamanashi Prefecture. *Annual Report of Rainfall and Water Level Observation of Yamanashi Prefecture (1957–1958)*; Yamanashi Prefecture: Kofu, Japan, 1962; p. 172. (In Japanese)
24. Yamanashi Prefecture. *Annual Report of Rainfall and Water Level Observation of Yamanashi Prefecture (1963–1964)*; Yamanashi Prefecture: Kofu, Japan, 1966; p. 127. (In Japanese)
25. Yamanashi Prefecture. *Annual Report of Rainfall and Water Level Observation of Yamanashi Prefecture (1965–1966)*; Yamanashi Prefecture: Kofu, Japan, 1967; p. 129. (In Japanese)
26. Yamanashi Prefecture. *Annual Report of Rainfall and Water Level Observation of Yamanashi Prefecture (1967–1968)*; Yamanashi Prefecture: Kofu, Japan, 1969; p. 119. (In Japanese)
27. Yamanashi Prefecture. *Water Level Monthly Report, 1981–1996*; Non-Published Data (viewed on 20 October 2020, pursuant to Article 12, Paragraph 1 of the Yamanashi Prefecture Information Disclosure Ordinance). (In Japanese)
28. Historical Water Levels in Fuji Five Lakes. Available online: https://www.pref.yamanashi.jp/chisui/113_006.html (accessed on 23 November 2021).
29. Tokyo Electric Power Company. *Water Level and Outflow Reports, 1972–2020*; Non-Published Data disclosed by the Governor of the Yamanashi Prefecture on 8 June 2022.
30. National Land Agency. *Sagami River Area Main Water System Survey*; National Land Agency: Tokyo, Japan, 2000; p. 232. (In Japanese)
31. Okai, K.; Fukushima, H.; Nakai, Y. A Study of Village Structure in Yamanaka District, Yamanakako Village, Minamitsuru-gun, Yamanashi Prefecture—Focusing on the Transformation of the Vertical Road, Landscape and Design Research Proceedings. *Landsc. Des. Res. Proc.* **2015**, *11*, 39–46. (In Japanese)
32. Takahashi, T.; Fukushima, H.; Nakai, Y. Transformation of Lakeside Landscape in Yamanakako Village—Focusing on Land Ownership and Livelihood, Landscape and Design Research Proceedings. *Landsc. Des. Res. Proc.* **2014**, *10*, 126–131. (In Japanese)
33. Takamura, T. Landslide disaster caused by "Typhoon No. 9" in 2010. *Sabo Chisui* **2012**, *44*, 40–43. (In Japanese)

 *water*

Article

Evaluation of Paired Watershed Runoff Relationships since Recovery from a Major Hurricane on a Coastal Forest—A Basis for Examining Effects of *Pinus palustris* Restoration on Water Yield

Devendra M. Amatya [1,*], Ssegane Herbert [2], Carl C. Trettin [1] and Mohammad Daud Hamidi [3]

1 Center for Forested Wetlands Research, USDA Forest Service, 3734 Highway 402, Cordesville, SC 29434, USA; carl.c.trettin@usda.gov
2 Oshkosh Corporation, 1917 Four Wheel Drive, Oshkosh, WI 54902, USA; hssegane@oshkoshcorp.com
3 Department of Earth Sciences, Durham University, Lower Mount Joy, South Rd., Durham DH1 3LE, UK; mohammad.d.hamidi@durham.ac.uk
* Correspondence: devendra.m.amatya@usda.gov

Abstract: The objective of this study was to test pre-treatment hydrologic calibration relationships between paired headwater watersheds (WS77 (treatment) and WS80 (control)) and explain the difference in flow, compared to earlier published data, using daily rainfall, runoff, and a water table measured during 2011–2019 in the Santee Experimental Forest in coastal South Carolina, USA. Mean monthly runoff difference between WS80 and WS77 of −6.80 mm for 2011–2019, excluding October 2015 with an extreme flow event, did not differ significantly from −8.57 mm ($p = 0.27$) for the 1969–1978 period or from −3.89 mm for 2004–2011, the post-Hurricane Hugo (1989) recovery period. Both the mean annual runoff coefficient and monthly runoff were non-significantly higher for WS77 than for WS80. The insignificant higher runoff by chance was attributed to WS77's three times smaller surface storage and higher hypsometrical integral than those of WS80, but not to rainfall. The 2011–2019 geometric mean regression-based monthly runoff calibration relationship, excluding the October 2015 runoff, did not differ from the relationship for the post-Hugo recovery period, indicating complete recovery of the forest stand by 2011. The 2011–2019 pre-treatment regression relationship, which was not affected by periodic prescribed burning on WS77, was significant and predictable, providing a basis for quantifying longleaf pine restoration effects on runoff later in the future. However, the relationship will have to be used cautiously when extrapolating for extremely large flow events that exceed its flow bounds.

Keywords: rainfall; runoff coefficient; water table; surface storage; soil water storage; evapotranspiration; calibration regression

Citation: Amatya, D.M.; Herbert, S.; Trettin, C.C.; Hamidi, M.D. Evaluation of Paired Watershed Runoff Relationships since Recovery from a Major Hurricane on a Coastal Forest—A Basis for Examining Effects of *Pinus palustris* Restoration on Water Yield. *Water* **2021**, *13*, 3121. https://doi.org/10.3390/w13213121

Academic Editor: Koichiro Kuraji

Received: 13 September 2021
Accepted: 1 November 2021
Published: 5 November 2021

Publisher's Note: MDPI stays neutral with regard to jurisdictional claims in published maps and institutional affiliations.

1. Introduction

Restoration of longleaf pine (LLP) (*Pinus palustris*) ecosystems is a public land management objective throughout the southeastern United States, and it is a principal goal in the Forest Plan for the Francis Marion National Forest in South Carolina, USA. While there have been numerous plot or stand-scale studies of LLP ecology, silviculture, and ecosystem services [1], there are uncertainties regarding the watershed-scale runoff effects of reestablishing longleaf pine communities due to the spatial heterogeneity of soil conditions, microtopography, slope, and understory vegetation, all of which affect soil water storage. In contrast to loblolly pine (*Pinus taeda* L.) (LP) stands managed for timber production, LLP stands managed for the open canopy with frequent prescribed fire have a much lower stocking, a longer period of open canopy, a sparse mid-story, and an understory generally dominated by grasses and sedges, potentially influencing soil moisture and evapotranspiration (ET) [2]. As a result of these differences in stand structure and composition, it may be expected that LLP stands will exhibit less leaf area, less interception loss and transpiration, more infiltration of rainfall recharging groundwater, and increasing runoff than stands managed for timber production, especially LP stands where fire is excluded.

Runoff generation in coastal watersheds with shallow water table (WT) (<2–3 m deep) soils with variable permeability and infiltration rates is dominated by saturation excess flow [3–6]. The runoff process is complicated by interactions of forest management and extreme events [7–11]. The near-surface or shallow WT, a surrogate of soil water storage regulated by ET [12–14], drives most streamflow (as shallow surface runoff and drainage) in these shallow coastal systems [15]. Furthermore, microtopography influencing both surface and subsurface storage [16–20], (dis)connectivity [21], and drainage network pathways [7,22] have been shown to be important factors affecting runoff and its timing. Thus, a careful examination of such spatial catchment characteristics, including the above-canopy and below-canopy leaf areas that regulate soil moisture and ET, is fundamental for an accurate interpretation of water yield [23].

A paired watershed approach, in which two neighboring watersheds (one reference or control and one treatment) are monitored concurrently during calibration (pre-treatment) and post-treatment periods [24–26], has been used extensively to assess the effects of water management and silvicultural practices on hydrologic variables (water yield, peak flow rate, ET, and the water table) and ecosystem services [27–30]. The control watershed accounts for year-to-year or seasonal climate variations and management practices and remains the same during the treatment period [31]. The basis for the paired watershed approach is that there is a significant and quantifiable relationship between the two watersheds that provides a basis for comparing whether a treatment alters that relationship. This approach has been used primarily on first-order headwater watersheds [28,32], although its applicability for predicting effects of flood events on larger systems has been challenged [33].

The aim of this study was to affirm a current pre-treatment (baseline) flow relationship, between a pair of headwater watersheds reported earlier [34,35] and discussed below, that is significant and predictable. The planned treatment will test the hypothesis that watershed-scale restoration to mature LLP stands will increase water yield, primarily due to reduced ET from the forest.

2. Baseline Paired Hydrology Relationship

The approach for this study has been to use the first-order paired watersheds (WS77 for treatment and WS80 for control) on the Santee Experimental Forest (SEF), located within the Francis Marion National Forest (Figure 1). This study has a long record that supported the comparative analyses, including a statistically significant relationship between monthly flows established between control and treatment watersheds [34], to evaluate effects of partial prescribed burning on streamflow for 1976–1980 [35,36]. However, the authors [35,36] found no significant difference in streamflow between the watersheds after partial prescribed burning. Monitoring of the paired watersheds, which was discontinued in early 1982, was restarted in 1990, soon after Hurricane Hugo (1989) significantly (>80%) damaged the forest canopy in the region [37]. Richter [36] found that the average annual streamflows from WS77 and WS80 were 28% and 20%, respectively, of precipitation. Richter also analyzed four possible explanations for this difference in water yield: (a) differences in deep seepage losses, (b) difference in vegetation, which influences ET and interception, (c) watershed boundaries, and (d) calibration errors in weir rating tables.

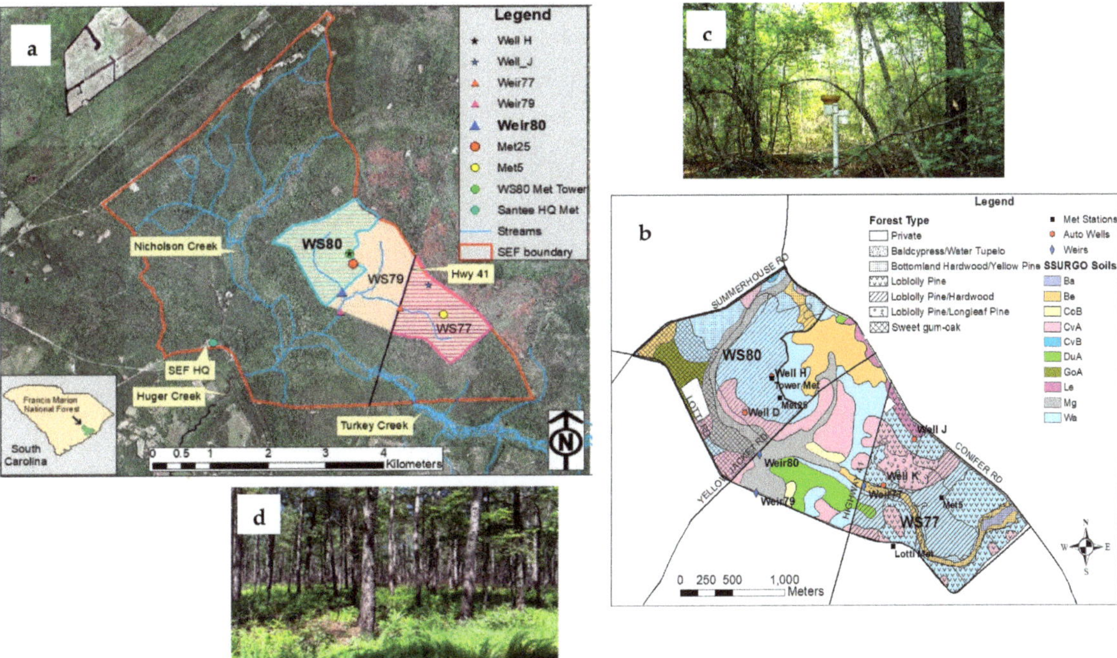

Figure 1. (**a**) Location map of the paired watersheds (WS77—treatment and WS80—control) and (**b**) their experimental layouts with existing monitoring stations, SSURGO soil types, and forest land cover types of both watersheds, and the forest stands of (**c**) WS80 and (**d**) WS77 within the Santee Experimental Forest (SEF) at Francis Marion National Forest, SC.

Richter [36] suggested negligible deep seepage losses for these poorly drained soils and found no evidence of weir leakage on either watershed. Based upon seasonal flows and vegetation composition analysis, the author also argued that differences in water yield cannot be explained by vegetational differences. His analysis also ruled out watershed boundary effects in these low-gradient systems, in which the watershed drainage areas are bounded by the elevated roads built with well compacted soils, minimizing any possible lateral seepage, except for the northeast corner of WS80, which is a watershed divide. However, Richter [36] suggested that because of the consistency in annual ET (rainfall–runoff) and predictability of runoff measured on WS77, differences were attributed possibly to WS80 runoff estimates, particularly for high flows. Nonetheless, the author also suggested a need for calibration of both stream gauges. In a long-term paired watershed study on grasslands in Uruguay, Chescheir et al. [38] also found similar inherent differences between the paired watersheds for the pre-treatment period, with higher runoff from the treatment than from the control, which was attributed to a higher baseflow from the treatment watershed, likely due to lower ET from its shallow soils or groundwater inflow from outside the watershed.

Interestingly, the paired pre-Hugo flow relationship (WS77 > WS80), reported by Richter [36] for 1969–1978, reversed (WS80 > WS77) four years after Hurricane Hugo's 1989 arrival for 10 years (1994–2003) before the relationship recovered to the pre-Hugo direction (WS77 > WS80) by 2004, as did the forest stands [24] (Figure S1). Jayakaran et al. [24], who analyzed pre- and post-Hugo monthly data over 2011, suggested that lowered vegetative water use likely increased outflows in both watersheds because trees were lost to the hurricane. However, WS77 recovered to its pre-hurricane runoff level by 1993, having an abundance of pine seedlings and saplings there compared to WS80, which recovered its flow pattern in 2003. Jayakaran et al. [24] noted that it seems likely that high rainfall in 2003 would have saturated soils in both watersheds, and 2004's drought-like conditions would have substantially drawn down the water table. The heavy rainfall and subsequent dry conditions in the following year might have somewhat confounded the exact timing and mechanisms responsible for the return of flow relationships to those pre-Hugo. Therefore, the authors cautioned that whether the 2003 wet and 2004 dry years accelerated the recovery to the pre-Hugo (WS77 > WS80) direction is an area for further study. Although the relationship was restored, the difference (WS77 > WS80) in the average magnitude in 2004 through to 2011 was 2–3 times smaller than before Hugo (Figure S1). Since these authors did not find any significant difference in the monthly rainfall totals between the two watersheds for any of the three periods, they attributed the relative periodical magnitude differences in paired streamflow to relative changes in ET dynamics between the watersheds.

Therefore, the objective of this study was to re-evaluate and re-establish the paired calibration relationship of watersheds, recovered since the 1989 hurricane, using climatic data for 2011–2019, which includes very large rainfall and dry events [39] (Table 1). This period was chosen because the stands reported by Jayakaran et al. [24] as being recovered by 2004, as stated above, were hypothesized to be fully recovered by 2011 (Figure S1) or 22 years after the hurricane. This hypothesis is consistent with studies reporting a recovery period of 7 to 25 years for the annual water yield in three forested watersheds in the northeastern US [8], as many as 20 years in the US and > 14 years in Japan [40], and 10 to 15 years for recovery of the water table and drainage outflow in managed pine forest watersheds in coastal North Carolina [26,41].

Table 1. Measured annual flow, with the average annual and standard deviation (StdDev), for paired watersheds WS77 and WS80 and the difference in flow between them for 2011–2019, pre-Hurricane Hugo (1969–1978), and post-Hugo (2004–2011), reported by Jayakaran et al. [24] (See also Figure S1).

Pre-Hurricane Hugo Period				Post-Hugo Recovery Period				Fully Recovered Baseline Period			
Year	WS77	WS80	Difference	Year	WS77	WS80	Difference	Year	WS77	WS80	Difference
	mm	mm	mm		Mm	mm	mm		mm	mm	mm
1969	441.0	259.9	181.1	2004	89.1	72.5	16.6	2011	57.5	31.0	26.5
1970	350.6	251.0	99.6	2005	351.0	276.1	75.0	2012	55.7	28.0	27.7
1971	734.9	494.5	240.4	2006	177.3	149.9	27.4	2013	334.4	219.0	115.4
1972	227.1	174.0	53.2	2007	105.4	69.9	35.5	2014	293.1	199.0	94.1
1973	404.5	315.0	89.5	2008	456.8	317.8	139.0	2015	949.9	967.0	−17.1
1974	305.2	229.0	76.2	2009	352.2	262.8	89.5	2016	633.0	562.0	71.0
1975	366.3	283.1	83.2	2010	271.4	307.0	−35.6	2017	391.9	217.0	174.9
1976	416.4	291.5	124.9	2011	57.5	31.2	26.3	2018	474.1	361.0	113.1
1977	179.7	140.8	38.9					2019	333.6	201.0	132.6
1978	187.7	146.4	41.2								
Average	**361.3**	**258.5**	**102.8**		**232.6**	**185.9**	**46.7**		**391.5**	**309.4**	**82.1**
StdDev	**161.7**	**102.8**	**64.3**		**146.7**	**118.1**	**53.1**		**295.0**	**277.8**	**60.5**

In addition, the choice of the 2011–2019 period as a pre-treatment baseline reference was supported by its closer agreement of the computed average annual flow difference of 82.1 mm between the treatment and control watersheds, than the 46.7 mm for 2004–2011, with the pre-Hugo average difference of 102.8 mm (Table 1). Furthermore, the StdDev of the flow difference for the baseline was closer to the pre-Hugo period than that of the post-Hugo, indicating their similar intra-annual variability. A similar approach was reported by Oda et al. [40] for testing disturbance effects using a paired watershed approach.

Regarding choosing a stable and sufficient record length for a baseline calibration period, Ssegane et al. [42] found statistically significant pre-treatment calibration relationships using only 762 days and 608 days, respectively, for two treatment watersheds from 2009 to 2012 that included some disturbances. Similarly, Bren and Lane [32] found a rapid increase in the quality of calibration relationship as the record length increased up to 3 years, but no increase was found beyond that, for all temporal scales of flow. The authors suggested that 5 years were adequate for most purposes, consistent with Clausen and Spooner [31], and the main advantage of longer periods was lower mean errors.

It was hypothesized, therefore, that the nine-year (2011–2019) record period, covering years with very low (2012) and very high (2015) runoff (Table 1), should be adequate for obtaining a stable pre-treatment (baseline) calibration relationship that is significant and quantifiable for future applications in treatment evaluations. This model would be applied using the measured flow from the control watershed to estimate expected flows for the WS77 treatment, assuming no disturbance, starting in 2020 when the harvesting and thinning treatments began for longleaf restoration. Next, the expected flow from the treatment watershed would be compared with actual measured flow. Deviations of the treated watershed's measured flow from expected values were considered to represent treatment effects if the deviations fell outside specified confidence intervals (95%) placed around the calibration regression line. In addition, the treatment regressions would also be evaluated against the pre-treatment baseline.

Various potential reasons, including rainfall and storm events, and understory prescribed burning implemented in 2013, 2016, and 2018 on the WS77, as shown by Richter et al. [35] and discussed above, were evaluated for the inherent differences in paired watershed flows. This study is novel in that no other studies, to the authors' knowledge, have reassessed the paired watershed calibration relationship after the reported recovery of forests following a major natural disturbance that altered the pre-disturbance flow regime between the watersheds.

Objective 1: Evaluate the annual rainfall, runoff coefficient, and ET (as the difference between rainfall and flow) in the paired watersheds for the pre-treatment baseline period and compare them with the 2004–2011 post-recovery period.

Hypothesis 1. *There will be no significant difference in the pre-treatment mean annual runoff coefficient (ROC) or in mean monthly rainfall between the paired watersheds, consistent with the post-recovery period, despite the effects of relatively very wet and dry years (defined, respectively, as 30% above and below the long-term average rainfall).*

Objective 2: Evaluate the difference in pre-treatment monthly runoff between the paired watersheds compared to the post-recovery period and the possible reasons for difference, if any, building upon earlier studies [24,36].

Hypothesis 2. *The difference in the monthly runoff response between the paired watersheds will be similar (WS77 > WS80) to that in the post-recovery period.*

Objective 3: Evaluate the pre-treatment monthly runoff calibration relationship between the watersheds compared to the post-recovery period.

Hypothesis 3. *The paired pre-treatment monthly runoff calibration relationship will not be different from the relationship for the post-recovery period and will be significant and quantifiable, with predictive capability.*

Hypothesis 4. *The periodic prescribed burning treatments in the pre-treatment period will not affect the change in paired runoff relationship between the watersheds.*

3. Materials and Methods

3.1. Site Description

The paired watersheds (WS77 and WS80) drain into Fox Gulley Creek and further down to Turkey Creek, a tributary of Huger Creek. These are parts of the headwaters of Huger Creek, a fourth-order stream and a major tributary of the East Branch of Cooper River, which drains into Charleston Harbor (Figure 1a). Basic characteristics of the watersheds are given in Table 2. The original WS80 watershed area was 206 ha when it was installed in 1968 [35], but on 6 November 2001, a culvert was installed to drain its north-eastern portion, thus reducing its area by 46 ha to 160 ha (Table 2). The vegetation in WS80 is a mixed hardwood-pine stand, regenerated since Hugo (Figure 1b,c). The vegetation in WS77 is dominated by loblolly pine (Figure 1b,d), planted for silvicultural research in the late 1970s. Soils in the watersheds are poorly to moderately well-drained sandy clay loam overlaying clay, typified by the Wahee and Craven soil series in the uplands and the Megget and Betheera soils in the riparian zones (Figure 1b). The control watershed is 48% wetlands compared to only 11% in WS77, as estimated from recent National Wetland Inventory data (Table 1). The mean surface depressional storage reported by Amoah et al. [17] for the WS80 watershed was nine times higher than for the WS77 (Table 1). The climate is warm-humid temperate, with an average daily temperature of 17.8 °C and annual rainfall of about 1370 mm [43]. Chronological activities of both watersheds are given in Table S1, and more details are described elsewhere [15,17,43].

Table 2. General characteristics of the paired watersheds.

Parameter	WS77 (Treatment)	WS80 (Control)
Location	33.14° N, 79.77° W	33.15° N, 79.8° W
Elevation (m a.m.s.l.)	4.9–10.4	3.5–10
Watershed size (ha)	155	206 until 2001; 160
Main channel length (km)	1.26	1.38
Drainage density (m^{-1}) *	0.0037	0.0023
Wetland area, %	11	48
Mean depressional storage capacity, mm	10 (±0.5)	93 (±2.7)

* Total stream length calculated using LiDAR based DEM analysis.

3.2. *Hydro-Meteorologic Monitoring*

Beginning in 2003, digital records of precipitation were collected using automatic tipping bucket gauges backed up by a manual gauge at the Met5 station in WS77 and at the Met25 station in WS80. Data from nearby gauges (Figure 1a,b) were also used to fill gaps [44,45]. Digital measurements of stage, also beginning in 2003, were recorded every 10 min by the Teledyne ISCO flowmeters installed upstream of both the WS77 and WS80 watershed weir outlet gauging stations (Figure 1a,b). These digital stage data were used with established rating curves for compound V-notch weirs for estimating streamflow rates [39,44,45]. Details of stream gauges, stage measurements, and estimates of flow rates and the quality control are given in [44,45]. The flow data have been recently verified and are of a high quality for the rating range they were developed for. Daily average weather parameters obtained from weather sensors installed on a 27-m tall tower (above the forest canopy) in WS80 in 2010 (Figure 1a,b) were used to estimate daily Penman–Monteith (P-M) [46] based potential evapotranspiration (PET) for the forest conditions following Amatya et al. [47] (Table 3). A 3 m weather station installed on open grass at the nearby SEF Headquarters (SHQ) (Figure 1a) [48] provided data to fill in some missing values for a few short periods [45].

Table 3. Measured annual rainfall, flow, ET (rainfall–flow), and ROC (runoff coefficient = flow/rainfall) and estimated PET for the WS77 and WS80 watersheds for 2011–2019.

Year	WS80	WS80	WS80	WS80	WS77	WS77	WS77	WS77	Forest
	Rainfall, mm	Flow, mm	ROC	ET mm	Rainfall, mm	Flow, mm	ROC	ET, mm	P-M PET, mm
2011	934	31	0.03	903	977	58	0.06	919	1351
2012	1174	28	0.02	1146	1148	55	0.05	1092	1239
2013	1433	220	0.15	1214	1502	334	0.22	1168	1017
2014	1375	199	0.15	1176	1340	293	0.22	1047	1123
2015	2171	968	0.45	1204	2146	950	0.44	1196	1098
2016	1743	562	0.32	1181	1709	633	0.37	1076	1197
2017	1443	217	0.15	1226	1555	392	0.25	1163	1177
2018	1633	361	0.22	1272	1661	474	0.29	1187	1146
2019	1381	201	0.15	1180	1429	334	0.23	1095	1200
Average	**1476**	**309**	**0.19**	**1167**	**1496**	**391**	**0.24**	**1105**	**1172**
Std Dev	**351.1**	**295.0**	**0.13**	**105.4**	**338.7**	**277.8**	**0.13**	**87.2**	**93.9**
COV	**0.24**	**0.95**	**0.71**	**0.09**	**0.23**	**0.71**	**0.54**	**0.08**	**0.08**

The WT in upland well H and riparian well D in WS80 and upland well J in WS77 have been measured hourly since 2004 using pressure transducers with a datalogger (Figure 1a,b). Well K near the riparian area in WS77 was installed in 2018 (Figure 1b). All wells were approximately 2.8 to 3 m deep. Plots of the daily water table depths of well J and well H for 2011–2019 are presented in Figure S2. Measurements of the leaf area index (LAI) were conducted every 2 or 3 weeks in 2019–2020 ($n = 9$) at three locations proposed for LLP treatment in WS77 (Figure 1b). The average LAI measured during 2008–2009 ($n = 40$) and reported by Dai et al. [43] for WS80 were used, assuming the LAI of the fully recovered stands on this control watershed remained unchanged. A comparison of WS77 LAI with WS80 LAI is shown in Figure S3. Details of all hydro-meteorologic measurements, including data quality control, can be found elsewhere [15,17,43,44,49].

3.3. Data and Statistical Analyses

Measured daily rainfall, streamflow or runoff (watershed area-based depth), and PET, estimated from the daily weather data for the 2011–2019 period, were used to obtain monthly and annual totals and to compute the annual rainfall normalized runoff coefficient (ROC). The number of daily rain events >25 mm in each year was also logged. Flow data were lost for both watersheds for Hurricane Joaquin (3–4 October 2015), while only WS77 lost some data for Hurricane Matthew (8 October 2016), because the measured stage exceeded the rating curve range of each watershed. The exceptionally high flow of October 4, 2015 was assumed to be an outlier, as discussed in the Results section below, so that month was excluded from both watersheds in the comparative monthly analysis. Data for WS77 for October 8, 2016 were constructed by assuming the maximum rating curve flow value for less than 9 h of the day when the measured stage exceeded the rating curve limit. Integration of all 10-min interval flow rates, including the peak rates for this day, yielded 242.2 mm of flow as a response to 204 mm rain on that day, preceded by 90.4 mm rain the day before with only 1.7 mm flow, indicating that most of the two-day rain contributed to this single day large event. This daily value of 242.2 mm, which was lower than the 187.6 mm observed for WS80, was used in the analyses. Daily flow data were used to derive the daily flow duration curves to identify differences in flow magnitudes, frequencies, and duration of daily runoff between the watersheds. Daily WT depths were obtained by integrating hourly data.

Monthly rainfall, as well as annual runoff and ROC, for both watersheds were statistically analyzed to test Hypothesis 1. Measured monthly runoff data were used to (a) compare the mean monthly difference in flow between the paired watersheds against the post-recovery period to test Hypothesis 2 and (b) develop a baseline calibration regression of the monthly flow between the paired watersheds to test Hypothesis 3. Finally, a MOSUM (moving sums of recursive residuals) approach was used to detect changes in the paired flow regime, if any, and also in the paired calibration relationship, due to the prescribed burning, to test Hypothesis 4.

The Shapiro–Wilk normality test [50] showed a non-normal distribution ($p < 0.001$) of monthly runoff. Therefore, the nonparametric Wilcoxon signed-rank test was used to assess the significance of differences in mean monthly runoff between the two watersheds measured for 108 months or nine (2011–2019) years. An ordinary least squares regression (OLS) was used to develop a calibration equation between the control and treatment watersheds and its significance test [51]. However, since the Durbin–Watson (DW) test [50] showed a positive autocorrelation of the monthly runoff of both watersheds (DW_WS77 = 0.054, $p < 0.0001$; DW_WS80 = 0.029, $p < 0.0001$), regression relationships using an OLS versus geometric mean (GM) regression were compared. Based on Ssegane et al. [42], the *ts* and *lmodel2* R statistical packages [52] were used to examine if the OLS was significantly different from the GM. The *ts* function is used to create time-series objects. These are vectors or matrices with a class of "*ts*" (and additional attributes), which represent data which have been sampled at equispaced points in time. In the matrix case, each column of the matrix data is assumed to contain a single (univariate) time series. Similarly, the *lmodel2* function computes model II simple linear regression using the following methods: ordinary least squares (OLS), major axis (MA), standard major axis (SMA), and reduced major axis (RMA) of the GM. The model accepts only one response and one explanatory variable. Model II regression should be used when the two variables in the regression equation are random, i.e., not controlled by the researcher. GM regression is a resampling technique that accounts for autocorrelation in the time series by resampling the original data in pre-determined blocks 1000 times to estimate regression coefficients. GM, also known as the reduced major axis (RMA) regression, is suited for paired watershed analysis, because it assumes errors are associated with both dependent (treatment watershed) and independent (control watershed) variables [53]. The coefficient of determination (R^2), Nash–Sutcliffe efficiency (NSE), and root mean squared error (RMSE) were used to evaluate the strength and significance of the regression. For R^2 ($0 \leq R^2 \leq 1.0$) and for NSE ($-\infty \leq \text{NSE} \leq 1.0$), a value of 1.0 indicates an optimal model. All statistical significance tests for similarity with no difference were conducted for the $\alpha = 0.05$ level.

An OLS-based MOSUM (moving sums of recursive residuals) approach using monthly flow data was conducted to detect the change in flow behavior, if any, between the watersheds due to prescribed burning of the WS77 watershed and potential effects on the monthly flow regression relationship. The null hypothesis (Hypothesis 4) tested by the MOSUM is that regression coefficients of a linear model are constant over time; the alternative hypothesis is that the coefficients change over time due to external factors [24,42].

A morphometric analysis, among other factors, was also used for explaining possible reasons for inherent differences in streamflow between the paired watersheds, with a higher, but insignificant, flow from the treatment than from the control watershed since the historic study [36]. A morphologic analysis was conducted by deriving the hypsometric curves and indices [54,55] to examine the effects of land morphologic characteristics on runoff generation for the paired watersheds. A system for automated geoscientific analysis (SAGA)-GIS [56] and LiDAR-based DEM were used to generate the hypsometric curves of WS77 and WS80. The hypsometric integral (HI), skewness (skew), and kurtosis of the hypsometric curves were computed using general formulations by Harlin [57] and Pérez-Peña, Azañón, and Azor [58].

4. Results

4.1. Annual Rainfall, Runoff, Runoff Coefficient (ROC), and ET

The first year (2011) of the pre-treatment (baseline) period was relatively dry, with rainfall below 32% of the long-term average (1370 mm) [43], and 2015 was relatively wet, with 58% above average rainfall. The nine-year average baseline period rainfall in WS77 was about 9% above the long-term average (Table 3). The nine-year baseline and the eight-year post-recovery periods yielded the highest and the lowest mean annual flow, respectively, for both watersheds (WS77 > WS80) (Table 1). An unusually high outflow in 2015 due to an extreme October event [59] might have caused the largest average flow in the baseline period. However, the mean annual ROC values, although almost consistently higher in WS77 (mean of 0.24) than in WS80 (0.19) (Table 3), were not statistically different ($p = 0.17$) between the pair and not different ($p > 0.80$) from those reported for the pre-Hugo period (1969–1978) (WS77 ROC = 0.25; WS80 ROC = 0.18) [36,41] and the 2004–2011 post-Hugo period ($p > 0.20$) (WS77 ROC = 0.18; WS80 ROC = 0.14). These results indicate consistency of the rainfall normalized flow (ROCs) between the paired watersheds in each of the three periods, supporting Hypothesis 1.

The annual ET, calculated as a difference between the annual rainfall and runoff, assuming no change in storage, varied from 903 mm in the relatively dry year of 2011 to as high as 1272 mm in the relatively wet year of 2018, with an average of 1167 mm for the control watershed (WS80) (Table 3). The annual ET was consistently lower in WS77, although not significantly so ($p = 0.07$), with a mean of only 1105 mm, primarily because it had a higher runoff than WS80. The mean annual ET for WS80 was very close to the estimated P-M PET, with no significant difference ($p = 0.46$), while the ET for WS77 was significantly lower ($p = 0.046$) than the PET, potentially indicating WS77's soil water limitations (Table 3). However, the annual ET increased insignificantly with rainfall, yielding a higher R^2 (0.71) for WS77 than for WS80 ($R^2 = 0.42$), indicating, again, WS77 as being more soil water limited than WS80.

4.2. Monthly Rainfall and Runoff between the Watersheds

A plot of the monthly rainfall averaged from each month of the 2009–2011 period is shown in Figure 2a for the paired watersheds with their standard deviations. Data show similar rainfall between the watersheds but higher values with larger variabilities in June-October, influenced by tropical storms/hurricanes, than in winter for both. For example, October 2015 yielded a very large rainfall of 667 mm for WS80 and 686 mm for WS77 due to an extreme two-day rainfall of nearly 500 mm on 3–4 October caused by Hurricane Joaquin [59], followed by the second large rainfall event of 296 mm for WS80 and 294 mm for WS77 in October of 2016 as a result of Hurricane Matthew (October 8). These data highly influenced the variability of rainfall in October (Figure 2a). However, the paired watershed monthly rainfall for this study period showed similar means (124.7 ± 93.4 mm for WS77 and 122.9 ± 90.8 mm for WS80) with no significant difference ($p = 0.89$), consistent with the earlier post-Hugo period reported by Jayakaran et al. [24]. However, the mean monthly rainfall for that period was insignificantly lower, by chance, than the baseline period. This was likely due to more than six times the average number of daily rain events > 25 mm during the 2011–2019 baseline period (not shown), compared to the earlier long-term (1946–2008) period reported by Dai et al. [43]. Moreover, the 2011–2019 period had eight rainfall events exceeding 100 mm in 24 h, induced by hurricanes and tropical depressions.

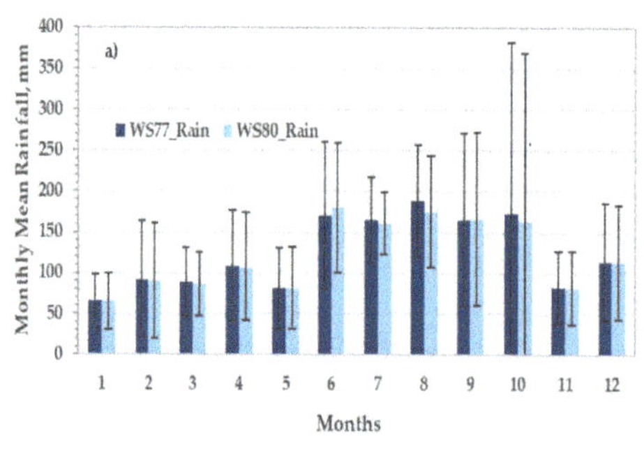

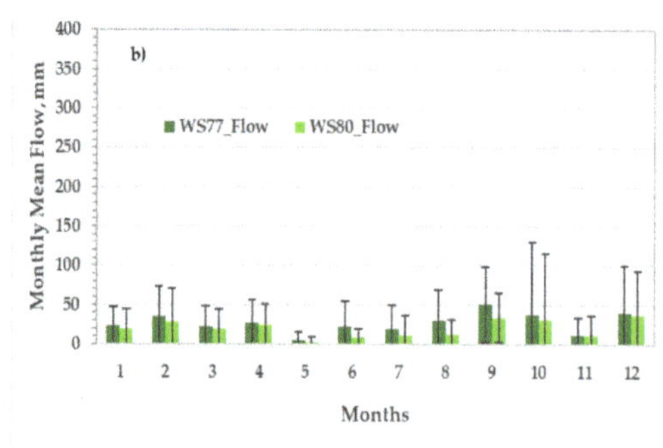

Figure 2. Distribution of (**a**) mean monthly rainfall and (**b**) mean monthly streamflow (runoff) of WS77 and WS80 for 2011–2019. Black vertical bars represent standard deviations.

The distribution of the monthly mean runoff (without October 2015) for the 2011–2019 baseline period depicted the lowest flow during May for both watersheds and the highest in September for WS77 (Figure 2b). The monthly mean runoff was consistently but not significantly ($p = 0.22$) higher for WS77 than WS80, with WS77 averaging 26.7 mm (0–210 mm) and WS80 averaging 20.4 mm (0–234 mm) without the October 2015 extreme event month (Figure S4a). Similarly, there was no difference in variance in the monthly flow between WS77 and WS80 ($p = 0.21$).

The effects of extreme rainfall on the water table influencing runoff for both watersheds during the October 2015 and 2016 hurricanes (Figure 2a,b) were similar due to fully saturated antecedent soil conditions (Figure S2). Monthly runoff responses to hurricanes in the following years, i.e., Irma (11–12 September 2017, with rainfall of 130 mm), Florence (14–15 September 2018, with rainfall of 110 mm), and Dorian (4–5 September 2019, with rainfall of 190 mm), were smaller than the two previous ones. The study period also experienced relatively drier months with more no-flow months for the control watershed than for WS77 (Figure 2b), with slightly more variability in monthly summer rainfall between the pair.

Data in Figure 3 for the monthly difference in runoff between the watersheds for 2011–2019 showed WS77 yielding somewhat higher flows (negative difference) than WS80, except for a few periods, consistent with the pre-Hugo (1969–1978) pattern [36] (Figure S1). The mean monthly runoff difference of −6.80 mm (±1.49 mm as standard error [SE]) between WS80 and WS77 for 2011–2019 (Figure 3) (without October 2015)) was 64% higher, although not significantly different ($p = 0.54$), than the −3.89 mm (±1.09 mm [SE]) obtained by Jayakaran et al. [24] for the 2004–2011 period, when the forest stands recovered. The difference for the 2011–2019 period was slightly, but not statistically ($p = 0.27$), lower than the pre-Hugo (1969–1978) mean of −8.57 mm (±1.65 mm [SE]) obtained by Richter [36], although the difference between the baseline and each of the pre- and post-Hugo periods was in the same direction. Thus, this result supports Hypothesis 2, confirming the validity of the 2011–2019 period data for pre-treatment calibration.

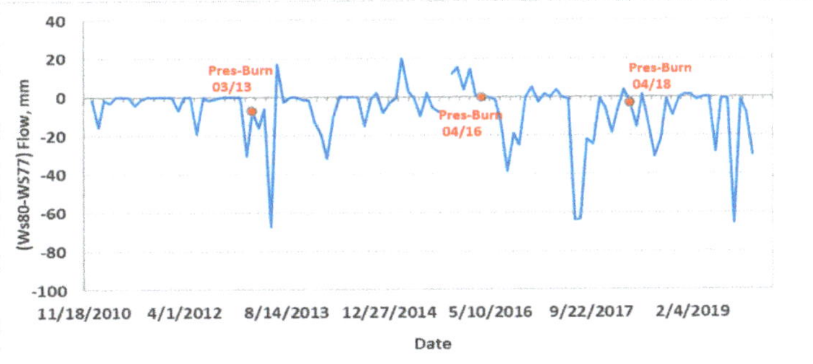

Figure 3. Difference in measured monthly flow (runoff) between the watersheds WS80 (control) and WS77 (treatment) for 2011–2019. The October 2015 data with an extreme event were omitted. Pres-Burn is prescribed burning.

4.3. Ordinary Least Squares Regression versus Geometric Mean Regression for Paired Monthly Runoff

The plot in Figure 4a compares the relationships of monthly runoff using ordinary least squares (OLS) and geometric mean (GM) regressions between the control watershed (WS80) and the treatment watershed (WS77) without October 2015 because of its extreme flow event. Results showed that the regression slope for the GM (WS77 = 1.15 × WS80 + 3.70; $R^2 = 0.87$) lay just at the border of the 95% confidence bounds (0.99–1.15) of the slope of the OLS regression (WS77 = 1.07 × WS80 + 5.39; $R^2 = 0.87$) (Figure 4a), and so it was barely statistically different ($p = 0.01$). Therefore, subsequent analysis focused only on the GM regression, which was also just within the bounds of the OLS slope for the recent post-Hugo regeneration (2004–2011) period (Figure 4b).

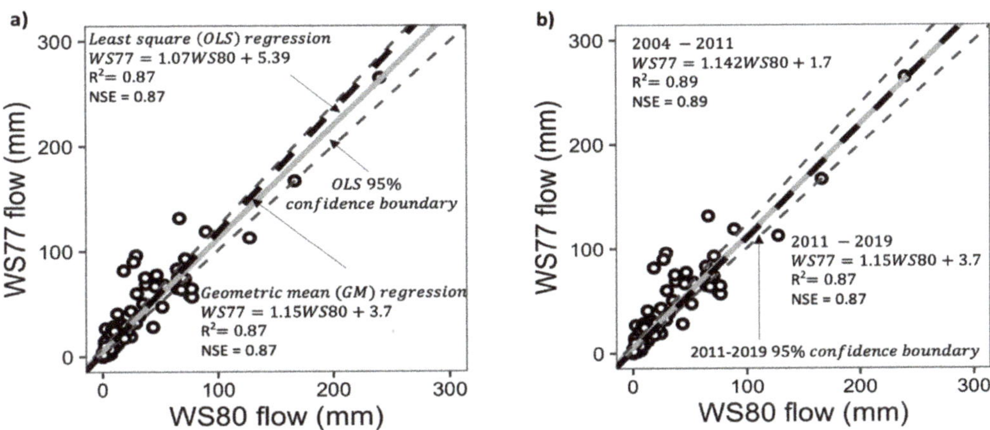

Figure 4. Comparison of regression lines for relationships between the monthly flow (runoff) for WS77 and WS80 using (**a**) OLS (solid line), with its 95% confidence intervals, and GM (dashed line) for 2011–2019, both without October 2015, and (**b**) GM (solid line) for the 2011–2019 period and GM for the 2004–2011 post-Hugo period, which was within the 95% confidence boundaries for the 2011–2019 GM mean.

4.4. Calibration Regression of Paired Monthly Flows

The plot in Figure 4a shows the regression relationships of measured monthly runoff between the control (WS80) and treatment (WS77) watersheds for the pre-treatment calibration period of 2011–2019, without October 2015 because of its extreme rainfall event.

The GM regression in Figure 4a yielded a significant monthly runoff relationship (WS77 = 1.15 × WS80 + 3.70; $R^2 = 0.87$) between the paired watersheds. Both the slope of 1.15 and an intercept of 3.7 mm were significant ($p < 0.0001$). This significance indicates that both the flow rate as well as the shift from the zero intercept could be attributed to the average difference in monthly flow, with WS77 insignificantly higher than WS80, as discussed above (Figure 3). The variability of flow around the 95% confidence limits of the regression line showed somewhat higher discharges for WS77 than for WS80 for most of the months for a flow of less than 100 mm. However, the regression with slope = 1.15 and intercept = 3.7 for the 2011–2019 pre-treatment period differed significantly from the 1969–1978 period with a slope = 1.43 and intercept of −0.68 [36] but not from the 2004–2011 post-Hugo period with 1.14 slope and 1.70 intercept [24] (Figure 4b). Thus, it both validates and invalidates Hypothesis 3. However, the relationship has to be cautiously interpreted as it includes one large event with flow >200 mm on October 8, 2016 (Hurricane Matthew), when a few hours of the unusable data during the peak flow of WS77 were estimated as explained above in Section 3.3, and thus it may have some uncertainties [60]. This event was included in this pre-treatment regression analysis because WS80 had good data, and frequencies of such large events are expected to increase in coming years [11]. For example, 17 out of 17 flow events >30 mm per day^{-1} occurred mostly because of hurricanes and tropical depressions within the 5 years since 2015 in this study (Figure 5).

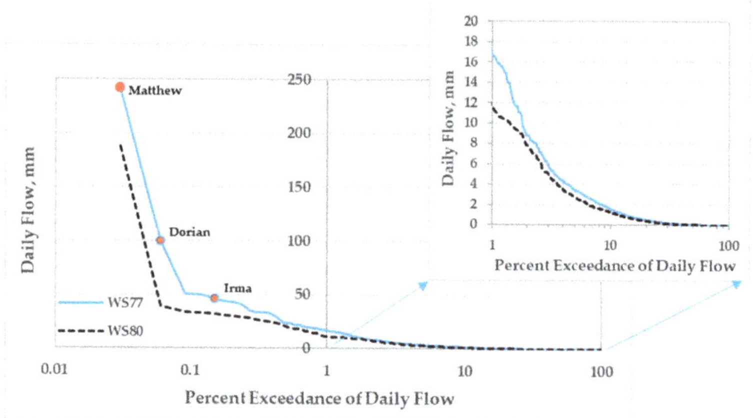

Figure 5. Daily flow duration curves for WS77 and WS80 for 2011–2019, without the three days (3–5 October) of an extreme event in 2015. Shown is an enlarged plot for a flow of less than 20 mm per day^{-1}.

4.5. Daily Flow (Runoff) Duration Curves

The watersheds' daily flow frequency duration curves for 2011–2020 are compared in Figure 5. Daily runoff was consistently lower from WS80 than from WS77, as in 1969–1978 [36]. The magnitude of daily runoff of 20 mm was exceeded 0.74% of the time for WS77 and 0.49% of the time for WS80 (Figure 5).

For 10% of the time, daily runoff exceeded 1.53 mm for WS77 and 1.23 mm for WS80. Similarly, WS77 had zero runoff 41.2% of the time, compared to 46.6% of the time for WS80 (Figure 5), which was somewhat similar to the 1969–1978 period, when WS77 had zero runoff 35.6% of the time and WS80 43.8% of the time (not shown). Furthermore, the difference in the percentage of zero runoff days between the watersheds was found to be smaller (5.5%) for the current period than the 8.2% for 1969–1978, although the two periods covered different numbers of days. Excluding the three days of the extreme event in October 2015, runoff exceeded nearly 100 mm day^{-1} for two hurricanes (Irma and Dorian, in September 2017 and September 2019, respectively), with a steeper slope for both, which indicates a flooding regime consistent with Amatya et al. [7]. Two other storm event days exceeded a 50 mm runoff, in August 2016 and July 2017, indicating an increasing pattern of large flow events during 2015–2019 for these low-gradient watersheds (Figure 5).

5. Discussion

The paired watersheds, besides being adjacent, are similar in many characteristics, including the area, topography, drainage, dominant soil types (Table 2), and mean Leaf Area Index (LAI) of the existing forest stands (Figure S3). Despite these similarities, the treatment watershed (WS77) yielded non-significantly, by chance, a higher monthly runoff than WS80 (Figure 3), except for a few periods with saturated soils (WT near the surface) in July 2013, March 2015, November–December 2015, and October 2016 (Figure S2), when WS80 runoff exceeded that of WS77 (Figure 3). The events in those periods resulted in large peak discharges of WS80, consistent with Harder et al. [15], who found exponentially increasing runoff as the WT in well H neared the surface or got ponded (Figure S2). The insignificantly higher monthly runoff (WS77 > WS80) (Figure 3), consistent with earlier studies [24,36], was also supported by the daily flow duration curves for 2011–2019 (Figure 5). Richter [36], in his study prior to Hugo (1989), ruled out the possible causes of groundwater seepage, drainage area, and vegetation effects for this difference in flow. However, he speculated some possible shortfalls in WS80 flow measurements for that period, particularly during high flow periods with a daily flow >5 mm. Since that historic study, however, several recent studies, including this one, have verified the flow measurements for WS80, as well as the drainage area and seepage [4,15,17,24,41,59,61]. Therefore, the flow measurements were consistent, except for the October 2015 hurricane [44], when flow rates exceeded the limits of the established rating curves for both watersheds and were estimated using theoretical equations for the WS80 outlet structure [59]. October 2015 was assumed to be an outlier and not used in the monthly analysis of this study.

Below, a few other potential factors were examined that may help explain the reasons why WS77 yielded slightly, but non-significantly, higher runoff than WS80. The mean annual rainfall, as a primary driver of runoff, was not significantly different ($p = 0.90$) between the watersheds, but both had more rainfall than the long-term average of 1370 mm reported by Dai et al. [43] for 1946–2008, meaning the study period was wetter. Total rainfall for each month was not significantly ($\alpha = 0.05$) different between the two watersheds, although WS77 experienced somewhat more (Table 3). This finding was consistent with Jayakaran et al. [24], who suggested that given the similarity of rainfall across the two watersheds, relative changes in streamflow are good indicators of relative changes in ET dynamics, as shown in Table 3. Data in Table 3 also show that WS77 is slightly more energy limited (ET/PET) than WS80, which is slightly more moisture limited (ET/Rain) than WS77 [7]. These observations are also supported by the annual rainfall–runoff relationships between the paired watersheds (Figure S4b), which yielded similar slopes, indicating similar rates of runoff response to rainfall. However, WS80 had a larger intercept, indicating more storage, potentially due to its greater ET than that of WS77 (Table 3). For example, although the annual ET through the baseline period was relatively constant, with a coefficient of variation (COV) <0.1 for both watersheds, WS80 yielded a somewhat higher mean value (1167 mm) than WS77 (1105 mm). These results, including the ROCs, are consistent with Boulet et al. [62], who also found such a hydrological difference between two paired Mediterranean headwater catchments with dissimilar land covers (*Pinis pinaster* and *Eucalyptus globulus*). In addition, the watersheds in this study differed in three important land use and management aspects, which were not addressed in earlier studies and are discussed below.

First, historic land use differed between WS80 and WS77. The lower reaches of WS80 were used for rice cultivation. As a result, the watershed has historical water management structures which WS77 does not have. LiDAR data analysis also revealed some depressions caused by legacy dikes on downstream riparian areas of WS80 [63]. More evidence comes from Amoah et al. [17], who found a nine-times higher mean overall surface depressional storage capacity (DSC) in WS80 than WS77, as well as more than four-times more wetland area in WS80 (Table 2). Thus, it is very likely that the high WS80 DSC values may have contributed to a higher water table during winter with lower ET demands and an increased ET with a lower water table during the summer growing season (Figure S2). Another likely cause of WS80's reduced streamflow is modulated peaks caused by the storage, as evidenced by WS80's flatter slope in the range of an approximately 10–40 mm daily flow (Figure 5), consistent with historical records [36]. The smaller flow rates of WS80 were further evidenced by the daily flow and 10-min hydrographs for two events of 8 June and 5 September in 2019, as an example (Figure 6a,b), in which the flow rates of WS77 with a much lower DSC were 3–4 times higher than that of WS80, consistent with some other years as well (not shown). These observations are consistent with other studies [18,19,64–67] that reported a water table position and microtopography influence storage, and they are critical factors that affect streamflow patterns, stormflow peaks, and volume on shallow coastal forests. For example, Rains et al. [66] noted that the cumulative effect of depressions (WS80 in our study) can play an important role in landscape-scale hydrology by regulating the frequency, magnitude, timing, duration, and rate of flows to downgradient waters along overland and groundwater flow paths. Similarly, Acreman and Holden's conclusions [64] on five characteristics (landscape location and configuration, topography, soil characteristics, soil moisture status, and drainage management) largely determined the influence on floods, consistent with our runoff observations for these two watersheds.

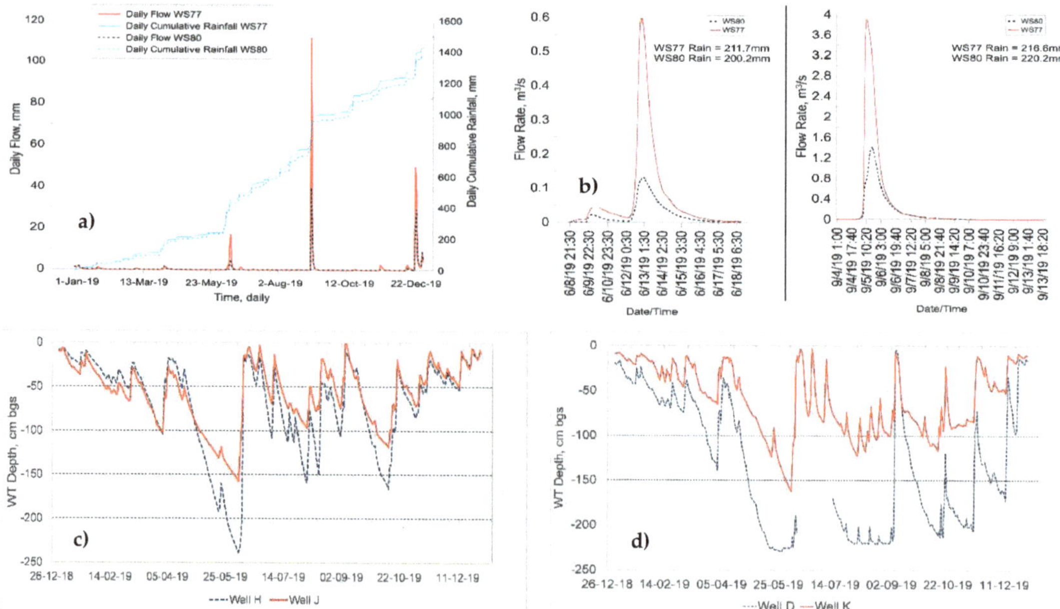

Figure 6. Measured (**a**) daily flow and daily cumulative rainfall and (**b**) hydrographs of June 8 and September 5 storm events for both WS77 and WS80, (**c**) daily water table depths of upland well H (WS80) and well J (WS77), and (**d**) daily water table depths for riparian well D (WS80) and well K (WS77) for 2019.

Secondly, a contemporary difference between the two watersheds is that WS80 has not received any forest management activities since it was established in 1968, while WS77 has been actively managed for loblolly pine silvicultural research using prescribed fire in a 2–3-year cycle (Figure S5a) for the past 20 years. There is a potential for flow to increase soon after fire [68,69]. For example, reduced understory vegetation (Figure S5b),

LAI, and ET caused by prescribed burning in March 2013, April 2016, and April 2018 (Figure 3) might have contributed to some temporary increased runoff in June 2013, August 2016, and August 2018. Accordingly, a detailed analysis using a MOSUM test in Figure 7 shows that the months immediately after prescribed burning captured a slight change in the relationship of paired flow, as shown by the upward or downward movement of the MOSUM curve, but not significantly impacting the linear regression coefficients (curve within the two red horizontal lines). Change was detected in June 2017, more than a year after prescribed burning. In addition, earlier studies [34,35] showed that prescribed fire had minimal or non-significant effects on soil properties, water quality, and water yield compared to the untreated reference for these watersheds. Furthermore, no heavy equipment, which may compact the soil, potentially reducing the conductivity, was used in this treatment. In addition, a rapid establishment of ground cover after the fire stabilizes the soil.

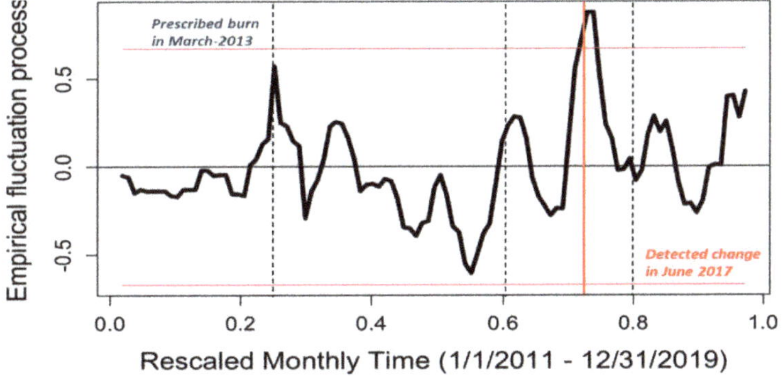

Figure 7. Ordinary least squares moving sums of recursive residuals (OLS-MOSUM) for monthly linear regression between WS77 (treatment) and WS80 (control) flow data. A shift of the MOSUM outside the 95% confidence intervals (horizontal red lines) is indicative of a structural break in the linear relationship. The vertical solid red line is the estimated structural break point. The three dotted vertical lines represent the months of the prescribed burning.

The fact that the mean monthly difference (WS80–WS77) in runoff (−6.8 mm) for the 2011–2019 period was found to be not different from the 1969–1978 (−8.6 mm) and 2004–2011 (−3.9 mm) periods further indicated the initiation of forest regeneration by 2004, with complete recovery by 2011 [24], as shown in Figure S1.

Third, active management of WS77 results in a stand that is predominately loblolly pine, in contrast to WS80, which is mixed hardwood-pine. Despite this difference in stand composition, the mean LAI of 2.31 m^2 m^{-2} (1.23 m^2 m^{-2}–3.36 m^2 m^{-2}), measured on the control watershed with pine-mixed-hardwood forest [43], was not significantly different (p = 0.34) from the mean of 2.54 m^2 m^{-2} (1.62 m^2 m^{-2}–2.92 m^2 m^{-2}) measured in 2019–2020 in WS77 with pine stands (Figure S3). Although the WS77 mean LAI was slightly higher than that of WS80, the growing season LAI, with a potential to influence ET, was higher (as high as 4–6 m^2 m^{-2}) in some plots during the growing season for WS80. Therefore, the lower WS80 runoff during growing season was partially attributed to a higher ET of the mixed hardwood-pine forest. For example, in the daily flow comparison for 2019 in Figure 6a, despite a lower WS77 total rainfall (235 mm) from Hurricane Dorian (September 5) than WS80's 242 mm, the WS77 daily flow was larger by 80 mm than WS80's. The larger flow of WS77 was likely due to its shallower WT, with less storage than that of WS80 in late August/early September (Figure 6c,d), resulting in an early initiation of flows. Accordingly, for WS80, a deeper WT and larger storage possibly caused larger ET loss than for WS77, with no flows for 54 days until this September 5 event, in contrast with only 2 days without flows for WS77. This pattern, with deeper WT depths and larger growing season deficits of

WS80 than of WS77, was also evident in the summers of 2011 to 2014, as well as briefly in 2016, 2017, and 2019 when WT fell below 100 cm (Figure S2), potentially contributing to higher ET and lower flows.

In addition, land morphology, defined by the hypsometric curve, also might have played a role in the runoff difference between the watersheds. The shape of the hypsometric curve is represented by a hypsometric integral, HI [70,71], with a value of 0.5 representing a threshold between the concave (HI < 0.5) and convex (HI $\geq$ 0.5) hypsometric forms. Vivoni et al. [71] found, keeping all other watershed variables constant (e.g., land use, land cover, rainfall), that modeled watersheds with a higher HI yielded higher runoff than those with a lower HI. Concave hypsometric curves for WS77 and WS80 (Figure 8a,b) clearly indicated that WS77 may be expected to yield more runoff than WS80 because of its higher HI of 0.405, compared to 0.285 for WS80 until 2001. The recomputed HI value of 0.313 after the drainage area of WS80 changed from 206 ha to 160 ha in 2001 was still < 0.405 (WS77). These results suggest that the shape of the basin hypsometry could be another reason for the difference in runoff between the two watersheds.

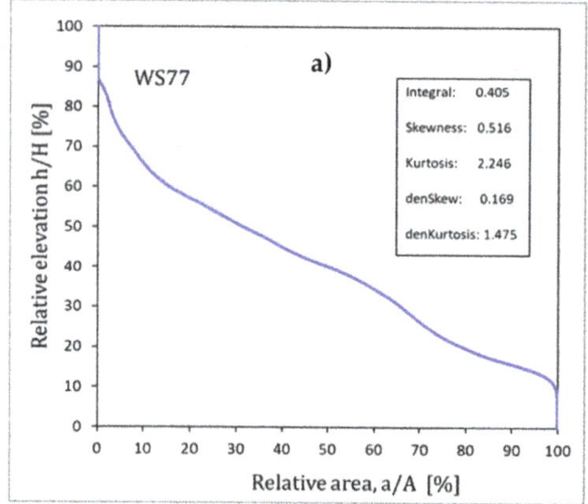

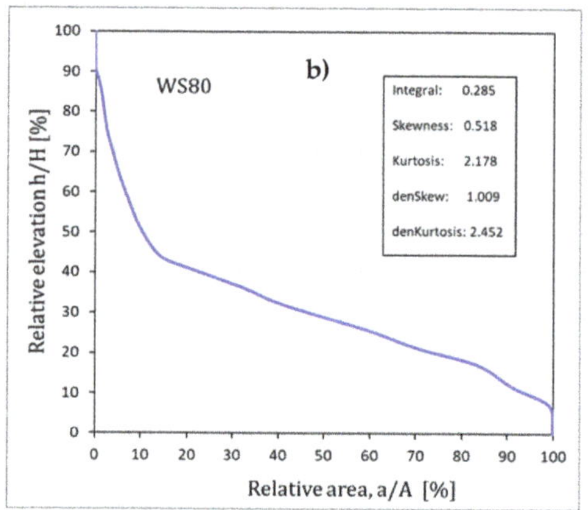

Figure 8. Hypsometric curves for watersheds (**a**) WS77 and (**b**) WS80. A hypsometric integral less than 0.5 and positive density skewness are characteristics of landforms dominated by surface runoff rather than subsurface drainage [71].

The pre-treatment monthly paired flow relationship without the October 2015 extreme runoff did not differ significantly from the 2004–2011 relationship (Figure 4b) reported by Jayakaran et al. [24]. The estimated monthly runoff of 609 mm was dominated by a one-day (4 October 2015) estimated extreme runoff of 311 mm on WS80. The peak flow rate for this hurricane event, estimated as 17.4 m^3 s^{-1} (10.9 m^3 s^{-1} km^{-2}), was assumed to have exceeded the 500-year flood [59,72], and therefore, this month, as an outlier, was omitted from the monthly analysis. However, the daily flow frequency duration analysis in this study also included the October 2015 month, except for the 3–5 October extreme flow days, and other large flow events caused by other hurricanes (Figure 5). These data may explain how the 2011–2019 calibration relationship might have been influenced by an increased number of high precipitation events. However, the fact that the 2004–2011 regression for the period with a reportedly recovered forest [24], but not the pre-Hugo 1969–1978, was like that of 2011–2019 may indicate the similar runoff response to rainfall during the two recent periods, dissimilar from 1969–1978. This similarity is potentially supported by the observations of Dai et al. [43], who reported more annual average storms > 50 mm in the 1982 to 2008 period than in 1946–2008, with even more storms by 2019 (not shown). We suggest, therefore, that the 2011–2019 relationship, which included some hurricane/tropical storm events, with no difference in either the mean monthly flow or regression relationship of the recent 2004–2011 period, is more justified than the pre-Hugo 1969–78 for its application in treatment effects evaluation of WS77 water yield later.

Our computed *p*-value, R^2, NSE, and RMSE statistics characterizing statistical significance and predictive quantifiable regression were also consistent with similar statistics (R^2 = 0.97, NSE = 0.97) for the paired daily flow relationships for 1988–1989 and higher than R^2 = 0.48 and NSE = 0.34 for the 2007–2008 calibration period reported by Ssegane [26] in their North Carolina pine forest study. Those values were also similar to R^2 = 0.83 and NSE = 0.82 and R^2 = 0.91 and NSE = 0.91 for two separate paired watersheds for the 2009–2012 calibration periods reported for studies in coastal North Carolina by Ssegane et al. [42].

Thus, the strong and significant 2011–2019 geometric regression-based pre-treatment baseline monthly runoff calibration relationship with given confidence limits (Figure 4b) could be used to compare the actual measured WS77 flow response with its expected flow response compared to WS80, within the bounds of data used in the calibration regression for quantifying the magnitude and significance of effects of longleaf pine restoration treatments in the near future. Nevertheless, it should still be cautiously interpreted and applied if frequencies of extreme events, like the October 2015 hurricane excluded from this study, continue to increase, as predicted by regional studies across the southeastern region [73]. This study also emphasizes a need to analyze long-term datasets, when available, to better understand the role of hydrological dynamics and their evolution and adaptation, including the paired watershed calibration for assessing treatment effects, in the context of a changing climate [74].

6. Conclusions

This study evaluated the seasonal rainfall and runoff response pattern and the flow calibration relationship using nine years (2011–2019) of hydro-meteorologic data for two long-term paired watersheds (155 ha, WS77 (treatment) and 160 ha, WS80 (control)) designated for a longleaf pine (LLP) restoration project at Santee Experimental Forest on the Atlantic Coastal Plain. The geometric mean regression-based monthly runoff relationship, proposed as a pre-treatment baseline, was compared to relationships reported earlier using 1969–1978 for pre-hurricane Hugo and 2004–2011 as post-Hugo recovery periods by Jayakaran et al. [24]. Other paired hydrologic metrics with a potential to influence the runoff were also used. Results revealed that the historical pattern in the runoff difference of WS77 > WS80 was maintained in the current baseline assessment. Furthermore, the difference in the mean monthly runoff between the two watersheds did not vary significantly (α = 0.05) from the pre-Hugo and post-Hugo periods, indicating a complete runoff recovery, as shown earlier by Jayakaran et al. [24]. The insignificantly higher, by chance, mean seasonal flow for WS77 than for WS80 was attributed to a lower surface storage (mean depressional storage capacity; Table 2) and higher hypsometric integral (a land morphological characteristic; Figure 8) for WS77 than for WS80, with a larger surface storage as well as subsurface storage indicated by a deeper average water table than that of WS77. In addition, the baseline monthly runoff calibration relationship, with multiple large flow events covering 2011–2019, except for an extreme of October 2015, did not differ from the 2004–2011 period but differed from 1969–1978, indicating a complete forest recovery and, possibly, a similarity in the climatic pattern of two recent periods. The baseline calibration relationship, found to be unaffected by periodic prescribed burning, was also significant (α = 0.05), predictable, and consistent, thereby providing a basis for quantifying post-treatment effects of the full LLP restoration on water yield later in the future. However, the relationship will have to be used cautiously when extrapolating for extremely large flow events, exceeding flow limits of the relationship as well as possibly exceeding the rating curve limits of the current gauging stations, otherwise equipped with well-defined compound weir control structures and dual sensors, including a backup for precise measurements of stage elevations.

Supplementary Materials: The following are available online at https://www.mdpi.com/article/10.3390/w13213121/s1, Figure S1: 12-month moving average of difference in monthly flow between WS80 and WS77 for 2011–2019 climatic conditions compared to those in 1969–1978 and 2004–2011 using whole long-term data with a large gap from 1982 to 1989, Figure S2: Measured daily water table depths on WS77 (Well J) and WS80 (Well H) watersheds for the years 2011 to 2019, Figure S3: LiCOR-2000 measured watershed averaged LAI for WS77 (top) and WS80 (bottom) for the 2019–2020 and 2008–2011 periods, respectively, Figure S4: (a) Box plot with median and interquartile range of monthly flow (MonFlow) measured on WS77 and WS80 watersheds in 2011–2019 without October 2015 as an outlier, and (b) Annual rainfall versus annual runoff between paired watersheds for 2011–2019, showing larger storage on WS80 than on WS77, Figure S5: Pictures of (a) operational prescribed burning and (b) post-burn land cover on WS77 treatment watershed, Table S1: Chronology of activities that took place on the WS77 and WS80 watersheds.

Author Contributions: Conceptualization, D.M.A. and C.C.T.; methodology, D.M.A. and S.H.; software, S.H. and M.D.H.; validation of formats, M.D.H.; formal analysis, D.M.A. and S.H.; investigation, D.M.A.; resources, D.M.A. and C.C.T.; writing—D.M.A.; writing—review and editing, D.M.A., S.H., C.C.T. and M.D.H. All authors have read and agreed to the published version of the manuscript.

Funding: This research received no external funding.

Institutional Review Board Statement: Not applicable.

Data Availability Statement: Data used in this paper can be accessed at https://doi.org/10.2737/RDS-2019-0033 for WS77 and https://doi.org/10.2737/RDS-2021-0043 for WS80.

Acknowledgments: The authors would like to acknowledge Andy Harrison and Julie Arnold of the Hydrologic Technician and Forestry Technician, respectively, at the US Forest Service Santee Experimental Forest for providing hydro-meteorologic data and Figure 1, respectively. The authors also thank the reviewers Ge Sun, USDA Forest Service, and Jami Nettles, Research Hydrologist at Weyerhaeuser, for their constructive comments and suggestions; Stephanie Siegel at Leading Solutions LLC for help with editing; and Priyanka Rao at Washington State University for help with providing plots and statistics from R-software. The opinions presented in this article are those of the authors and should not be construed to represent any official USDA or U.S. Government determination or policy.

Conflicts of Interest: The authors declare no conflict of interest.

References

1. Samuelson, L.J.; Stokes, T.A.; Johnsen, K.H. Ecophysiological comparison of 50-year-old longleaf pine, slash pine and loblolly pine. *For. Ecol. Manag.* **2012**, *274*, 108–115. [CrossRef]
2. Brantley, S.T.; Vose, J.M.; Wear, D.N.; Band, L. Potential of longleaf pine restoration to mitigate water scarcity and sustain carbon sequestration: Planning for an uncertain future. In *Proceedings of the Ecological Restoration and Management of Longleaf Pine Forests;* CRC Press: Boca Raton, FL, USA, 2018; pp. 291–310.
3. Eshleman, K.N.; Pollard, J.S.; O'Brien, A.K. Interactions between groundwater and surface water in a Virginia coastal plain watershed. 1. Hydrological flowpaths. *Hydrol. Process.* **1994**, *8*, 389–410. [CrossRef]
4. Griffin, M.P.; Callahan, T.J.; Vulava, V.M.; Williams, T.M. Storm-event flow pathways in lower coastal plain forested watersheds of the southeastern United States. *Water Resour. Res.* **2014**, *50*, 8265–8280. [CrossRef]
5. Slattery, M.C.; Gares, P.A.; Phillips, J.D. Multiple modes of storm runoff generation in a North Carolina coastal plain watershed. *Hydrol. Process.* **2006**, *20*, 2953–2969. [CrossRef]
6. Williams, T.M. Evidence of runoff production mechanisms in low gradient coastal forested watersheds. In Proceedings of the 2007 Minneapolis, St. Joseph, MI, USA, 17–20 June 2007; Paper # 072228. American Society of Agricultural and Biological Engineers: St. Joseph, MI, USA, 2007; Volume 5, pp. 1–13.
7. Amatya, D.M.; Williams, T.M.; Nettles, J.E.; Skaggs, R.W.; Trettin, C.C. Comparison of hydrology of two Atlantic coastal plain forests. *Trans. ASABE* **2019**, *62*, 1509–1529. [CrossRef]
8. Hornbeck, J.W.; Adams, M.B.; Corbett, E.S.; Verry, E.S.; Lynch, J.A. *Long-Term Impacts of Forest Treatments on Water Yield: A Summary for Northeastern USA*; Scientific Research Publishing: Wuhan, China, 1993; Volume 150, pp. 323–344. [CrossRef]
9. Kelly, C.N.; McGuire, K.J.; Miniat, C.F.; Vose, J.M. Streamflow response to increasing precipitation extremes altered by forest management. *Geophys. Res. Lett.* **2016**, *43*, 3727–3736. [CrossRef]
10. Shelby, J.D.; Chescheir, G.M.; Skaggs, R.W.; Amatya, D.M. Hydrologic and Water-Quality Response of Forested and Agricultural Lands During The 1999 Extreme Weather Conditions in Eastern North Carolina. *Am. Soc. Agric. Eng.* **2006**, *48*, 2179–2188. [CrossRef]
11. Kundzewicz, Z.W.; Kanae, S.; Seneviratne, S.I.; Handmer, J.; Nicholls, N.; Peduzzi, P.; Mechler, R.; Bouwer, L.M.; Arnell, N.; Mach, K.; et al. Flood risk and climate change: Global and regional perspectives. *Hydrol. Sci. J.* **2014**, *59*, 1–28. [CrossRef]
12. Acharya, S.; Jawitz, J.W.; Mylavarapu, R.S. Analytical expressions for drainable and fillable porosity of phreatic aquifers under vertical fluxes from evapotranspiration and recharge. *Water Resour. Res.* **2012**, *48*, 11526. [CrossRef]

13. Amatya, D.M.; Skaggs, R.W.; Gregory, J.D. Effects of controlled drainage on the hydrology of drained pine plantations in the North Carolina coastal plain. *J. Hydrol.* **1996**, *181*, 211–232. [CrossRef]
14. Loheide, S.P.; Butler, J.J.; Gorelick, S.M. Estimation of groundwater consumption by phreatophytes using diurnal water table fluctuations: A saturated-unsaturated flow assessment. *Water Resour. Res.* **2005**, *41*, 1–14. [CrossRef]
15. Harder, S.V.; Amatya, D.M.; Callahan, T.J.; Trettin, C.C.; Hakkila, J. Hydrology and Water Budget for a Forested Atlantic Coastal Plain Watershed, South Carolina. *J. Am. Water Resour. Assoc.* **2007**, *43*, 563–575. [CrossRef]
16. Evaristo, J.; McDonnell, J.J. Global analysis of streamflow response to forest management. *Nature* **2019**, *570*, 455–461. [CrossRef] [PubMed]
17. Amoah, J.K.O.; Amatya, D.M.; Nnaji, S. Quantifying watershed surface depression storage: Determination and application in a hydrologic model. *Hydrol. Process.* **2013**, *27*, 2401–2413. [CrossRef]
18. Hu, L.; Bao, W.; Shi, P.; Wang, J.; Lu, M. Simulation of overland flow considering the influence of topographic depressions. *Sci. Rep.* **2020**, *10*, 1–14. [CrossRef]
19. Walega, A.; Amatya, D.M.; Caldwell, P.; Marion, D.; Panda, S. Assessment of storm direct runoff and peak flow rates using improved SCS-CN models for selected forested watersheds in the Southeastern United States. *J. Hydrol. Reg. Stud.* **2020**, *27*, 100645. [CrossRef]
20. Wu, K.; Johnston, C.A. Hydrologic comparison between a forested and a wetland/lake dominated watershed using SWAT. *Hydrol. Process.* **2008**, *22*, 1431–1442. [CrossRef]
21. Ares, M.G.; Varni, M.; Chagas, C. Runoff response of a small agricultural basin in the argentine Pampas considering connectivity aspects. *Hydrol. Process.* **2020**, *34*, 3102–3119. [CrossRef]
22. Todd, A.K.; Buttle, J.M.; Taylor, C.H. Hydrologic dynamics and linkages in a wetland-dominated basin. *J. Hydrol.* **2006**, *319*, 15–35. [CrossRef]
23. Trettin, C.C.; Amatya, D.M.; Gaskins, A.H.; Miniat, C.F.; Chow, A.; Callahan, T. Watershed response to longleaf pine restoration–application of paired watersheds on the Santee Experimental Forest. In Proceedings of the 6th Interagency Conference on Research in Watershed, Shepherdstown, WV, USA, 23–26 July 2018; Volume 2018, pp. 194–201.
24. Jayakaran, A.D.; Williams, T.M.; Ssegane, H.; Amatya, D.M.; Song, B.; Trettin, C.C. Hurricane impacts on a pair of coastal forested watersheds: Implications of selective hurricane damage to forest structure and streamflow dynamics. *Hydrol. Earth Syst. Sci.* **2014**, *18*, 1151–1164. [CrossRef]
25. Loftis, J.C.; MacDonald, L.H.; Streett, S.; Iyer, H.K.; Bunte, K. Detecting cumulative watershed effects: The statistical power of pairing. *J. Hydrol.* **2001**, *251*, 49–64. [CrossRef]
26. Ssegane, H.; Amatya, D.M.; Chescheir, G.M.; Skaggs, W.R.; Tollner, E.W.; Nettles, J.E. Consistency of Hydrologic Relationships of a Paired Watershed Approach. *Am. J. Clim. Chang.* **2013**, *2*, 147–164. [CrossRef]
27. Amatya, D.M.; Gregory, J.D.; Skaggs, R.W. Effects of Controlled Drainage on Storm Event Hydrology in A Loblolly Pine Plantation. *J. Am. Water Resour. Assoc.* **2000**, *36*, 175–190. [CrossRef]
28. Bosch, J.M.; Hewlett, J.D. A review of catchment experiments to determine the effect of vegetation changes on water yield and evapotranspiration. *J. Hydrol.* **1982**, *55*, 3–23. [CrossRef]
29. Brown, A.E.; Zhang, L.; McMahon, T.A.; Western, A.W.; Vertessy, R.A. A review of paired catchment studies for determining changes in water yield resulting from alterations in vegetation. *J. Hydrol.* **2005**, *310*, 28–61. [CrossRef]
30. Tomer, M.D.; Schilling, K.E. A simple approach to distinguish land-use and climate-change effects on watershed hydrology. *J. Hydrol.* **2009**, *376*, 24–33. [CrossRef]
31. Clausen, J.; Spooner, J. *Paired Watershed Study Design*; Technical Report R841-F-93-009; United States Environmental Protection Agency, Office of Water: Washington, DC, USA, 1993.
32. Bren, L.J.; Lane, P.N.J. Optimal development of calibration equations for paired catchment projects. *J. Hydrol.* **2014**, *519*, 720–731. [CrossRef]
33. Alila, Y.; Kuraś, P.K.; Schnorbus, M.; Hudson, R. Forests and floods: A new paradigm sheds light on age-old controversies. *Water Resour. Res.* **2009**, *45*, 8416. [CrossRef]
34. Binstock, D.A. Effects of a Prescribed Winter Burn on Anion Nutrient Budgets in the Santee Experimental Forest Watershed Ecosystem. Ph.D. Thesis, Duke University, Durham, NC, USA, 1978.
35. Richter, D.D.; Ralston, C.W.; Harms, W.R. Prescribed fire: Effects on water quality and forest nutrient cycling. *Science* **1982**, *215*, 661–663. [CrossRef]
36. Richter, D. Effects of Water Quality and Nutrient Cycling in Forested Watersheds of the Santee Experimental Forest in South Carolina. Ph.D. Thesis, Duke University, Durham, NC, USA, 1980.
37. Hook, D.D.; Buford, M.A.; Williams, T.M. Impact of Hurricane Hugo on the South Carolina Coastal Plain. *J. Coast. Res.* **1991**, 291–300. [CrossRef]
38. Chescheir, G.M.; Skaggs, R.W.; Amatya, D.M. Quantifying the Hydrologic Impacts of Afforestation in Uruguay: A Paired Watershed Study. In Proceedings of the XIII World Forestry Congress, Buenos Aires, Argentina, 18–23 October 2009.
39. Amatya, D.M.; Trettin, C.C. Long-Term Ecohydrologic Monitoring: A Case Study from the Santee Experimental Forest, South Carolina. *J. South Carolina Water Resour.* **2020**, *6*, 46–55. [CrossRef]

40. Oda, T.; Green, M.B.; Urakawa, R.; Scanlon, T.M.; Sebestyen, S.D.; McGuire, K.J.; Katsuyama, M.; Fukuzawa, K.; Adams, M.B.; Ohte, N. Stream Runoff and Nitrate Recovery Times After Forest Disturbance in the USA and Japan. *Water Resour. Res.* **2018**, *54*, 6042–6054. [CrossRef]
41. Amatya, D.M.; Miwa, M.; Harrison, C.A.; Trettin, C.C.; Sun, G. *Hydrology and Water Quality of Two First Order Forested Watersheds in Coastal South Carolina*; 2006, Paper # 062182; American Society of Agricultural and Biological Engineers: Portland, OR, USA, 2006.
42. Ssegane, H.; Amatya, D.M.; Muwamba, A.; Chescheir, G.M.; Appelboom, T.; Tollner, E.W.; Nettles, J.E.; Youssef, M.A.; Birgand, F.; Skaggs, R.W. Calibration of paired watersheds: Utility of moving sums in presence of externalities. *Hydrol. Process.* **2017**, *31*, 3458–3471. [CrossRef]
43. Dai, Z.; Trettin, C.C.; Amatya, D.M. *Effects of Climate Variability on Forest Hydrology and Carbon Sequestration on the Santee Experimental Forest in Coastal South Carolina*; General Technical Report 172; United States Department of Agriculture, Forest Service, Southern Research Station: Asheville, NC, USA, 2013.
44. Amatya, D.M.; Trettin, C.C. *Santee Experimental Forest, Watershed 77: Streamflow, Water Chemistry, Water Table, and Weather Data*; Forest Service Research Data Archive: Fort Collins, CO, USA, 2019; Updated 9 June 2020. [CrossRef]
45. Amatya, D.M.; Trettin, C.C. *Santee Experimental Forest, Watershed 80: Streamflow, Water Chemistry, Water Table, and Weather Data*; Forest Service Research Data Archive: Fort Collins, CO, USA, 2021. [CrossRef]
46. Monteith, J.L. Evaporation and environment. *Symp. Soc. Exp. Biol.* **1965**, *19*, 205–234. [CrossRef]
47. Amatya, D.M.; Harrison, C.A. Grass and Forest Potential Evapotranspiration Comparison Using Five Methods in the Atlantic Coastal Plain. *J. Hydrol. Eng.* **2016**, *21*, 05016007. [CrossRef]
48. Amatya, D.M.; Trettin, C.C. *Santee Experimental Forest, Headquarters: Climate Data*; Forest Service Research Data Archive: Fort Collins, CO, USA, 2020. [CrossRef]
49. Amatya, D.M.; Trettin, C.C. Development of watershed hydrologic studies at Santee Experimental Forest, South Carolina. In *Advancing the Fundamental Sciences, Proceedings of the Forest Service National Earth Sciences Conference, San Diego, CA, 18–22 October 2004*; Furniss, M.J., Clifton, C.F., Ronnenberg, K.L., Eds.; Pacific Northwest Research Station: Portland, OR, USA, 2007; pp. 180–190.
50. *SAS Base SAS 9.1.3 Procedures Guide*; SAS Institute, Inc.: Cary, NC, USA, 2006.
51. Warton, D.I.; Wright, I.J.; Falster, D.S.; Westoby, M. Bivariate line-fitting methods for allometry. *Biol. Rev.* **2006**, *81*, 259. [CrossRef]
52. R Development Core Team. *A Language and Environment for Statistical Computing*; R Foundation for Statistical Computing: Vienna, Austria, 2015.
53. Friedman, J.; Bohonak, A.J.; Levine, R.A. When are two pieces better than one: Fitting and testing OLS and RMA regressions. *Environmetrics* **2013**, *24*, 306–316. [CrossRef]
54. Krug, J.A.; Wrather, W.E.; Langbein, W.B. Topographic characteristics of drainage basins. *Water Supply Pap.* **1947**, 125–157. [CrossRef]
55. Strahler, A.N. Hypsometric (Area-Altitude) Analysis of Erosional Topography. *GSA Bull.* **1952**, *63*, 1117–1142. [CrossRef]
56. Olaya, V.; Conrad, O. Chapter 12 Geomorphometry in SAGA. *Dev. Soil Sci.* **2009**, *33*, 298–308.
57. Harlin, J.M. Statistical moments of the hypsometric curve and its density function. *J. Int. Assoc. Math. Geol.* **1978**, *10*, 59–72. [CrossRef]
58. Pérez-Peña, J.V.; Azañón, J.M.; Azor, A. CalHypso: An ArcGIS extension to calculate hypsometric curves and their statistical moments. Applications to drainage basin analysis in SE Spain. *Comput. Geosci.* **2009**, *35*, 1214–1223. [CrossRef]
59. Amatya, D.M.; Harrison, C.A.; Trettin, C.C. Hydro-meteorologic Assessment of October 2015 Extreme Precipitation Event on Santee Experimental Forest Watersheds, South Carolina. *J. South Carol. Water Resour.* **2016**, *3*, 12.
60. Saleh, F.; Ramaswamy, V.; Georgas, N.; Blumberg, A.F.; Pullen, J. A retrospective streamflow ensemble forecast for an extreme hydrologic event: A case study of Hurricane Irene and on the Hudson River basin. *Hydrol. Earth Syst. Sci.* **2016**, *20*, 2649–2667. [CrossRef]
61. Callahan, T.J.; Vulava, V.M.; Passarello, M.C.; Garrett, C.G. Estimating groundwater recharge in lowland watersheds. *Hydrol. Proc.* **2012**, *26*, 2845–2855. [CrossRef]
62. Boulet, A.-K.; Rial-Rivas, M.E.; Ferreira, C.; Coelho, C.O.A.; Kalantari, Z.; Keizer, J.J.; Ferreira, A.J.D. Hydrological Processes in Eucalypt and Pine Forested Headwater Catchments within Mediterranean Region. *Water* **2021**, *13*, 1418. [CrossRef]
63. Amatya, D.; Trettin, C.; Panda, S.; Ssegane, H. Application of LiDAR Data for Hydrologic Assessments of Low-Gradient Coastal Watershed Drainage Characteristics. *J. Geogr. Inf. Syst.* **2013**, *5*, 175–191. [CrossRef]
64. Acreman, M.; Holden, J. How wetlands affect floods. *Wetlands* **2013**, *33*, 773–786. [CrossRef]
65. Amatya, D.M.; Chescheir, G.M.; Skaggs, R.W. Hydrologic effects of the location and size of a natural wetland in an agricultural landscape. In *Proceedings of the AWRA/ASAE International Conference on "Versatility of Wetlands in the Agricultural Landscape"*; American Society of Agricultural Engineers: Tampa, FL, USA, 1995; pp. 477–488.
66. Rains, M.C.; Leibowitz, S.G.; Cohen, M.J.; Creed, I.F.; Golden, H.E.; Jawitz, J.W.; Kalla, P.; Lane, C.R.; Lang, M.W.; Mclaughlin, D.L. Geographically isolated wetlands are part of the hydrological landscape. *Hydrol. Process.* **2016**, *30*, 153–160. [CrossRef]
67. Vogel, R.M.; Fennessey, N.M. Flow Duration Curves II: A Review of Applications in Water Resources Planning. *J. Am. Water Resour. Assoc.* **1995**, *31*, 1029–1039. [CrossRef]
68. Ebel, B.A.; Moody, J.A.; Martin, D.A. Hydrologic conditions controlling runoff generation immediately after wildfire. *Water Resour. Res.* **2012**, *48*, 3529. [CrossRef]

69. Robichaud, P.R. Fire effects on infiltration rates after prescribed fire in Northern Rocky Mountain forests, USA. *J. Hydrol.* **2000**, *231*, 220–229. [CrossRef]
70. Luo, W. Quantifying groundwater-sapping landforms with a hypsometric technique. *J. Geophys. Res. Planets* **2000**, *105*, 1685–1694. [CrossRef]
71. Vivoni, E.R.; Di Benedetto, F.; Grimaldi, S.; Eltahir, E.A.B. Hypsometric control on surface and subsurface runoff. *Water Resour. Res.* **2008**, *44*, 12502. [CrossRef]
72. Amatya, D.M.; Radecki-Pawlik, A.A. Flow Dynamics of Three Experimental Forested Watersheds in Coastal South Carolina (USA). *ACTA Sci. Pol. Form. Circumiectus* **2007**, *6*, 3–17.
73. Ingram, K.T.; Dow, K.; Carter, L.; Anderson, J.A. *Climate of the Southeast United States: Variability, Change, Impacts, and Vulnerability*; Island Press-Center for Resource Economics: Washington, DC, USA, 2013; ISBN 9781610915090.
74. Juez, C.; Pena-Angulo, D.; Khorchani, M.; Regues, D.; Nadal-Romero, E. 20-Years of hindsight into hydrological dynamics of a mountain forest catchment in the Central Spanish Pyrenees. *Sci. Total Environ.* **2021**, *766*. [CrossRef] [PubMed]

Article

Characterizing the Interception Capacity of Floor Litter with Rainfall Simulation Experiments

Qiwen Li [1], Ye Eun Lee [2] and Sangjun Im [1,3,*]

1 Department of Agriculture, Forestry and Bioresources, Seoul National University, Seoul 08826, Korea; gimunlee@snu.ac.kr

2 Division of Forest Disaster Management, National Institute of Forest Science, Seoul 02455, Korea; dldpdms109@korea.kr

3 Department of Agriculture, Forestry and Bioresources, Research Institute of Agriculture and Life Sciences, Seoul National University, Seoul 08826, Korea

* Correspondence: junie@snu.ac.kr; Tel.: +82-2-880-4759

Received: 11 September 2020; Accepted: 8 November 2020; Published: 10 November 2020

Abstract: Floor litter can reduce the amount of water reaching the soil layer through rainfall interception. The rainfall interception capacity of floor litter varies with the physical features of the litter and rainfall characteristics. This study aimed to define the maximum and minimum interception storages (C_{mx}, C_{mn}) of litter layers using rainfall simulation experiments, and examine the effects of litter type and rainfall characteristics on rainfall retention and drainage processes that occur in the litter layer. Different types of needle-leaf and broadleaf litters were used: *Abies holophylla*, *Pinus strobus*, *Pinus rigida*, *Quercus acutissima*, *Quercus variabilis*, and *Sorbus alnifolia*. Our results indicate a wide variation in interception storage values of needle leaf litter, regardless of the rainfall intensity and duration. The *A. holophylla* needle-leaf litter showed the highest C_{mx} and C_{mn} values owing to its short length and low porosity. Conversely, the lowest interception storage values were determined for the *P. strobus* needle leaf litter. No significant differences in interception storage were established for the broadleaf litter. Moreover, except for *A. holophylla* litter, the broadleaf litter retained more water than the needle leaf litter. An increase in the intensity or duration of rainfall events leads to an increase in the water retention storage of litter. However, these factors do not influence the litter's drainage capacity, which depends primarily on the force of gravity.

Keywords: floor litter; rainfall interception capacity; rainfall simulation experiment; litter drainage

---

1. Introduction

Rainfall interception is recognized at present as one of the most underrated and underpriced processes in forest hydrology. In places where floor litter has developed on a near-ground surface, rainfall that falls on forest cover is intercepted by the litter layer and subsequently evaporates back into the atmosphere [1–3]. This retention and redistribution process profoundly influences the water budget of forest areas, altering the amount of water available to percolate into the uppermost layer of forest soils [4–6]. Compared to tree canopy interception, floor litter interception has received less attention because it is often regarded as a minor component of the water cycle [7]. Forest litter refers to recently fallen and partially decomposed tree leaves, twigs, and small branches, distinct from humus, resting on the upper surface of soils [8]. It forms a porous barrier that retains a small portion of the incident rainfall.

Researchers have been investigating floor litter interception since the middle of the 20th century; however, only recently has it received a significant amount of research attention [2,3]. Taking accurate measurements of litter interception capacity is challenging [3,7]. Field experiments can provide a reliable estimation of the interception loss but are inaccurate in their estimation of the interception capacity over time and space [9–13]. Thus, laboratory experiments are often employed to assess litter interception loss [4,7,14–19]. Regardless of the approach, water intercepted by floor litter is of similar magnitude or sometimes higher than that intercepted by the canopy. A wide range of litter interception loss has been observed in previous studies, ranging from 1–2% to 50–70% of gross rainfall [2,3,10,14,20,21], although inconsistencies in measurement exist. The reason for these inconsistencies can be attributed to the variation in the litter type, litter thickness, and rainfall characteristics [4,7,11,15,18–21].

Floor litter is highly heterogeneous in terms of its physical structure and its accumulation is not evenly distributed over the soil surface. Furthermore, the interception capacity of floor litter can vary across tree species and depends entirely on the litter's physiological and morphological characteristics [22]. Broadleaf litter is generally large and curved in shape and thus can easily capture rainwater, while needle-leaf litter is well compacted and can block flow paths that run through the litter [7,19]. Broadleaf litter individually stores more water than needle litter during storm periods [4,7,14]. In contrast, the loosely packed, flat litter layer in deciduous species intercepts less water than the clumped, needle litter accumulation in coniferous species [16,19]. The absolute amount of water retained is likely to depend on the thickness of the accumulated litter, where a thicker litter layer retains more water [10,15,18].

Water retention is not only affected by the inherent nature of litter but also by rainfall characteristics, which influence the water storage capacity of floor litter. Although the amount of water stored is proportional to rainfall intensity, litter storage capacity shows a poor relationship with rainfall intensity [16,23]. This inverse relationship exists because more water is needed to saturate the deeper litter accumulation thoroughly [10,24]. Therefore, floor litter can retain a smaller rainfall percentage during a short, intense storm event than in a long, less intense event [25,26]. Li et al. [25] and others [17] found that the effects of rainfall intensity on interception capacity are not apparent, and no linear relationship exists across litter types.

The influence of rainfall on the extent of interception capacity has not yet been revealed as raindrop size, intensity, and the pattern of natural rainfall vary in space and time. Although research on litter interception is limited, several studies have attempted to quantify it under natural or artificial rainfall events. The variance and inhomogeneity of natural rainfall make it difficult to identify the fingerprint of rainfall on litter interception storage under in situ experiments [25–27]. Over the past few decades, rainfall simulation experiments have been used in numerous hydrologic studies to quantitatively demonstrate the influence of rainfall characteristics on experimental variables [4,14,23]. Using simulated rainfall enables a greater control of the rainfall variables and simplifies data collection during the experiment.

Although the interception capacity of rainfall by floor litter has been widely investigated, studies on the precise nature of hydrologic processes occurring in the litter layer during rainfall have not been conducted. Furthermore, the effects of litter and rainfall characteristics on water retention and drainage have not been adequately examined. In this regard, the present study conducts a laboratory investigation of retention and drainage processes as to whether or not the physical features of litter and rainfall characteristics would affect the hydrologic function of litter. The primary aim of the experiment is to quantitatively estimate the rainfall interception storage of various litter types under controlled conditions of rainfall and litter. The secondary aim is to examine the hydrologic processes of the litter layer over the short period of time by relating the retention and drainage processes to the litter's physical traits and the rainfall characteristics.

2. Materials and Methods

2.1. Litter Collection and Characterization

Two contrasting types of litter were used in this study: needle-leaf litter and broadleaf litter. All litter samples were collected at Mt. Gwanak, which is approximately 1500 ha and is managed by the College of Agriculture and Life Sciences, Seoul National University. This forest is located in the southern part of Seoul in Korea and is at a height of 632 m a.s.l., consisting of young-growth and mixed-coniferous type stands. Its topography is relatively steep, and the dominant soil type is sandy loam.

Undecomposed litter was taken from six different sites: three litter samples beneath coniferous trees, and three beneath deciduous trees. Coniferous tree species included *Abies holophylla* Maxim., *Pinus strobus*, *Pinus rigida* Mill., while *Quercus acutissima* Carruth., *Quercus variabilis* Blume, and *Sorbus alnifolia* (Siebold & Zucc.) C. Koch represented broadleaf litter. Fallen leaves from the current year were manually collected from the ground surface beneath the mature trees. We also removed leaves with obvious symptoms of pathogen or herbivore attack. Most of the litter samples were collected in the period of October to November 2016, while some needle litter samples were supplemented in April 2017. All the collected litter samples were placed in sealed plastic bags and immediately transported to the laboratory for analysis.

The morphological characteristics of litter may vary with tree species. Apparent features, such as length and diameter, of the needle-leaf litter were measured manually and the measurements were repeated for 200 needles for each species, thereby accounting for the natural variations in its physical shape. The measurements were performed using a digital caliper and rounded to the nearest 0.1 mm. The needle's all-side surface area and volume were estimated from the litter length and diameter, under the assumption that the pine-type needle has a cylindrical form with the terminal area neglected [4]. The projected areas of 50 randomly selected broadleaf litters were measured with an LI-3000C leaf-area meter (LiCOR Inc., Lincoln, NE, USA) and used to determine the total surface area of individual litter. The leaf volume of the broadleaf litter was estimated by multiplying the surface area of the individual litter by its thickness. Leaf thickness was averaged from five measurement points that were randomly made with a digital caliper.

Because of the wide variation in the leaf morphology between tree species, this study used two reliable parameters, specific surface area (SSA) and surface-area-to-volume ratio (SAV), to distinguish the physical properties of the litter. A finite amount of water adheres to the litter surface as a result of surface tension. Therefore, SSA has been used extensively to examine the influences of litter traits on rainfall retention [3,28]. SSA is a geometric estimator representing the total surface area per unit mass of litter, thereby indicating the extent to which the litter surface interacts with its surroundings. The litter was dried in laboratory at nearly 23 °C temperature, 40% relative humidity for at least seven days and then weighed to determine its dry mass. SAV is a crucial litter parameter that best describes litter geometry and its relative dimensions. It refers to the ratio between the surface area and its volume [29]. The SAV value of individual litter can be determined by separate measurements of the surface area and volume. Large SAVs increase the rates of energy and mass exchange in the gaseous phase, implying that litter can quickly become wet or dry in response to changes in the surrounding conditions.

2.2. Rainfall Simulation Experiment

A rainfall simulation experiment was conducted to quantitatively demonstrate the effects of rainfall characteristics on litter interception storage. The experimental apparatus consisted of a portable rainfall simulator, a litter container, a drainage collector, and electronic balances (Figure 1).

A portable rainfall simulator with a 0.25 m × 0.25 m sprinkling area (Eijkelkamp®, Giesbeek, Netherlands) supplied simulated rainfall of assigned rainfall intensity and duration. The rainfall simulation pours water droplets with a mass of 0.106 g and a diameter of 5.9 mm, which is similar to the canopy drip of throughfall [30].

Litter mass influences water storage and drainage [14,25]. Kang et al. [31] observed a litter accumulation of 944 ± 512 g/m^2 in a deciduous forest in Korea, which is similar to a previous study [15]. Therefore, a weight of 60 g (on average 960 g/m^2) of litter was used in the rainfall simulation experiments. Litter was placed in a 0.25 m × 0.25 m rectangular container beneath the portable simulator. The thickness of the litter layer varied with the density and consolidation of the layer.

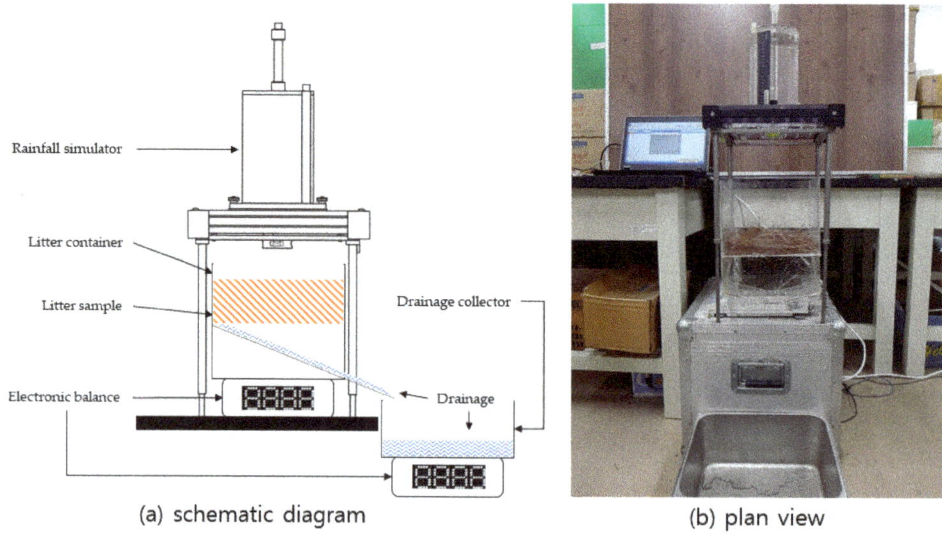

Figure 1. Rainfall simulation apparatus: (**a**) schematic diagram, and (**b**) plan view.

Water moved along the macropore channels in the litter layer during the experiment. Assuming that floor litters are hygroscopic, the distribution of pore space influences the vertical or lateral movement of water. Therefore, porosity was calculated to enable a direct comparison with the extent of pore space in the litter layer [32]:

$$\varepsilon = 1 - \frac{W_l}{\rho_l \times V_l} \tag{1}$$

where ε is the litter layer's porosity, W_l is the weight of the litter layer, V_l is the total volume of the litter layer, and ρ_l is litter density.

The litter container was placed on a CUX-4200H electronic balance (CAS®, Yangju, Korea), connected to a personal computer by a communication cable. When the rain began, the weight of the container was continuously recorded every 5 s. Part of the incident rainfall percolated through the pore channel of litter and consequently reached the drainage collector. The drainage rate was also measured at 5 s intervals from the drainage collector on the CUX-4200H electronic balance. Varying intensities of 50, 75, and 100 mm/h were designated as rainfall intensity parameters, and the rainfall simulator was operated at these intensities for 10, 20, 30, and 40 min. An intensity of 100 mm/h is approximately equivalent to a 100-year, 1-h design storm event in the Seoul region. The rainfall simulation experiment was repeated five times for each litter type under specified rainfall conditions. In total, 180 experiments were conducted in this study.

2.3. Interception Storage Measurement

Interception storage capacity is a predictive variable that quantitatively represents the retention ability of floor litter. It can be defined as the depth of water stored or retained on the litter surface and within the macropores of the litter layer [27]. There are two types of interception storage [15]. The maximum interception storage capacity (C_{mx}) is the amount of rainwater retained by litter temporarily before the rain stops. It represents the water stored by litter that can flow down to the uppermost soil layer and is therefore essential for understanding the soil water relationship. The minimum interception storage capacity (C_{mn}) refers to the thin film of detained water held on the litter surface that is removed by evaporation only. C_{mn} is a more crucial hydrological parameter than C_{mx} because it indicates the ultimate retention capacity of floor litter [14,23].

Through rainfall simulation experiments, the above-mentioned two storages were determined from the litter weight curve. C_{mx} was experimentally obtained as the amount of detained water at an asymptotic stable line during rainfall and C_{mn} was determined to be the water stored at the completion of post-rainfall drainage, as shown in Figure 2.

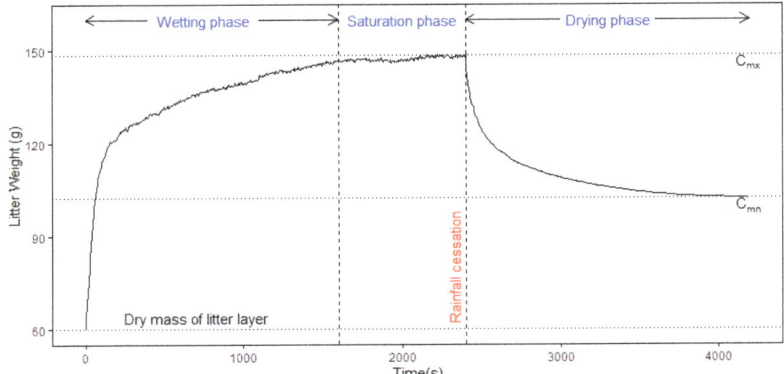

Figure 2. Graphical representation of interception storage capacity in the litter weight curve.

2.4. Litter Drainage Estimation

Water retention and drainage processes occurring in the litter layer displayed three phases in terms of the timing and flux of water: wetting, saturation, and drying. As shown in Figure 2, the amount of retained water increased rapidly during the wetting phase. This phase usually took place during the first stages of the rain over a short time. As the rain continued for a certain period, the water retained on the litter plateaued and reached the saturation phase. The saturation phase was relatively stable because the litter was too wet to retain more water and, consequently, excess water drained into the collector. When the rain stopped, a gravitational flow was produced in the pores until the surface tension and gravitation force established an equilibrium. During the drying phase, the drainage rate tended to decrease exponentially, similar to the infiltration curve.

The wetting phase consists of the initial abstraction (lag time), percolation, and gravitational flow (Figure 3). When rain begins to fall, all the rain can be stored for a short period as the initial abstraction. As shown in Figure 3, the initial abstraction is commonly referred to as the lag time, which is the elapsed time for producing litter drainage. In succession to the lag time, percolation occurred through the litter layer [33]. The percolation period was estimated from the time-varying flux of rainfall and drainage, $\frac{dD}{dR}$, where dD and dR are the deviations of drainage and rainfall rates for a 5-s interval, respectively. We assume that the value of $\frac{dD}{dR}$ increases linearly for the period of percolation. Although the increase in rainfall is unbounded, drainage rate is bounded by the maximum interception storage and asymptotically approaches a constant value of $\frac{dD}{dR}$, as shown in Figure 3. This asymptotic convergence occurs because a certain amount of liquid is gradually absorbed throughout the amorphous parts in cell walls of dead leaves [3]. The gravitation flow is referred to as the dominant hydrologic process occurring in the litter layer during this period.

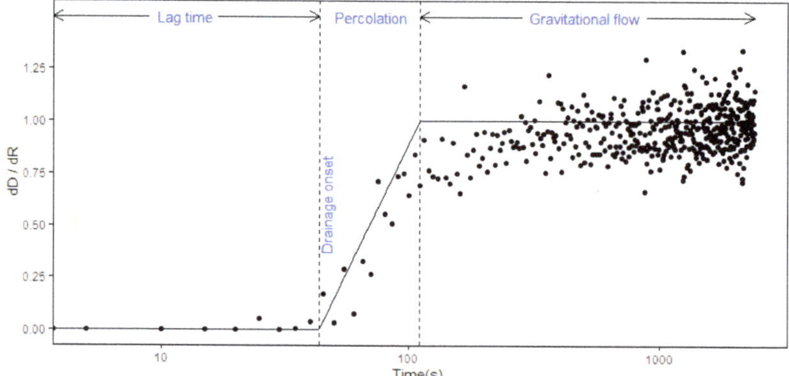

Figure 3. Deviations of drainage and rainfall rates (*dD/dR*) for 5-s intervals.

When the rain stops, water retained in the litter surface begins to evaporate. If evaporation is negligible, the post-rainfall retention of litter is determined by the competition of the available water ($C_{mx} - C_{mn}$) subject to the gravitational force and the cohesive and/or adhesive forces of water. When the rain ceased, the retained water exponentially decreased as follows:

$$C_t = C_{mn} + (C_{mx} - C_{mn})e^{-kt} \quad (2)$$

where C_t is the water retained in the litter layer at t-min after rain cessation, C_{mx} and C_{mn} are the maximum and minimum interception storage of the litter layer, and k is the recession coefficient of litter drainage. The parameter k represents the time-dependent decline of water retained in the litter layer.

2.5. Statistical Analysis

Statistical differences among the groups were evaluated by one-way ANOVA or Kruskal-Wallis test (one-way ANOVA on ranks). If the treatments satisfied the assumptions of the ANOVA, i.e., the independence, normality and homogeneity of variances, the one-way ANOVA method was used. Otherwise, the Kruskal–Wallis test, a nonparametric method, was used.

In addition, a post hoc test was conducted when the results showed significant differences ($p < 0.05$). A Tukey's test was used for the post-hoc analysis of one-way ANOVA test. Meanwhile, a Mann-Whitney U test was used for that of the Kruskal-Wallis test. This nonparametric test allows two groups to be compared without assuming that the values are normally distributed [34].

Thus, both Kruskal-Wallis and Mann-Whitney U tests were conducted for a comparison of physical traits and the interception storage capacity. The one-way ANOVA and Tukey's tests were used for comparing the recession coefficient of litter drainage.

All statistical analyses were performed using R 4.0.2 (The R Foundation for Statistical Computing, Vienna, Austria) and Python 3.7.8 (The Python Software Foundation, Wilmington, DE, United States).

3. Results

3.1. Litter Physical Characteristics

Leaf litter samples taken from three deciduous tree species were used in this study. The litters of *Q. variabilis* and *Q. acutissima* were characterized by a longer length and narrower width than the oval-shaped litter of *S. alnifolia*. The projected surface area of the *Q. variabilis* litter was significantly larger than that of the *S. alnifolia* and *Q. acutissima* litters. As shown in Figure 4a, small litters with areas of less than 100 cm^2 contributed the most to the *Q. acutissima* and *S. alnifolia* litters but contributed only 38% to the *Q. variabilis* litter.

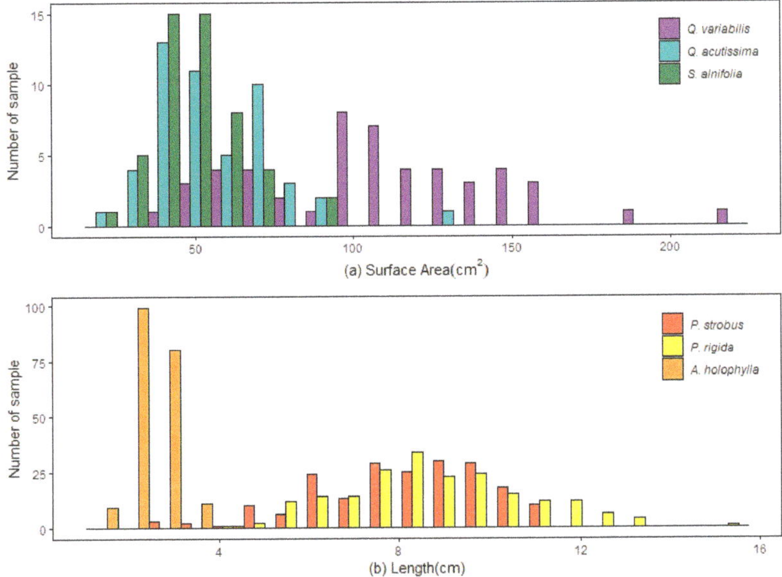

Figure 4. Distribution of (**a**) the surface area of the broadleaf litter, and (**b**) the length of the needle-leaf litter.

Figure 4b shows the distinct difference in litter length distribution between *A. holophylla* and pine litter (*P. rigida* and *P. strobus*). The majority of *A. holophylla* litters were 2–3 cm in length. The needle length of pine trees ranged mostly from 4 cm to 13 cm, with an average length of 8.89 cm in *P. rigida* and 8.19 cm in *P. strobus*.

The mean and standard deviation values of density, SSA, and SAV in various litter types are presented in Table 1. Significant differences were detected in the SSAs and SAVs between the broadleaf and needle-leaf litters ($p < 0.01$); however, the difference was less significant in term of the litter density ($p < 0.05$). Between the broadleaf or needle litter types, interspecific differences in the physical properties were observed ($p < 0.01$), which are in partial agreement with the findings reported in previous studies [2,4,35].

Table 1. Physical characteristics of individual litter samples.

Litter Type	Species	Sample Size	Density (g/cm^3) *	Specific Surface Area (SSA) (cm^2/g) *	Surface-Area-to-Volume Ratio (SAV) (cm^2/cm^3) *
Broadleaf litter	*Q. variabilis*	50	0.590 ± 0.046	167.83 ± 26.56	98.53 ± 14.03
	Q. acutissima	50	0.535 ± 0.065	221.36 ± 49.13	116.20 ± 15.71
	S. alnifolia	50	0.482 ± 0.070	283.88 ± 77.91	135.02 ± 34.47
Needle-leaf litter	*P. strobus*	200	0.650 ± 0.028	176.87 ± 33.05	115.11 ± 23.97
	P. rigida	200	0.661 ± 0.113	96.68 ± 14.04	62.94 ± 7.76
	A. holophylla	200	0.393 ± 0.036	30.64 ± 9.72	11.96 ± 3.70

* indicates the mean ± standard deviation.

The *P. rigida* litter is considered to be the heaviest floor litter (0.661 g/cm^3), while the density of the *A. holophylla* litter is 0.393 g/cm^3, which is the lightest among the six litter types. Leaf surface area was significantly higher in the broadleaf litter than in the needle-leaf litter. The SSAs of the broadleaf litter varied from 167.83 cm^2/g (*Q. variabilis*) to 283.88 cm^2/g (*S. alnifolia*), while they ranged from 30.64 cm^2/g (*A. holophylla*) to 176.87 cm^2/g (*P. strobus*) in the case of needle-leaf litter. A similar variation was observed in the SAV values, which were higher in the order of *S. alnifolia*, *Q. acutissima*, and *Q. variabilis* for the broadleaf litter, and *P. strobus*, *P. rigida*, and *A. holophylla* for the needle-leaf litter.

Simulated rainfall was poured over the litter accumulation in the litter container. The litter layer's characteristics, such as layer thickness and porosity, can affect the water retention capacity. Table 2 displays the thickness and porosity of the litter layer for all the experiments. Litter thickness varied with litter density and porosity. The *A. holophylla* litter layer had a lower thickness (1.37 cm) than the pine litter layers. The broadleaf litter layer thickness ranged from 10.84 cm in *Q. variabilis* to 13.16 cm in *Q. acutissima*. Porosities of the broadleaf litter layer were higher than those of the needle litter layer.

Table 2. Physical characteristics of litter layers used for rainfall simulation experiments.

Litter Type	Species	Litter Layer	
		Thickness(cm) *	Porosity (%) *
Broadleaf litter	*Q. variabilis*	10.84 ± 0.70	98.48 ± 0.10
	Q. acutissima	13.16 ± 0.45	98.61 ± 0.05
	S. alnifolia	12.75 ± 0.45	98.40 ± 0.06
Needle-leaf litter	*P. strobus*	7.13 ± 0.62	97.92 ± 0.19
	P. rigida	5.80 ± 0.53	97.46 ± 0.23
	A. holophylla	1.37 ± 0.22	81.49 ± 3.67

* indicates the mean ± standard deviation for five experiments.

3.2. Litter Interception Storage

Litter interception storage is given by the equivalent depth of rainfall per unit thickness of the litter layer (Table 3). For needle-leaf litter, the average C_{mn} value per unit thickness of the litter layer was the highest in *A. holophylla* (1.146 mm/cm), followed by *P. rigida* (0.173 mm/cm) and *P. strobus* (0.097 mm/cm). The water stored in the broadleaf litter layer ranged from 0.088 mm in *Q. acutissima* to 0.098 mm in *Q. variabilis* per effective depth of the layer, depending on species and rainfall intensity, which were considerably lower than those of the needle litters.

Table 3. Variations of C_{mx} and C_{mn} across litter type and rainfall.

Intensity (mm/h)	Duration (min)	Broadleaf Litter			Needle-Leaf Litter		
		Q. variabilis	*Q. acutissima*	*S. alnifolia*	*P. strobus*	*P. rigida*	*A. holophylla*
				(a) C_{mx} (mm/cm) *			
50	10	0.118 ± 0.014	0.128 ± 0.016	0.126 ± 0.007	0.174 ± 0.007	0.242 ± 0.029	1.447 ± 0.322
	20	0.145 ± 0.008	0.138 ± 0.011	0.135 ± 0.015	0.196 ± 0.016	0.290 ± 0.018	2.075 ± 0.467
	30	0.159 ± 0.017	0.133 ± 0.008	0.144 ± 0.020	0.219 ± 0.012	0.282 ± 0.035	1.611 ± 0.069
	40	0.157 ± 0.025	0.143 ± 0.016	0.142 ± 0.009	0.214 ± 0.021	0.286 ± 0.021	1.601 ± 0.118
75	20	0.171 ± 0.031	0.151 ± 0.009	0.140 ± 0.008	0.237 ± 0.017	0.322 ± 0.030	1.738 ± 0.344
100	20	0.191 ± 0.019	0.152 ± 0.011	0.164 ± 0.011	0.260 ± 0.029	0.331 ± 0.031	1.623 ± 0.224
				(b) C_{mn} (mm/cm) *			
50	10	0.074 ± 0.014	0.079 ± 0.013	0.083 ± 0.006	0.088 ± 0.003	0.156 ± 0.027	1.101 ± 0.189
	20	0.090 ± 0.006	0.088 ± 0.008	0.088 ± 0.010	0.095 ± 0.010	0.171 ± 0.017	1.420 ± 0.309
	30	0.103 ± 0.014	0.086 ± 0.005	0.094 ± 0.013	0.095 ± 0.010	0.167 ± 0.017	1.089 ± 0.079
	40	0.103 ± 0.018	0.093 ± 0.011	0.097 ± 0.009	0.096 ± 0.019	0.174 ± 0.014	1.043 ± 0.108
75	20	0.105 ± 0.021	0.093 ± 0.006	0.084 ± 0.005	0.097 ± 0.006	0.184 ± 0.021	1.203 ± 0.177
100	20	0.111 ± 0.014	0.087 ± 0.007	0.101 ± 0.007	0.110 ± 0.015	0.188 ± 0.026	1.022 ± 0.194

* indicates the mean ± standard deviation for five experiments.

Figure 5 shows how the interception storage capacities (C_{mx}, C_{mn}) vary with litter type and rainfall duration at a constant intensity of 50 mm/h for a given period. As shown, the interception storage responses to rainfall duration varied among the litter types. The influence of needle-leaf litter was more evident than that of broadleaf litter. Both the C_{mx} and C_{mn} values were highest in the *A. holophylla* litter and lowest in the *P. strobus* litter for the needle-leaf litters, regardless of rainfall duration.

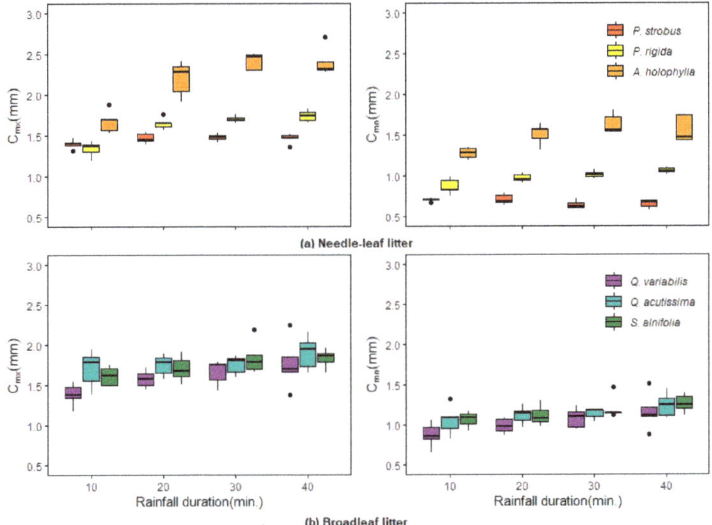

Figure 5. Variations in litter interception storage capacity at various durations and an intensity of 50 mm/h rainfall. (**a**) needle-leaf litter, and (**b**) broadleaf litter.

The averaged C_{mx} values in the *A. holophylla* litter varied from 1.67 mm for a 10-min duration to 2.42 mm for a 30-min duration, while the C_{mn} values ranged from 1.28 mm to 1.63 mm under the same rainfall characteristics. There were significant differences in the C_{mx} and C_{mn} values between needle-leaf litters ($p < 0.05$). However, unlike the needle-leaf litters, the values of C_{mx} and C_{mn} in the *Q. acutissima* and *S. alnifolia* litters occurred in the same group ($p \geq 0.05$), and those of the *Q. variabilis* litter differed from other broadleaf litters. Slight increasing trends were observed in the C_{mx} and C_{mn} values with increasing rainfall for all the broadleaf litters.

The influence of rainfall intensity on interception storage was also examined. Rainfall intensities of 50, 75, and 100 mm/h were poured for 20 min over various litter covers. As shown in Figure 6, both C_{mx} and C_{mn} increased marginally with increasing rainfall intensity. However, the effect of rainfall intensity on C_{mx} was more apparent. As mentioned previously, C_{mx} is the sum of C_{mn} and gravitational flow. Thus, several studies have reported a directly proportional relationship between C_{mn} and rainfall intensity [14–16]. The results of this study indicate that a two-fold increase in rainfall intensity (50 mm/h to 100 mm/h) caused C_{mn} to increase by 6.5% for the needle litter and 12.7% for the broadleaf litter.

The effects of rainfall characteristics on litter interception storage were substantial. The extent of litter interception capacity is mainly dependent on the rainfall amount and its intensity and duration. The current study demonstrated that a higher intensity or longer duration of rainfall leads to an increase in the interception storage of the litter layer, which is similar to the findings of Sato et al. [14], Putuhena and Cordery [15], and others [16,25].

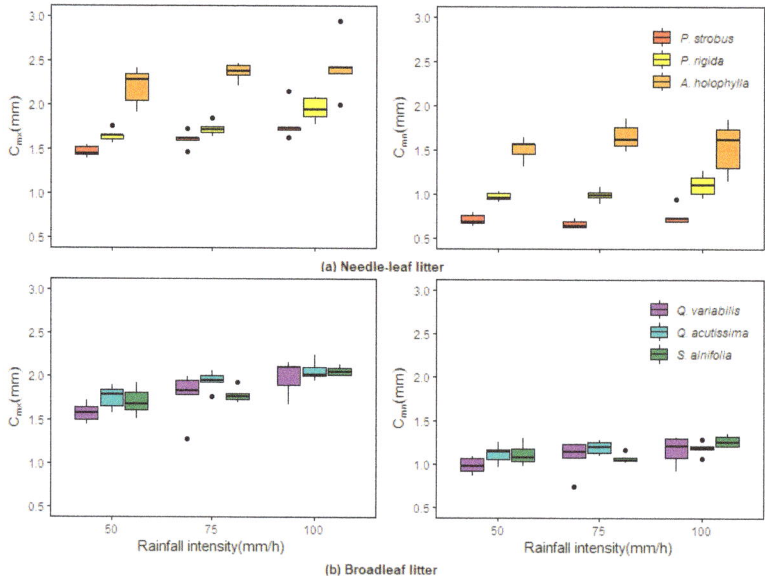

Figure 6. Variations in litter interception storage with rainfall intensity. (**a**) needle-leaf litter, and (**b**) broadleaf litter.

3.3. Litter Drainage

The amount of water retained in the litter is the result of the throughfall, evaporation and litter drainage. When assuming little or no evaporation loss for the experiment's short duration, rainfall that exceeds the litter's interception storage may begin to move downward, under gravitational forces, through the litter layer as litter drainage. During the experiment, litter drainage reached a stable condition (equivalent to the rainfall rate) after approximately 20–30 min, depending on the litter type and rainfall characteristics.

Table 4 presents the portion of the litter drainage corresponding to the rainfall amount. When the litter layer reached its saturation point, a nearly constant amount of water was drained. Litter drainage, expressed as a percentage of the total rainfall, for broadleaf litter with 50 mm/h of rainfall varied from 0.865% after 10 min to 0.966% after 40 min. At a rainfall intensity of 100 mm/h, litter drainage increased by approximately 0.026% at a 50 mm/h intensity for the same 20 min duration. Therefore, the drainage percentage increased with a longer duration or higher intensity of rainfall. However, for the rainfall-simulated experiments, the rainfall amount was found to be a crucial variable for producing litter drainage [14,19]. When compared with events producing the same rainfall amount, such as 25-mm rainfall events (30 min, 50 mm/h and 20 min, 75 mm/h) and 33.3-mm rainfall events (40 min, 50 mm/h and 20 min, 100 mm/h), the drainage of each litter was not significantly different. Table 4 also indicates that the effects of litter type on drainage were not apparent.

3.4. Water Retention and Drainage Process of the Litter Layer

Figure 7a depicts the lag time for drainage onset under rainfall simulation experiments. The drainage onset occurred earliest with pine litters and was delayed longest by the *A. holophylla* litter. Unlike the needle leaf litter, no significant differences in lag time existed for the broadleaf litter. For the rainfall experiments with an intensity of 50 mm/h and a 20-min duration, an average lag time of 53.0 s (equivalent to 0.74 mm) was required for the drainage onset in the *A. holophylla* litter, followed by the *S. alnifolia* litter (51.4 s), *Q. variabilis* (48.3 s), *Q. acutissima* litter (48.0 s), *P. strobus* litter (37.0 s), and *P. rigida* litter (35.0 s). As shown in Figure 7a, the lag time varied with litter type, showing a slight decrease with an increase in rainfall intensity. Generally, fast drainage responses were typically associated with needle litter, while broadleaf litter was slower to respond. This phenomenon was also reported by Zhao et al. [19] and Li et al. [25]. They demonstrated that the broadleaf litter's larger surface depression functions as a rainwater harvester at the beginning of a rainfall event.

Table 4. Variation in litter drainage with litter type and rainfall (expressed as a percentage of the total rainfall).

Litter Type	Species	Total Litter Drainage Ratio					
		50 mm/h				75 mm/h	100 mm/h
		10 min	20 min	30 min	40 min	20 min	20 min
Broadleaf litter	*Q. variabilis*	0.887	0.938	0.955	0.963	0.954	0.963
	Q. acutissima	0.855	0.927	0.951	0.961	0.950	0.962
	S. alnifolia	0.852	0.928	0.949	0.960	0.957	0.959
Needle-leaf litter	*P. strobus*	0.914	0.956	0.973	0.979	0.973	0.976
	P. rigida	0.892	0.939	0.958	0.968	0.960	0.965
	A. holophylla	0.851	0.907	0.931	0.952	0.932	0.952

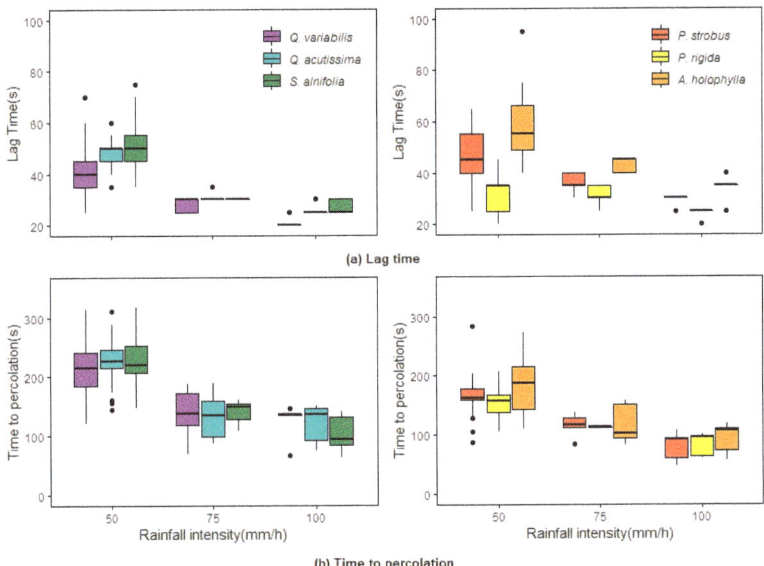

Figure 7. Variation in lag time and time to percolation with litter type and rainfall intensity. (**a**) Lag time, (**b**) Time to percolation.

The fraction of rainfall that moved through the litter layer as percolation and the litter container's weight gradually increased with an increasing rainfall rate. The percolation period was defined as the time at which the $\frac{dD}{dR}$ curve meets the threshold line. The percolation period lasted for 1.7–4.1 min for the broadleaf litter, and 1.3–3.3 min for the needle litter (Figure 7b). This period indicates that most rainfall can be intercepted and retained by the litter layer, although a small portion of rainfall drips off.

The recession coefficient (k) in Equation (2) was analyzed using the interception storage and drainage relationships. Table 5 displays the best relationships of drainage recession limbs for each litter type, showing a sufficiently good agreement. The recession coefficients varied from 0.00202 in the *A. holophylla* litter to 0.00236 in the *S. alnifolia* litter, indicating no significant difference between litter types. Therefore, we propose a practical use recession coefficient of 0.00224 to estimate drainage flow from floor litter. Table 5 indicates the absence of a significant difference in the post-rainfall drainage recession coefficient. This suggests that post-rainfall drainage from the saturated litter layer is governed by the force of gravity and not litter type.

Table 5. Recession coefficient (k) values of water retention functions.

Litter Type	Species	k		R^2
		Mean	SD *	
Broadleaf litter	*Q. variabilis*	0.00229	0.000237	0.893
	Q. acutissima	0.00222	0.000163	0.935
	S. alnifolia	0.00236	0.000202	0.939
Needle-leaf litter	*P. strobus*	0.00232	0.000205	0.911
	P. rigida	0.00222	0.000137	0.955
	A. holophylla	0.00202	0.000168	0.911

* indicates the standard deviation.

4. Discussion

4.1. The Influence of Litter Type on Interception Capacity

The amount of intercepted water can be largely attributed to the physical differences across litter types. Morphologically, broadleaf litters have either oval or elliptical forms and form well-arranged stacks on the ground surface. Leaf bundles or fascicles are a distinct characteristic of needle-leaf litters. The number of needles per fascicle differs among tree species. *A. holophylla* litters comprise flattened needles that are singly attached around the twig. The *P. rigida* litter has fascicles of three needles, while *P. strobus* litters have five-needle fascicles.

Under the rainfall simulation experiments, C_{mx} is controlled by the porosity and volume of litter, and the ability of water to penetrate through the exterior of the litter into the inner pores. C_{mn} is related to litter's physical features such as surface area, arrangement and litter surface [4]. Sato et al. [14] and others [4] suggested that broadleaf litter stores substantially more water than needle-leaf litter. No significant differences in the interception storage were established for broadleaf litter. This is due to the fact that rainwater is mostly retained in the macropores of broadleaf litter when rain pours for a short period, which is not enough to saturate the litter tissue [10]. Moreover, except for the *A. holophylla* litter, broadleaf litter retained more water than needle leaf litter. As shown in Table 2, the lower porosity of *A. holophylla* litter contributes to higher storage retention due to the greater resistance of the packed litter to the vertical movement of water [36,37].

The *A. holophylla* litter has the advantage of retaining water within the layer due to its exceptionally accumulated litter structure with a relatively smaller length and lower porosity compared to pine litter. Thus, the adhesion and surface tension of water molecules are strong on the *A. holophylla* layer. Rainfall applied to the *A. holophylla* litter layer did not disperse sufficiently [14], and the dense litter mat restricted water movement into the litter layer [38]. The SAV value was approximately 1.6 times greater in *A. holophylla* litter than in the pine litters. Litter interception storage, in general, was inversely related to SSA and SAV for the needle-leaf litter. In contrast, SSA and SAV's influences were not significant for the broadleaf litter because water was stored on surface pits or leave concavities. The interception storage was also affected by the thickness of the litter layer [10,15]. This indicates that interception storage in needle litters is more strongly dependent on physical properties in comparison to broadleaf litters.

Additionally, the individual litter's orientation or arrangement can affect the amount of water retained in the litter layer. For needle litter, the horizontally oriented layer retains more water than the curved or suspended layer. In this study, litter was horizontally piled in the litter container to ignore the effect of litter orientation.

4.2. Litter Drainage Characteristics

It is expected that no drainage occurs until rainfall exceeds the litter layer's interception storage capacity. However, drainage from the litter container was observed even at the beginning of the rainfall (less than 1 min), which is most likely due to quick flow such as the preferential flow through macropores and bypass flow on the litter container walls. Some rainfall passed quickly through the litter layer's macropores or flowed down along the container surface. However, the initial abstraction process was not fully investigated in this study because the experiment was conducted with a small, thin-layered litter layer. Nevertheless, initial abstraction is thought to be more considerable in the forest cover water cycle and thereby warrants further research.

When the rain continued to fall, the rainfall occupied the macropores of the litter layer until the litter reached saturation. During this period, drainage tended to increase logarithmically because the litter layer became partly saturated, and gravitational flow occurred under the force of gravity. Moreover, the timespan for gravitational flow generation is affected by the rainfall intensity and litter type. In the experiment, the drainage rate was always below the rainfall rate during the wetting phase.

When litter can be completely saturated under long rainfall durations, the average drainage rate is approximately equal to the rainfall rate. In most cases, the constant drainage rate did not attain a long duration of rainfall because water absorption into leaf tissues may have accounted for some of the gradual rainfall loss. The maximum drainage rate was nevertheless defined, as the water was discharged immediately before rain cessation. This made it possible to define the potential amount of water that reached the soil layer under various rainfall conditions.

4.3. Limitation and Future Application

We noted several limitations in this study. The rainfall intensities used in the experiments were relatively high. This led to a decrease in the litter's retention ability compared to that exposed to lower-intensity natural rainfall [14]. Furthermore, short rainfall duration caused a lower retention capacity [25,26], particularly in the *A. holophylla* litter, because all needles of the litter layer may not have been wetted. Moreover, experiments were also conducted with a homogeneous litter, yet under natural conditions, floor litter consists of leaves, branches, and fruits of several trees and plants, rather than a single species. In natural conditions, lateral flow on inclined surfaces may occur, thereby affecting percolation and retention [19,23]. However, slope was not considered in the experiments and no measurements were made of the lateral movement of water. Retention or drainage processes occurring in the litter layer may be affected by these factors. The litter's interception capacity is relatively constant for a specific tree species, but it can vary with geographical location, tree age, or canopy position. Additional precision can be achieved by enlarging the number of observations to compensate for variations in both the litter's physical properties and its water retention ability.

The presence of litter layer on the ground is a characteristic feature of forest soils. Rainfall interception of litter layer is one of the hydrologic processes in forest watersheds. The changes in water stored within the floor litter is determined by summing direct rainfall as throughfall, drip from the foliage, stems, and branches, lateral flow out of the layer, and percolation from the litter layer into the upper soil layer. In addition, C_{mn} corresponds to the threshold value of percolation through the litter layer, which controls the litter drainage during rainfall events. Leaf litter, characterized by the interception storage and drainage capacity here, are important in determining what happens in forest floors and examining how the vegetation influences the hydrologic cycle of forested watersheds. However, there have been few attempts to develop the hydrologic models regarding the interception storage-drainage relationship of floor litter [33,37] because the knowledge of the process has not yet been fully explored [27]. Here, we presented the simplified rainfall-interception relationship of floor litter with the rainfall simulation experiments, but more research is needed to overcome the above-mentioned drawbacks.

5. Conclusions

Floor litter plays an essential role in forest hydrological cycles by capturing a fraction of rainfall and evaporating it into the atmosphere. During a short period of rainfall events, litter's water retention ability may be affected by the physical properties of the litter, amount of litter, and rainfall characteristics. In this study, rainfall simulation experiments were conducted to experimentally explore the influences of litter type and rainfall patterns on water retention and drainage in the litter layer. Three types of broadleaf and needle-leaf litters were investigated to demonstrate how the litter's physical features can influence the rainfall interception process.

The interception storage capacity of needle litters varied significantly with the litter type. The *A. holophylla* litter showed the highest values for the maximum and minimum interception storage regardless of the rainfall intensity and duration. Moreover, small *A. holophylla* litter retained more water than the other litters owing to its much lower thickness and porosity. No significant differences in interception storage existed across the broadleaf litter types. Rainfall interception of broadleaf litter occurred on the surface storage of horizontal or sub-horizontal leaves, particularly in concavities. It was also revealed that the amount of water retained by litter is a function of the intensity and duration of rainfall. A higher intensity or longer duration of rainfall events can increase the interception storage capacity of litter on all broadleaf and needle-leaf litters.

The amount of rainfall available for initial abstraction and percolation in the litter layer varied according to litter type and rainfall pattern. However, after rainfall cessation, the recession limbs did not vary significantly according to litter type and rainfall characteristics. This implies that the force of gravity is the primary factor governing post-rainfall drainage from the saturated litter layer.

Author Contributions: Conceptualization, Q.L. and S.I.; methodology, Q.L. and Y.E.L.; software, Q.L.; validation, Q.L., Y.E.L. and S.I.; formal analysis, Q.L.; investigation, Y.E.L.; data curation, Q.L.; writing—original draft preparation, S.I.; writing—review and editing, Q.L. and S.I.; visualization, Q.L.; supervision, S.I. All authors have read and agreed to the published version of the manuscript.

Funding: This research received no external funding.

Acknowledgments: This study was carried out with the support of R&D Program for Forest Science Technology (Project No. 2020185B10-2022-AA02) provided by Korea Forest Service (Korea Forestry Promotion Institute).

Conflicts of Interest: The authors declare no conflict of interest.

References

1. Horton, R.E. Rainfall Interception. *Mon. Weather. Rev.* **1919**, *47*, 603–623. [CrossRef]
2. Gentilli, J.; Kittredge, J. Forest Influences: The Effects of Woody Vegetation on Climate, Water, and Soil, with Applications to the Conservation of Water and the Control of Floods and Erosion. *Geogr. Rev.* **1949**, *39*, 164. [CrossRef]
3. Gerrits, A.M.J.; Savenije, H.H.G. Forest floor interception. In *Forest Hydrology and Biogeochemistry: Synthesis of Past Research and Future Direction*, 1st ed.; Levia, D.F., Carlyle-Moses, D., Tanaka, T., Eds.; Springer: Heidelberg, Germany, 2011; pp. 445–454.
4. Walsh, R.P.D.; Voigt, P.J. Vegetation Litter: An Underestimated Variable in Hydrology and Geomorphology. *J. Biogeogr.* **1977**, *4*, 253. [CrossRef]
5. Savennije, H.H.G. The importance of interception and why we should delete the term evapotranspiration from our vocabulary. *Hydrol. Process.* **2004**, *18*, 1507–1511. [CrossRef]
6. Crockford, R.; Richardson, D. Partitioning of rainfall into throughfall, stemflow and interception: Effect of forest type, ground cover and climate. *Hydrol. Process.* **2000**, *14*, 2903–2920. [CrossRef]
7. Li, X.; Xiao, Q.; Niu, J.; Dymond, S.; McPherson, E.G.; Van Doorn, N.; Yu, X.; Xie, B.; Zhang, K.; Li, J. Rainfall interception by tree crown and leaf litter: An interactive process. *Hydrol. Process.* **2017**, *31*, 3533–3542. [CrossRef]
8. Hoover, M.D.; Lunt, H.A. A Key for the Classification of Forest Humus Types. *Soil Sci. Soc. Am. J.* **1952**, *16*, 368. [CrossRef]

9. Helvey, J.D. Rainfall interception by hardwood forest litter in the southern Appalachians. In *US Forest Service Research Paper SE-8*; USDA Southeastern Forest Experimental Station: Asheville, NC, USA, 1964.
10. Helvey, J.D.; Patric, J.H. Canopy and litter interception of rainfall by hardwoods of eastern United States. *Water Resour. Res.* **1965**, *1*, 193–206. [CrossRef]
11. Tsiko, C.; Makurira, H.; Gerrits, A.M.J.; Savenije, H.H.G. Measuring forest floor and canopy interception in a savannah ecosystem. *Phys. Chem. Earth Parts A/B/C* **2012**, *47*, 122–127. [CrossRef]
12. Naeth, M.A.; Bailey, A.W.; Chanasyk, D.S.; Pluth, D.J. Water Holding Capacity of Litter and Soil Organic Matter in Mixed Prairie and Fescue Grassland Ecosystems of Alberta. *J. Range Manag.* **1991**, *44*, 13. [CrossRef]
13. Reynolds, J.F.; Knight, D.H. Magnitude of snowmelt and rainfall interception by litter in lodgepole pine and spruce-fir forests in Wyoming. *Northwest Sci.* **1973**, *47*, 50–60.
14. Sato, Y.; Kumagai, T.; Kume, A.; Otsuki, K.; Ogawa, S. Experimental analysis of moisture dynamics of litter layers?the effects of rainfall conditions and leaf shapes. *Hydrol. Process.* **2004**, *18*, 3007–3018. [CrossRef]
15. Putuhena, W.M.; Cordery, I. Estimation of interception capacity of the forest floor. *J. Hydrol.* **1996**, *180*, 283–299. [CrossRef]
16. Keim, R.F.; Skaugset, A.; Weiler, M. Storage of water on vegetation under simulated rainfall of varying intensity. *Adv. Water Resour.* **2006**, *29*, 974–986. [CrossRef]
17. Rosalem, L.M.P.; Wendland, E.; Anache, J.A.A. Understanding the water dynamics on a tropical forest litter using a new device for interception measurement. *Ecohydrology* **2018**, *12*, e2058. [CrossRef]
18. Bulcock, H.H.; Jewitt, G. Field data collection and analysis of canopy and litter interception in commercial forest plantations in the KwaZulu-Natal Midlands, South Africa. *Hydrol. Earth Syst. Sci.* **2012**, *16*, 3717–3728. [CrossRef]
19. Zhao, L.; Hou, R.; Fang, Q. Differences in interception storage capacities of undecomposed broad-leaf and needle-leaf litter under simulated rainfall conditions. *For. Ecol. Manag.* **2019**, *446*, 135–142. [CrossRef]
20. Brye, K.R.; Norman, J.M.; Bundy, L.G.; Gower, S.T. Water-Budget Evaluation of Prairie and Maize Ecosystems. *Soil Sci. Soc. Am. J.* **2000**, *64*, 715–724. [CrossRef]
21. Price, J.; Rochefort, L.; Quinty, F. Energy and moisture considerations on cutover peatlands: Surface microtopography, mulch cover and Sphagnum regeneration. *Ecol. Eng.* **1998**, *10*, 293–312. [CrossRef]
22. Talhelm, A.F.; Smith, A.M. Litter moisture adsorption is tied to tissue structure, chemistry, and energy concentration. *Ecosphere* **2018**, *9*, 02198. [CrossRef]
23. Guevara-Escobar, A.; González-Sosa, E.; Véliz-Chávez, C.; Ventura-Ramos, E.; Ramos-Salinas, M. Rainfall interception and distribution patterns of gross precipitation around an isolated Ficus benjamina tree in an urban area. *J. Hydrol.* **2007**, *333*, 532–541. [CrossRef]
24. Bernard, J.M. Forest Floor Moisture Capacity of the New Jersey Pine Barrens. *Ecology* **1963**, *44*, 574–576. [CrossRef]
25. Li, X.; Niu, J.; Xie, B. Study on Hydrological Functions of Litter Layers in North China. *PLoS ONE* **2013**, *8*, e70328. [CrossRef] [PubMed]
26. Xiao, Q.; McPherson, E.G. Surface Water Storage Capacity of Twenty Tree Species in Davis, California. *J. Environ. Qual.* **2016**, *45*, 188–198. [CrossRef]
27. Guevara-Escobar, A.; Gonzalez-Sosa, E.; Ramos-Salinas, M.; Hernandez-Delgado, G.D. Experimental analysis of drainage and water storage of litter layers. *Hydrol. Earth Syst. Sci.* **2007**, *11*, 1703–1716. [CrossRef]
28. Carlyle-Moses, D.E.; Gash, J.H.C. Rainfall interception loss by forest canopies. In *Forest Hydrology and Biogeochemistry: Synthesis of Past Research and Future Directions*; Levia, D.F., Carlyle-Moses, D.E., Tanaka, T., Eds.; Springer: New York, NY, USA, 2011; pp. 407–423.
29. Fernandes, P.; Rego, F. A New Method to Estimate Fuel Surface Area-to-Volume Ratio Using Water Immersion. *Int. J. Wildland Fire* **1998**, *8*, 121–128. [CrossRef]
30. Levia, D.F.; Hudson, S.A.; Llorens, P.; Nanko, K. Throughfall drop size distributions: A review and prospectus for future research. *Wiley Interdiscip. Rev. Water* **2017**, *4*, e1225. [CrossRef]
31. Kang, M.-S.; Hong, J.-W.; Bong, H.-Y.; Jang, H.-M.; Choi, M.; Jang, Y.-H.; Cheon, J.-H.; Kim, J. On Estimating Interception Storage Capacity of Litter Layer at Gwangneung Deciduous Forest. *Korean J. Agric. For. Meteorol.* **2011**, *13*, 87–92. [CrossRef]
32. Wang, H.; Van Eyk, P.J.; Medwell, P.R.; Birzer, C.H.; Tian, Z.F.; Possell, M.; Huang, X. Air Permeability of the Litter Layer in Broadleaf Forests. *Front. Mech. Eng.* **2019**, *5*, 53. [CrossRef]

33. Dunkerley, D. Percolation through leaf litter: What happens during rainfall events of varying intensity? *J. Hydrol.* **2015**, *525*, 737–746. [CrossRef]
34. Fay, M.P.; Proschan, M.A. Wilcoxon-Mann-Whitney or t-test? On assumptions for hypothesis tests and multiple interpretations of decision rules. *Stat. Surv.* **2010**, *4*, 1–39. [CrossRef]
35. Zagyvai-Kiss, K.A.; Kalicz, P.; Szilagyi, J.; Gribovszki, Z. On the specific water holding capacity of litter for three forest ecosystems in the eastern foothills of the Alps. *Agric. For. Meteorol.* **2019**, *278*, 107656. [CrossRef]
36. Bristow, K.; Campbell, G.; Papendick, R.; Elliott, L. Simulation of heat and moisture transfer through a surface residue—Soil system. *Agric. For. Meteorol.* **1986**, *36*, 193–214. [CrossRef]
37. Bussière, F.; Cellier, P. Modification of the soil temperature and water content regimes by a crop residue mulch: Experiment and modelling. *Agric. For. Meteorol.* **1994**, *68*, 1–28. [CrossRef]
38. Warning, R.H.; Rogers, J.J.; Swank, W.T. Water relation and hydrologic cycles. In *Dynamic Properties of Forest Ecosystem*; Reichle, D.E., Ed.; Cambridge University Press: New York, NY, USA, 1982; pp. 205–264.

Article

To What Extent Can a Sediment Yield Model Be Trusted? A Case Study from the Passaúna Catchment, Brazil

Klajdi Sotiri [1,*], Stephan Hilgert [1], Matheus Duraes [2], Robson André Armindo [3], Nils Wolf [4], Mauricio Bergamini Scheer [5], Regina Kishi [6], Kian Pakzad [4] and Stephan Fuchs [1]

1 Department of Aquatic Environmental Engineering, Institute for Water and River Basin Management, Karlsruhe Institute of Technology, 76131 Karlsruhe, Germany; stephan.hilgert@kit.edu (S.H.); stephan.fuchs@kit.edu (S.F.)
2 Institute of Geography, Federal University of Uberlandia, Uberlandia 38400-902, Brazil; duraes@ufu.br
3 Department of Physics, Institute of Natural Sciences, Federal University of Lavras, Lavras 37200-900, Brazil; robson.armindo@ufla.br
4 EFTAS Fernerkundung Technologietransfer GmbH, 48145 Münster, Germany; nils.wolf@eftas.com (N.W.); kian.pakzad@eftas.com (K.P.)
5 Sanitation Company of Paraná SANEPAR, Curitiba 80215-900, Brazil; mauriciobs@sanepar.com.br
6 Department of Hydraulics and Sanitation, Federal University of Paraná, Curitiba 80050-540, Brazil; rtkishi@gmail.com
* Correspondence: klajdi.sotiri@kit.edu; Tel.: +49-721-608-44111

Citation: Sotiri, K.; Hilgert, S.; Duraes, M.; Armindo, R.A.; Wolf, N.; Scheer, M.B.; Kishi, R.; Pakzad, K.; Fuchs, S. To What Extent Can a Sediment Yield Model Be Trusted? A Case Study from the Passaúna Catchment, Brazil. *Water* **2021**, *13*, 1045. https://doi.org/10.3390/w13081045

Academic Editor: Koichiro Kuraji

Received: 30 March 2021
Accepted: 8 April 2021
Published: 10 April 2021

Publisher's Note: MDPI stays neutral with regard to jurisdictional claims in published maps and institutional affiliations.

Abstract: Soil degradation and reservoir siltation are two of the major actual environmental, scientific, and engineering challenges. With the actual trend of world population increase, further pressure is expected on both water and soil systems around the world. Soil degradation and reservoir siltation are, however, strongly interlinked with the erosion processes that take place in the hydrological catchments, as both are consequences of these processes. Due to the spatial scale and duration of erosion events, the installation and operation of monitoring systems are rather cost- and time-consuming. Modeling is a feasible alternative for assessing the soil loss adequately. In this study, the possibility of adopting reservoir sediment stock as a validation measure for a monthly time-step sediment input model was investigated. For the assessment of sediment stock in the reservoir, the commercial free-fall penetrometer GraviProbe (GP) was used, while the calculation of sediment yield was calculated by combining a revised universal soil loss equation (RUSLE)-based model with a sediment delivery ratio model based on the connectivity approach. For the RUSLE factors, a combination of remote sensing, literature review, and conventional sampling was used. For calculation of the C Factor, satellite imagery from the Sentinel-2 platform was used. The C Factor was derived from an empirical approach by combining the normalized difference vegetation index (NDVI), the degree of soil sealing, and land-use/land-cover data. The key research objective of this study was to examine to what extent a reservoir can be used to validate a long-term erosion model, and to find out the limiting factors in this regard. Another focus was to assess the potential improvements in erosion modeling from the use of Sentinel-2 data. The use of such data showed good potential to improve the overall spatial and temporal performance of the model and also dictated further opportunities for using such types of model as reliable decision support systems for sustainable catchment management and reservoir protection measures.

Keywords: sediment yield; RUSLE; Sentinel-2; reservoir siltation; penetrometer; sediment balance

1. Introduction

Soil is a dynamic system that is highly dependent on the variations of the surrounding environment. Erosion-induced changes are the dominant processes in terms of landscape and terrain shaping [1]. Erosion has multiple environmental and economic impacts. The first and most obvious impact is the degradation and productivity loss of fertile soils.

Population growth goes hand in hand with a growth in food demand. The removal of the natural vegetation, deforestation, and the intensification of crop cultivation have increased the vulnerability of soil towards erosion [2,3]. Based on the results by [4], only during the last century, the per-capita removed soil has increased by around 400%. In comparison to 2000 years ago, the per-capita removed amount of soil today is around 2000% higher. In contrast, soil formation is extremely slow. Under tropical and temperate agricultural conditions, 200 to 1000 years are needed for the creation of 340 t ha^{-1} of soil. The yearly renewal rate is around 0.2–2 t ha^{-1} a^{-1}, while the soil loss in intense agricultural regions fluctuates from 10 to 100 t ha^{-1} a^{-1} [5]. A recent review study of [6] suggests a crop yield loss of up to 10% is to be expected by the year 2050 if the actual rates of soil loss continue. With such high differences between soil loss and renewal rates and also the high impact that soil loss has on the global food availability soil conservation practices become essential concerning the global food economy.

Water is the main natural erosive agent, as it is responsible for 80% of soil erosion worldwide [7]. Erosion has severe impacts on the aquatic ecosystems and water budget in reservoir systems. Sediment input and related nutrient flux due to erosion are the main factors deteriorating the water quality, threatening aquatic biodiversity, and reducing the lifetime of river impoundments. Therefore, soil loss is not just an issue concerning only food scarcity, but also water scarcity.

The cross-scale characteristics of the erosion phenomenon with its high spatial-temporal variation may cause high costs for the adequate quantification of soil loss by monitoring programs. Hence, alternatives like modeling are often considered for quantification of soil loss and localization of hotspots. A vast range of model types (physical, stochastic, or empirical) has been developed, but these models are normally specific to local or regional environmental conditions, and the performance varies based on the data availability and quality [8].

The aim of this study is to validate sediment input modeling by using validation measurements of sediment stock from the long-term siltation estimate in a reservoir. Large river impoundments represent the perfect opportunity as they often have trapping efficiencies >95% and consequently may serve as validation points for transported material [9–11]. For this study the sediment stock was assessed by high-resolution sediment magnitude measurements in the reservoir via a dynamic penetrometer [12]. For calculation of the revised universal soil loss equation (RUSLE) factors, satellite data were used for the land-use and land-cover (LULC) assessment, two soil sampling campaigns to define the soil properties of the area, and available datasets from the literature.

2. Materials and Methods

2.1. Study Area

The Passaúna Reservoir catchment (152.6 km^2; 25°31′43″ S and 49°23′37″ W; 25°18′15″ S and 49°21′03″ W) is located near the Metropolitan Region of Curitiba and is part of a water supply system that provides water for more than three million people. About 30% of the population of the Metropolitan Region of Curitiba is supplied by this catchment. In 2001, the Environmental Protection Area of Passaúna was established, comprising 16,060 ha of territory Even so, anthropic pressure on the catchment has continued over the years (Figure 1). The Passaúna river composes 65.6% of the contribution area of the reservoir, followed by the contribution of the small sub-basins < 1 km^2 (8.4%), the Ferraria river (6.9%), the reservoir area (5.9%), the runoff lands around the reservoir (4.0%), the Eneas river (3.6%), and two other unnamed sub-basins with 3.2% and 2.6%, respectively [13].

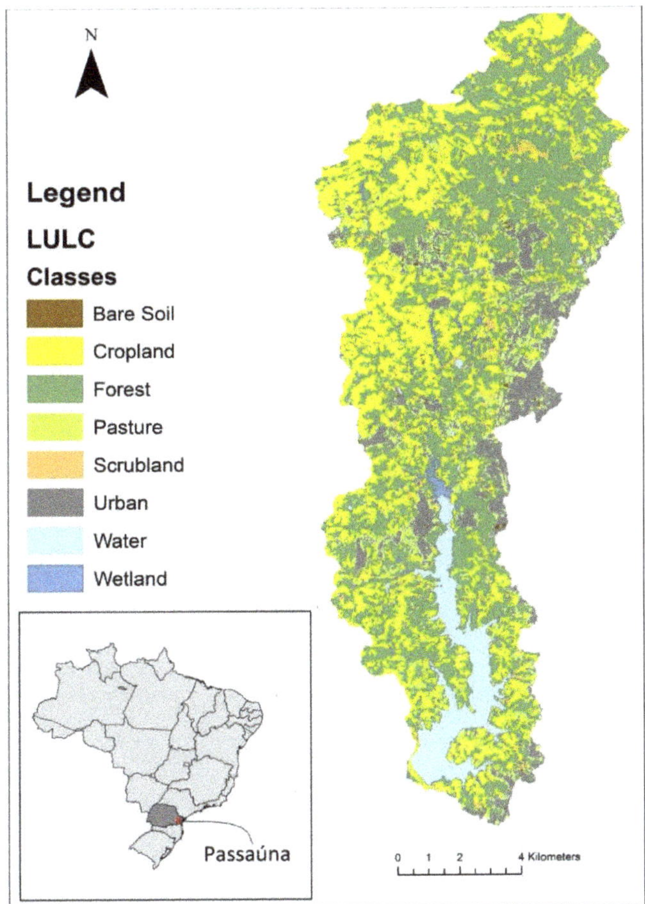

Figure 1. Location of Passaúna catchment in the lower left corner of the graph, and the land use/land cover of the catchment.

Passaúna reservoir initiated operation in 1989. The total water surface area is 895 ha and the reservoir has an actual volume of 69.3 hm^3, considering the spillway level at 887.2 masl. The intake is located approximately 3 km upstream of the dam. The impoundment structure is a 1200 m long and 17 m high rock-fill dam with a clayey core.

2.2. Sediment Yield Model

The sediment input (or sediment yield) is calculated as a product of soil loss from the hillslopes and a sediment delivery ratio:

$$SI = A \cdot SDR \tag{1}$$

where SI stands for sediment input (or sediment yield), A for soil loss, and SDR for sediment delivery ratio.

As mentioned, the soil loss is calculated based on the RUSLE model. The universal soil loss equation (USLE) originated from [14] to assess the soil erosion in US agricultural land. Research for quantifying the soil loss started in 1940 in the Corn Belt and ended with the final publication by [14], where figures and relations were added for calculating each of the parameters. The next development in USLE happened in 1997, when [15] published the revised form of the Universal Soil Loss Equation (RUSLE). In the new version of RUSLE, the core philosophy of USLE was retained, even though significant changes in the calculation of the single parameters were included. The idea of USLE/RUSLE consists in the parametrization of the factors that affect erosion (terrain geometry, soil physical properties, rain characteristics, land use/land cover, and conservation practices).

In this study, due to the adequate data availability, a model in a monthly time resolution was used. Mathematically, RUSLE is presented in the following form:

$$A = L \cdot S \cdot R \cdot C \cdot K \cdot P \tag{2}$$

where

- A is the soil loss at the investigated area
- L is the slope length factor
- S is the slope steepness factor
- R is the rainfall-runoff erosivity factor
- C is the cover management factor
- K is the soil erodibility factor
- P is the support practice factor

2.2.1. Topographic Factor LS

LS expresses the expected ratio of soil loss per unit area from a field slope to that from a 72.6 ft (22.13 m) length of uniform 9% slope under otherwise identical conditions [14]. The relation was adapted by [16], especially for the L Factor. The basis for calculation of the pixel-based topographic factor was a digital elevation model (DEM) of an accuracy of 10 m available from TanDEM-X service (Figure 2). For calculation of the LS Factor, the open source platform inVEST was used [17]. The LS Factor was calculated as follows:

$$LSi = (((A_{i\text{-}in} + D^2)\hat{}m + 1 - A_{i\text{-}in}\,\hat{}m + 1)/(D\,\hat{}m + 2 \cdot x_i\,\hat{}m \cdot 22.13\,\hat{}m)) \cdot S_i \tag{3}$$

where:

- S_i is the slope factor calculated from terrain slope θ in radians as showed below $S = 10.8 \sin\theta + 0.03$ when $\theta < 9\%$ $S = 16.8 \sin\theta - 0.50$ when $\theta > 9\%$
- D is the gridcell dimension
- $A_{i\text{-}in}$ is the contributing area (m^2) at the inlet of a grid cell which is computed from the d-infinity flow direction method
- $x_i = |\sin a_i| + |\cos a_i|$ when $\theta > 9\%$ and a_i is the aspect direction for grid cell i
- m is the length exponent factor (Table 1)

Table 1. Values of dimensionless factor m.

Slope % [s]	m
s < 1	0.2
1 < s < 3.5	0.3
3.5 < s < 5	0.4
5 < 9	0.5
s > 9	m = β/((1 + β)) [1]

[1] $\beta = ((\sin\theta/0.0986))/((3 \cdot \sin\theta\,\hat{}0.8 + 0.56))$.

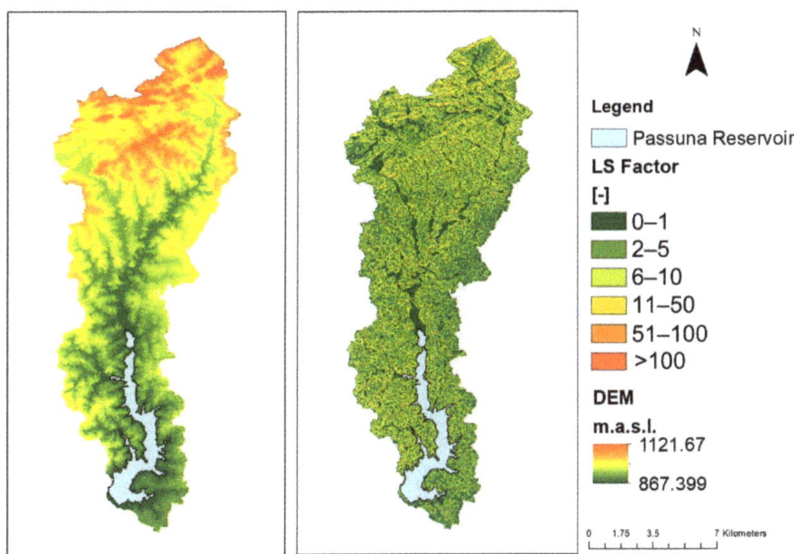

Figure 2. (**Left**). Digital elevation model; (**Right**) spatial distribution of LS Factor.

2.2.2. Soil Erodibility Factor K

The K Factor corresponds to the soil erodibility or the soil's intrinsic susceptibility to erosion, which reflects the spatial variability of possible soil erosion depending on its structural and compositional characteristics [18]. This factor can be determined through experiments, and carried out in field plots using a specific measurement setup [19]. Alternatively, it may be obtained from predefined estimates based on the soil classes documented in the published literature reporting soil erodibility values for soil classes observed in different regions of Brazil (Table 2).

Table 2. K Factor values from literature data base.

Soil Class	K Factor Value (t h MJ^{-1} mm^{-1})	Soil Class
Haplic Inceptisol	0.03	[20]
Humic Inceptisol	0.0175	[21]
Oxisol	0.018	[22]

In order to determine the K Factor, two soil sampling campaigns were organized in the Passaúna catchment with a total of 22 soil samples (Figure 3Left). The texture (silt, clay, and sand fractions) and loss on ignition at 550 °C (LOI550) were defined for each sample. For each point, three subsamples were taken as replicates within a radius of 5 m. Disturbed material was dried and sieved in 2 mm mesh, and the texture analysis was undertaken by the Bouyoucos hydrometer method [23] based on the classification by the United States Department of Agriculture (USDA), which addresses that the particle sizes between 0.05–2 mm are sand, between 0.002–0.05 mm are silt, and smaller than 0.002 mm are clay. For the samples of the first campaign, also some physical parameters of the soil were measured. All soil samples were used to calculate the K Factor at each location. For this study, Equation (4), proposed by [24] for the sample points collected covering Ultisol, Red Oxisol, and Typic Eutraquox classes, was used.

$$K = ((SAN + SIL)/CLA)/100 \tag{4}$$

where SAN, SIL, and CLA are sand, silt, and clay fraction in percentage, respectively.

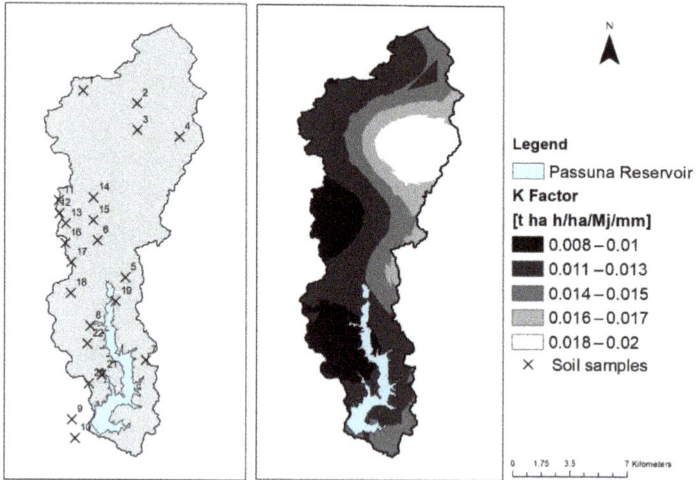

Figure 3. (**Left**) Location of soil samples; (**Right**). Interpolated map of K Factor.

Afterwards, the values were interpolated using the inverse distance weighting (IDW) approach in order to obtain the information for the full coverage of the watershed.

2.2.3. Rain Erosivity Factor R

Based on the availability of data, two approaches to calculate the R Factor were investigated.

1. Based on literature findings

For lack of 10–20 min frequency precipitation data for the Passaúna catchment, initially literature findings were used to determine the rainfall erosivity in the catchment [25] studied extensively the relations between the rain erosivity calculated from pluviographic and pluviometric data. Optimally, the rain erosivity is calculated by using long-term pluviographic (disdrometric) data, even though this type of data is mostly unavailable. The pluviometric data is often more easy to access but has a major disadvantage as it gives no information about the duration of the rain [25]; derived three different equations for three different locations in Parana to relate the erosivity calculated from the pluviometric data (R_{Pm}) with the erosivity calculated from the pluviographic data (R_{Pg}). Based on the aforementioned research [26], calculated the erosivity factor for the whole state of Paraná in a monthly resolution (Figure 4). In their research [26], integrated data from 114 pluviometric and pluviographic stations with more than 20 years of data (1986–2008).

The values used for this study were extracted from the monthly erosivity maps for the area of Curitiba. For the whole catchment with its 150 km^2, a constant value of R was used for each month.

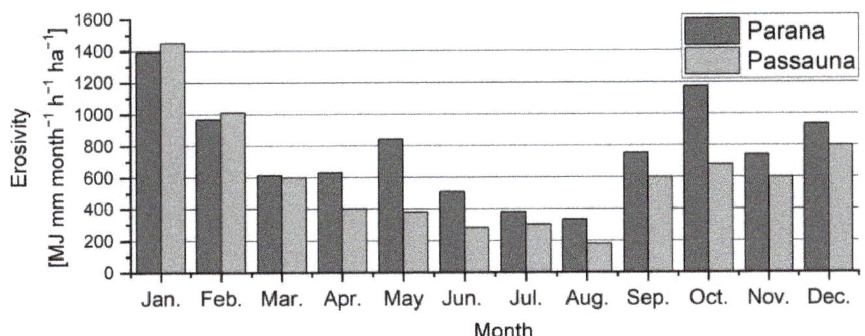

Figure 4. Erosivity in Passaúna and Paraná after [26].

2. Based on Pluviometric Data of Daily Frequency

For calculation of the R Factor using the second approach, the data of two pluviometric stations in the catchment was used. The stations are part of the hydrological information system of Instituto das Águas do Paraná. The station of Colonia Dom Pedro was located in the central part of the catchment while the other station Barragem Sanepar (Dam), is located in the Southern part of the catchment near the dam (Figure 5). For both stations, precipitation data from 2000 until 2018 were available with a daily resolution.

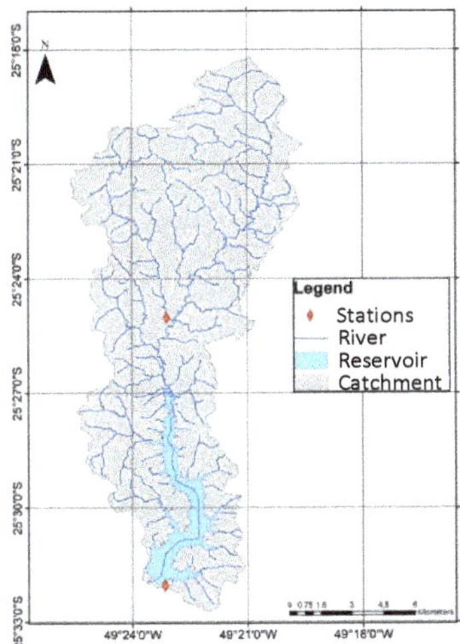

Figure 5. Location of pluviometric stations in the Passaúna catchment.

For calculation of EI, thus the R Factor, the approach by [27] (Equation (5)) was applied. The precipitation patterns at both locations are similar; therefore, only one value of erosivity factor was used for the whole catchment (Figure 6).

$$EI = 68.730 \cdot (C_c)^{\wedge}0.841 \quad (5)$$

$$C_c = (p^{\wedge}2)/P \quad (6)$$

where p is the average monthly precipitation in mm and P is the yearly average precipitation in mm.

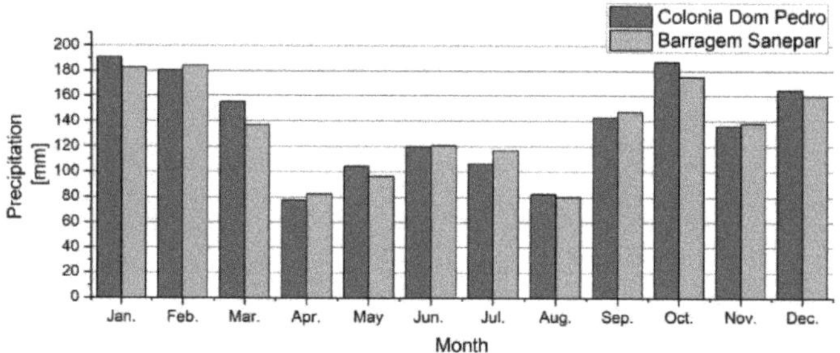

Figure 6. Monthly precipitation for the two locations in the Passaúna catchment.

2.2.4. Cover and Management Factor C

The land-cover factor C is one of the most important factors, when it comes to what causes the highest inconsistencies in the outputs of a RUSLE-based model [28–30]. Optimally, the C Factor is determined from experimental soil erosion plots under natural rainfall conditions [31,32]. This type of data is often expensive to produce and in most cases, C Factors are derived from the literature. One of the most important drawbacks for the use of constant C Factors is the high variability of values for the same land-cover class among different literature sources. A literature review by [33] showed that the C Factors among the same class could differ by up to a factor of 100 (Table 3). Another major disadvantage of constant C Factor values is the inability to capture the spatial and temporal variability of the factor values among the same LULC class. With the developments in satellite-based earth observation systems and the increase in data availability during the last decade, more scientists base their approaches on remote-sensing data [34–36].

Table 3. C-factor values for five land-use/land-cover (LULC) classes in Brazil from a literature review by [33].

Land Use	C_{max}	C_{min}	C_{min}
Bare soil	1.000	0.696	0.100
Impervious areas	1.000	0.257	0.000
High vegetation	0.090	0.008	0.00004
Low vegetation	0.630	0.099	0.008
Water	0.000	0.000	0.000

For calculation of the C Factor in this study, the Sentinel-2 data was processed, and spatial information about LULC, urban soil sealing, and NDVI was derived.

For generation of the LULC maps, the Random Forest algorithm was used for pixel-wise labeling of a Sentinel-2 time series raster stack [37]. The scenes were selected based on image quality criteria and with the aim of representing different phenological phases. Train and test sample data were collected through visual interpretation of aerial images and field work. The estimate of overall accuracy based on a hold-out test set is 84%.

NDVI was the core parameter derived from the Sentinel-2 dataset. The use of NDVI values for calculation of the C Factors enabled a model setup with a monthly temporal resolution. NDVI is related to the vegetation density, biomass, and productivity [38]. It was calculated based on the 10 m red (Band 4) and near-infrared (Band 8) bands of Sentinel-2. An automated processing chain was established comprising the download, preprocessing (atmospheric correction), optimized cloud masking, scene selection, and processing of land surface variables. The automated processing was focused not only on NDVI but also on other variables like the degree of soil sealing or LULC.

Degree of soil sealing or imperviousness is defined as the fractional coverage of artificially sealed ground which impedes water from infiltrating into the ground. The calculation of imperviousness is based on a strong inverse relationship between vegetation cover and impervious surface as well as on the idea that an urban landscape can be linearly decomposed into vegetation, impervious layer, and soil [39,40]. The imperviousness layer was calculated based on a min-max rescaling of the NDVI derived from satellite acquisitions between the maturity and senescence onsets. The rescaling was guided by a visual comparison of results with submeter resolution aerial images as well as findings by [40], who studied the linear relationship between NDVI and imperviousness across several European cities.

Two NDVI-based approaches were considered for calculation of the C Factor in this study: [34,41]. As shown in [32], for the conditions prevailing in Brazil, the methodology derived from [34] (Equation (7)) produces more reliable results. Therefore, this approach was used for calculation of the C Factor.

$$C = (-\text{NDVI} + 1)/2 \tag{7}$$

The previously mentioned satellite-derived data was used to calculate the C Factor also in non-sealed urban areas. As can be seen from Figure 7, in the urban areas, the NDVI is in the range of 0.25, which would result in a C Factor of 0.35–0.40, which corresponds to the C Factor values of arable land. Therefore, a filter was applied to the data with the simple logical condition that if a pixel in the urban areas had more than 60% soil sealing, the NDVI at the same location was set to 0.999, as it was assumed that no or very little sediment can occur from sealed areas.

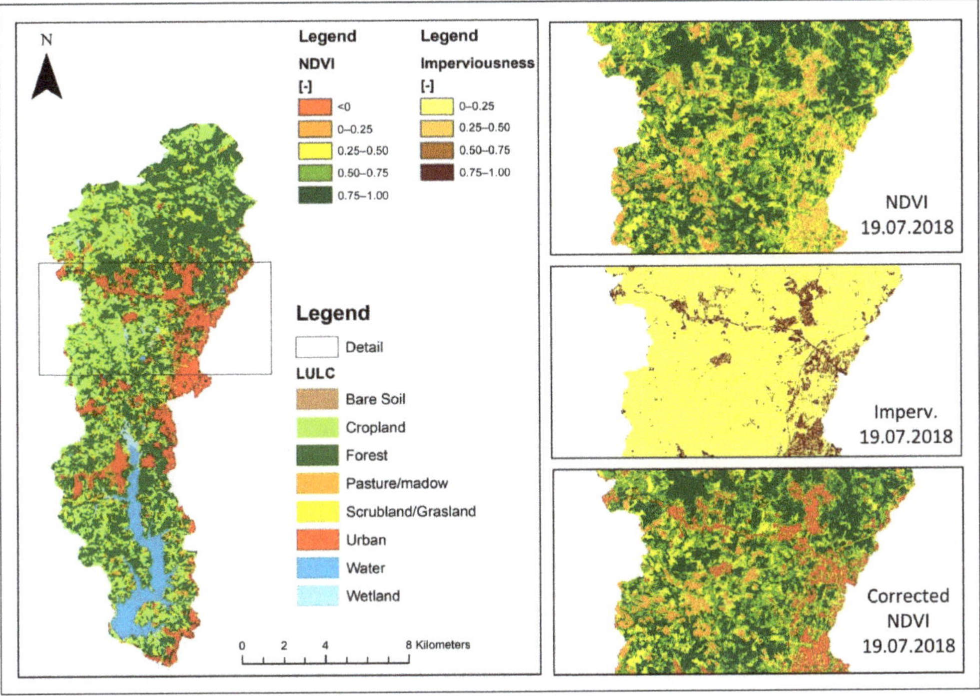

Figure 7. Correction of normalized difference vegetation index (NDVI) values for urban areas.

2.2.5. Conservation Practices Factor P

During several field trips in the Passaúna catchment, many agricultural properties were visited. Support practice was observed at almost none of them (Figure 8). Therefore the P Factor was set to a constant value of $P = 1$.

Figure 8. Typical arable land in Passaúna catchment.

2.2.6. Sediment Delivery Model

The sediment delivery ratio plays a crucial role in the outcome of the final sediment amount reaching the river, as it is directly related to a large number of factors (amount of soil displacement, geometry of the transporting paths, land cover of the surrounding area, or amount of surface runoff) [42]. For this study, the SDR was calculated based on the flow connectivity approach by [43]. Hydrological connectivity is a term often used to describe the linkages between runoff and sediment generation in the upper parts of catchments and the receiving waters [44]. The use of the connectivity index as an input parameter for SDR has shown satisfying results globally [45–48]. For this study, the calculation of the connectivity index, and subsequently SDR, was implemented in ArcMap 10.5 as described in [49] based on the following formula:

$$SDR_k = SDR_{max}/(1 + \exp((IC_{0,k} - IC_k)/K_{IC,k})) \quad (8)$$

where SDR_{max} is is the maximum attainable SDR coefficient at kth cell, set to 1, as soil in the Passaúna catchment has a high percentage of clay (0.002 mm), silt (0.002–0.05 mm), and fine sand (0.05–0.25 mm). IC_k is the index of connectivity as explained in [49], $IC_{0,k}$ is a calibration parameter with a value of 0.5 [43,50] and $K_{IC,k}$ is a calibration parameter with a value of 2.0 [43,50].

2.3. *Sediment in the Reservoir*

RUSLE results represent averaged long-term (mostly over decades) soil loss and sediment input (when RUSLE is combined with the SDR) from the catchment. When dealing with the results, the challenges are encountered mainly in having a reliable assessment of the sediment input in cases where no monitoring station is available. Due to their high trapping efficiency, large reservoirs with long residence times act as sinks for the incoming sediment. Therefore, they can be used as suitable long-term validation points for sediment input modeling. Several studies were conducted in this regard [9,51–53]. In order to acquire fast and accurate sediment information in areas where no previous data are available, a so-called portable free-fall penetrometer (PFFP) was used to assess the sediment distribution in the reservoir. PFFPs are not new in marine research (mostly sediment management in harbors), while their application in freshwater is still limited [54–63].

For this study, the commercial system GraviProbe (GP) produced by the Belgian company dotOcean (Figure 9) was used. Often, the spatial distribution of sediment thickness in a reservoir can be determined by bathymetric surveys, if precise pre-impoundment bathymetry is available [9,64,65]. Most of the time, this information is missing, and even when it is accessible, often the accuracy of the data does not allow for proper sediment stock estimation. One of the main advantages of the GP is its independence from previous data. The GP can deliver rapid results on penetration depth, cone penetration resistance, and shear strength of the sediment for each deployment. The GP was deployed at 134 points in the Passaúna reservoir (Figure 10). To determine the sediment magnitude, the information from the dynamic cone penetration resistance (DCPR) at some locations in the reservoir was related to the sediment magnitude data from core samples [12,66]. Characteristic changes in the sediment density/composition were identified and then related to the changes in the DCPR from the GP. The sediment thickness could be derived for all other points, where GP data was available because the relation between DCPR and sediment-pre-impoundment soil interface was established. This was possible due to the fact that the share of sand and coarse particles in the Passaúna sediment is smaller than 5% at most locations and a full penetration could be achieved. From the relation between core samples and GP information, it was observed that a DCPR of 200 kPa is the threshold between the sediment and the pre-impoundment soil. More information about the method can be found in [12].

Figure 9. Picture of GraviProbe (GP) after deployment.

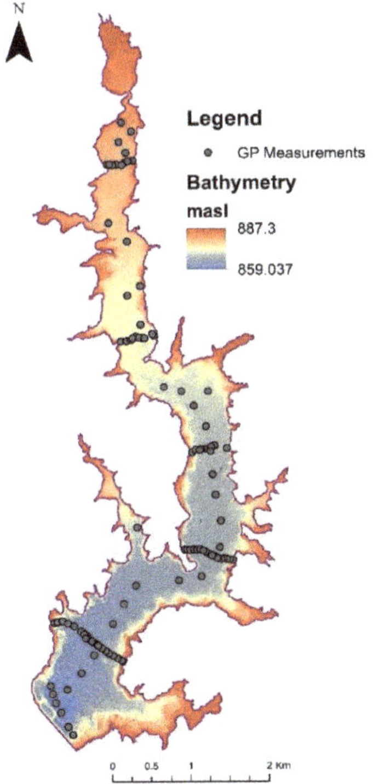

Figure 10. Location of all GP measurements.

3. Results

3.1. C Factor

For each of the available NDVI maps, the C factor was computed. The highest seasonal changes in the C Factor values were observed for the cropland (Figure 11). Between January and February, which is harvesting time, and November, which is seeding time, there is a change of almost 100% in the average C Factor of the catchment (from 0.15 to 0.28). A high interannual change in the C Factor was also observed in the scrubland/grassland areas. Winter and spring are characterized by a low vegetation coverage, whereas summer and partially autumn reveal a high vegetation coverage. Forests showed moderate changes mainly because a small percentage of the trees in humid subtropical regions lose their leaves in winter. The seasonal change in the forest average C Factor (change of <0.05 between maximum and minimum average C Factor) can also be related to the misclassification of certain areas with other LULC into forest class. Pasture and meadow follow also a similar land cover pattern. In summer and autumn, the vegetation cover is high, and in winter and spring, it diminishes. Bare soil has the smallest changes of all classes. There is a seasonal change of a maximum of 0.05 among the months and this can be attributed to the errors of the LULC classification process.

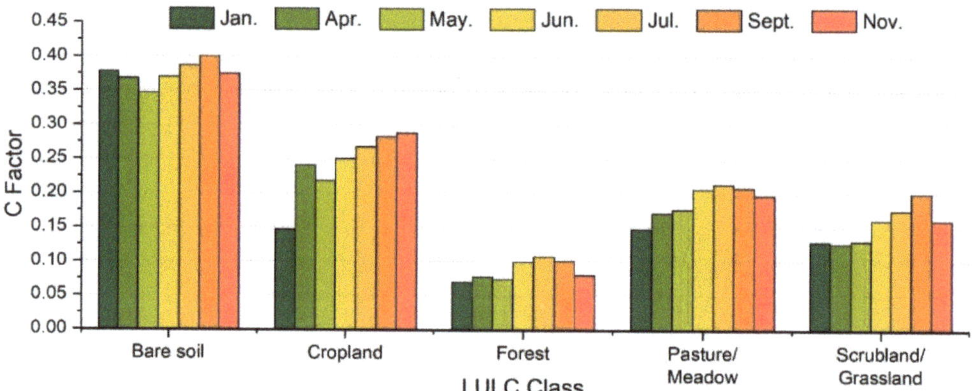

Figure 11. Mean C Factor for each land-use/land-cover (LULC) class for all months with available NDVI data.

The difference among the seasons can be clearly observed also in the spatial distribution of the C Factor (Figure 12). The western area of the catchment, where most of the agriculture activity is located, shows higher values in July than in January. In July (winter period), the soil is mostly uncovered and has an average C Factor greater than 0.3. While in January (summer and wet season), vegetation covers most of the catchment area. Only sporadic parts of the agricultural areas, which were not seeded, had high C Factor values also in January.

3.2. R Factor

The R factor computed based on precipitation data showed results different from those calculated by [26]. The largest differences are observed in January, April, and October. The data by [26] shows high differences among the months and overestimates substantially for the month of January. The precipitation data from both pluviometric stations show a more uniform distribution than what is suggested by [26] (Figure 13). The calculations by [26] also include a margin of error due to the low density of weather stations. In certain regions, these coarser maps cannot represent accurately the rain erosivity when brought in a mesoscale plot. Therefore, for the final calculation of erosion and sediment input, the R Factor calculated from the pluviometric data in the Passaúna catchment was used.

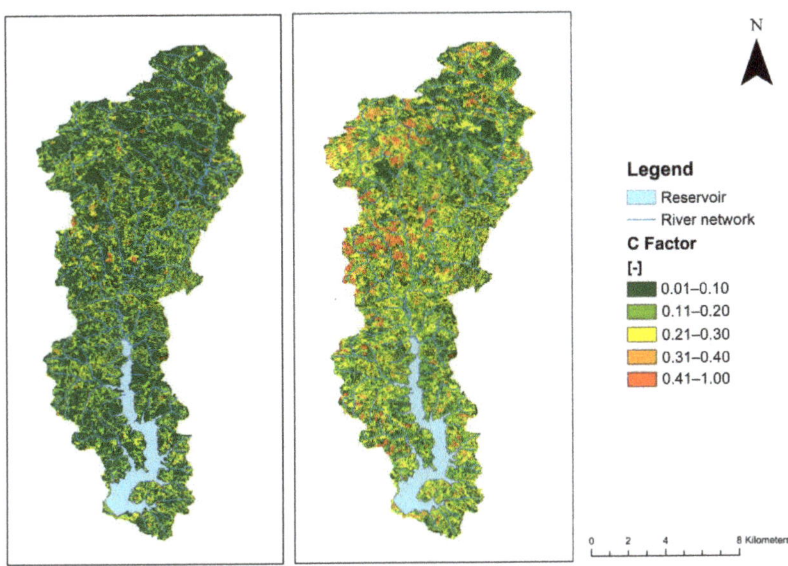

Figure 12. Spatial distribution of C factor for January (**left**) and July (**right**).

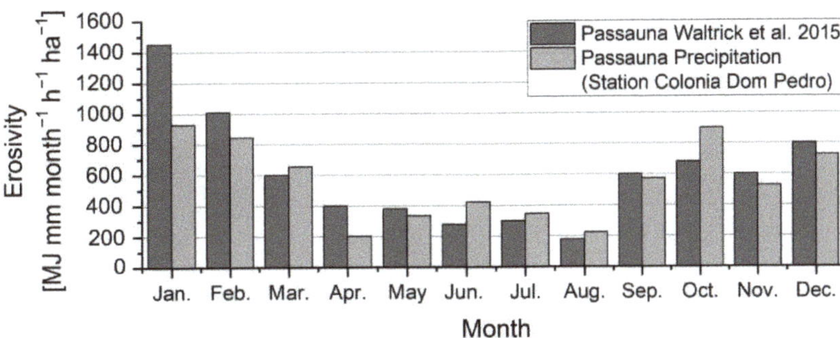

Figure 13. R Factor from two approaches.

3.3. K Factor

In general, the soil in the Passaúna catchment shows a low erodibility factor (<0.02 t ha h ha^{-1} MJ^{-1} mm^{-1}). The most erodible soils, dominated by Distrofic Latossol (Oxisol), are located in the northern part of the catchment, according to the soil map provided by the Brazilian Agricultural Corporation (EMBRAPA) [67]. The results are also aligned with further literature values in that geographic area, which also assessed that the K Factor for Latossol (Oxisol) is in the range of 0.019–0.026 t ha h ha^{-1} MJ^{-1} mm^{-1} [68–71]. The western part of the catchment, which is also dominated by Oxisol, showed low soil erodibility with values reaching up to 0.013 t ha h ha^{-1} MJ^{-1} mm^{-1}.

In general, as shown in Figure 14, the soil has a similar texture pattern throughout the catchment area. The silt-clay content of the samples was always larger than 50%. The sand content in the soil is also relatively high (reaching up to 50% at some locations). Most of the catchment is covered in sandy clay, which has a low to average erodibility.

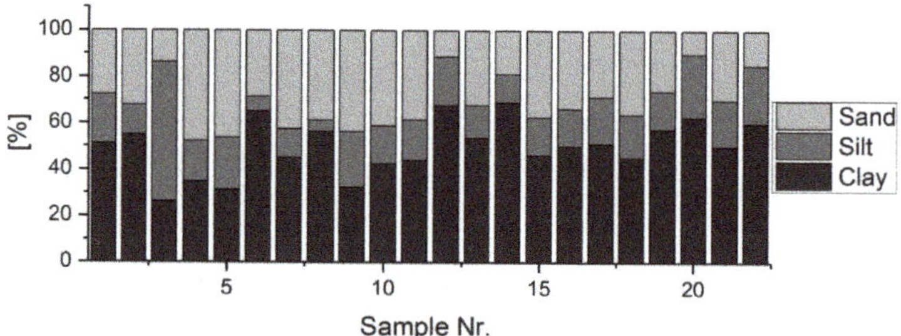

Figure 14. Texture of soil samples (sample number as in Figure 3 Left).

3.4. Sediment Delivery Ratio

Based on the physiographic characteristics of the Passaúna catchment, the SDR was calculated for each of the investigated months by applying the approach developed by [43] (Figure 15). In general, the calculated SDR could reach values of up to 0.15 in the dry months and rarely in some locations of above 0.15. The interannual vegetation cover which characterizes the region contributes to having low SDR values throughout the catchment. The highest SDR was observed in unprotected soil areas near the river stretches and at high slopes. The largest part of the catchment has SDR values lower than 7.5% in both dry and wet seasons. As explained in [49], connectivity, thus SDR, varies in both time and space. To define the change in the spatial patterns, the mean SDR was computed for each of the months. The results show low differences between the months. The mean SDR for the dry month of July was calculated to be 6%, while for the wet month of January, it was 5%.

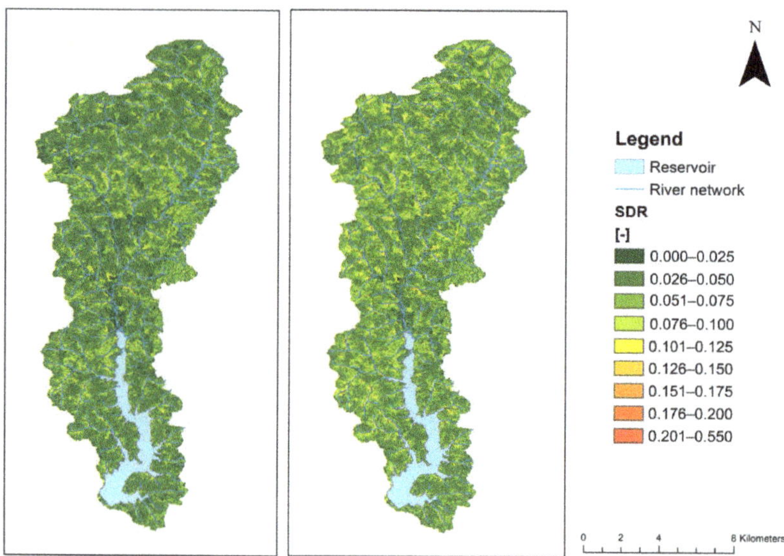

Figure 15. Sediment delivery ratio (SDR) for the months of July 2017 (**right**) and January 2018 (**left**).

3.5. Sediment Input—Initial Model Run

The results show high sediment input in all of the wet months. The highest sediment input happens in January with 14,000 t, even though the vegetation coverage of the catchment is rather high. In the month of August, despite the low vegetation cover, the overall sediment input from the catchment is the lowest (Figure 16). By comparing the soil loss distribution to the LULC map (Figure 17), it can be observed that high sediment input occurs from the forested areas. Even in the months of winter, when precipitation is low, there is significant sediment input from the forested areas. By comparing the mean value of the calculated C Factor for the forest areas in the Passaúna catchment with literature values, it was found that the assessed C Factor is significantly overestimated (Figure 18) from the method used in this paper. The average C Factor found by [33] (Table 3) is between 10 to 20 times lower than the calculated C Factor values during the months of July and October (Figure 18) from this study. The values of Max. and Average in Figure 18 refer to the maximum C Factor found in the literature review (*Min* = 0). The approach developed by [34] seems to overestimate the C Factor in forested areas, even though it is not sure from their research whether the empirical approach they developed can be used in forested areas. Therefore, an arbitrary correction factor of 0.05 was applied to the C Factor by multiplication. This value of 0.05 was chosen, as the calculated values of C Factor were 10 to 20 times higher than the average C Factor of forest areas in that region. Furthermore, the new R Factor, calculated from the pluviometric data, was also included in the new equation for giving the final results shown in Figures 19 and 20.

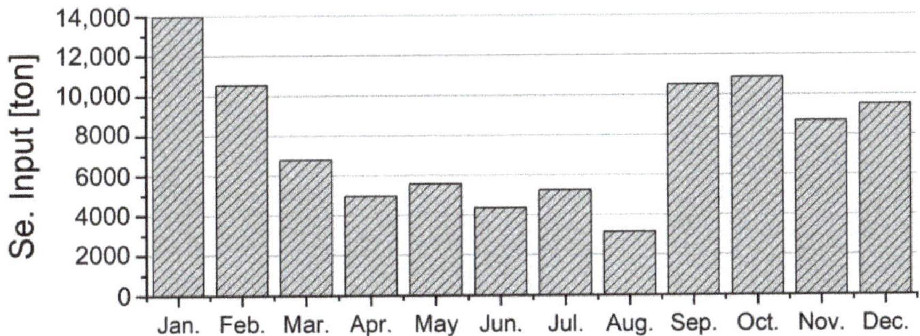

Figure 16. Monthly distribution of sediment input from the initial model run.

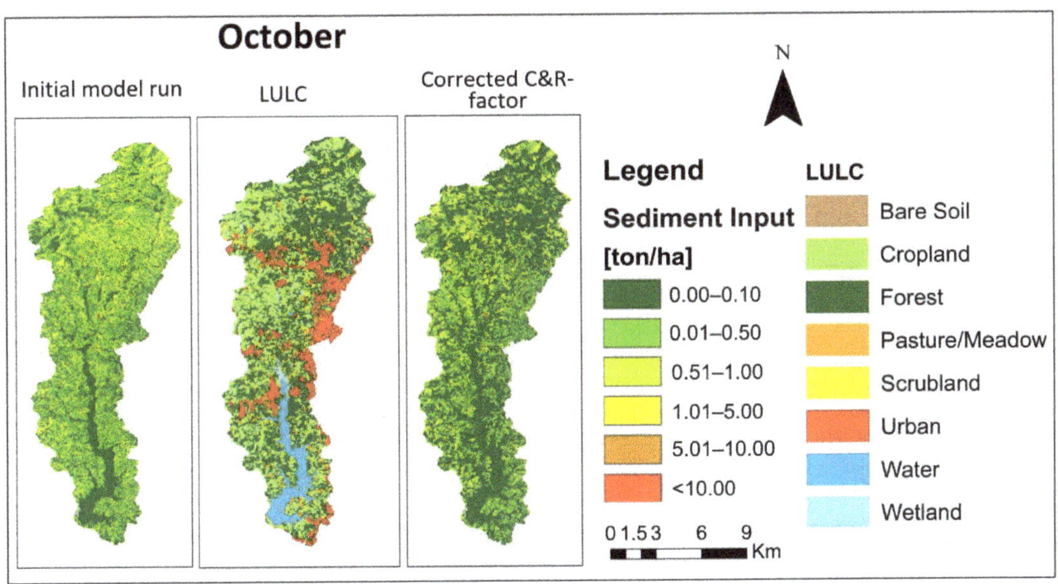

Figure 17. Comparison between initial model run and LULC.

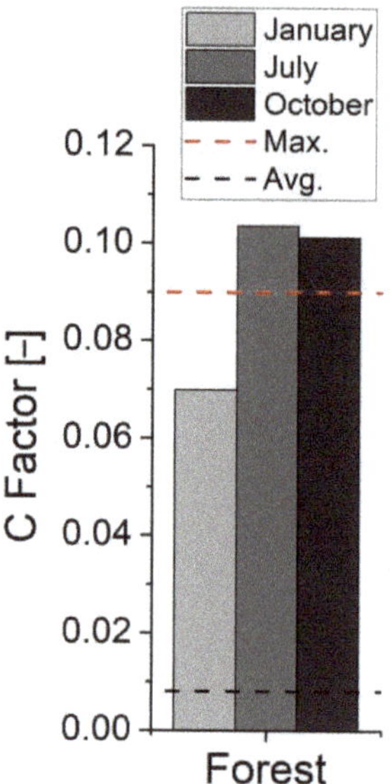

Figure 18. Comparison of the average C Factor to maximum and average values found in the literature.

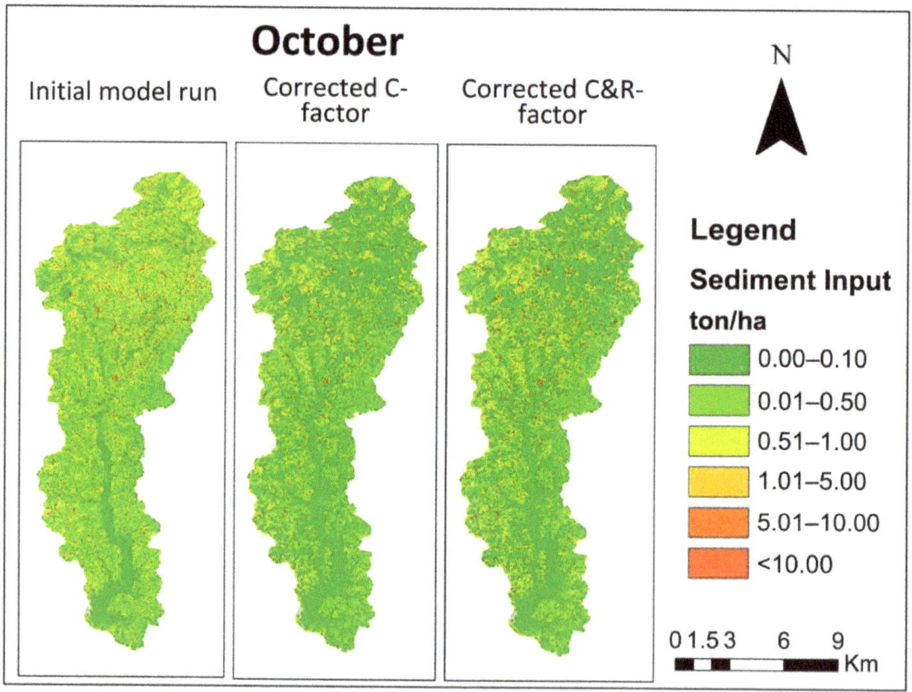

Figure 19. Final distribution of sediment input after C and R Factor correction.

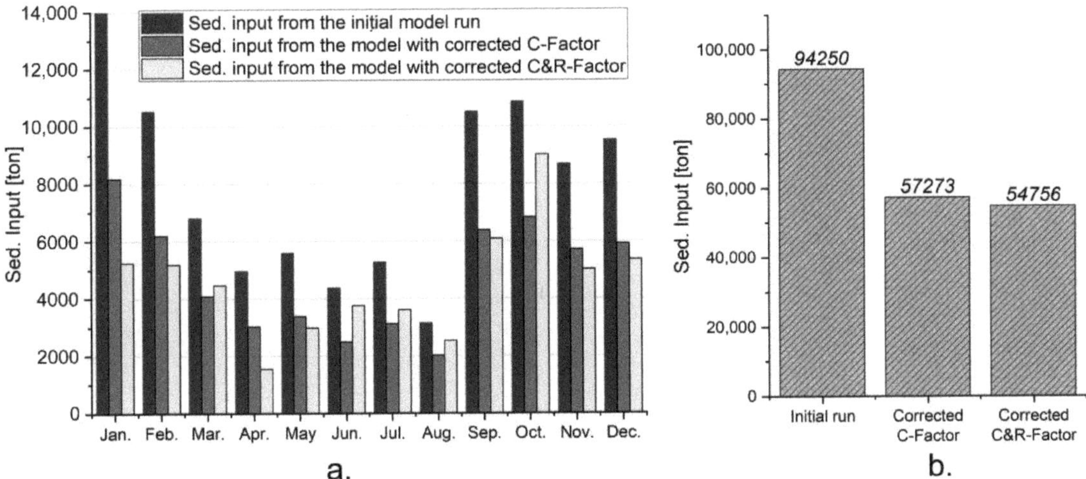

Figure 20. (**a**) Comparison of the interannual dynamics of the system (**b**). Comparison of the yearly sediment input.

The C Factor correction decreased the overall amount of sediment input into the Passaúna reservoir by 30% from an initial 94,300 t a^{-1} to 57,300 t a^{-1}. After the inclusion of the new R Factor calculated from the daily precipitation data, the sediment input decreased by a further 5% to 54,800 t a^{-1} (Figure 20b). The use of the new R Factor shifted also the seasonal dynamics of the sediment input. The final model indicates that the most important month in terms of sediment yield is not January but October (Figure 20a). The month with the lowest input is April and not August, as suggested by the initial model results. The spatial distribution of sediment input changes significantly between the final model run (C and R Factor correction) and the initial model run (Figure 19). The sediment input from forested areas is reduced substantially in the final model run to less than 0.1 t ha^{-1}. For the overall operational time of the Passaúna reservoir (30 years), according to the modeling results, the accumulated sediment stock should be approximately 1.6×10^6 t.

3.6. Reservoir Sediment Stock

As explained in [12], a DCPR of 200 kPa is defined as the vertical consolidation threshold between the sediment overlay and pre-impoundment soil. The sediment in the reservoir showed a high spatial heterogeneity of the siltation patterns. Even points at a horizontal distance of 10 m showed different sediment thickness values, mainly because of the bottom topography. This underlines the need for large numbers of measurement points, to obtain representative estimates of the accumulated sediment. Near the deepest part of the reservoir, a sediment thickness of up to 1.8 m could be observed. The areas with the highest sediment accumulation are located near the dam and near the inflow (Figure 21). Also, the sidearm located in the southwestern part of the reservoir showed high sedimentation rates compared to the northern areas of the reservoir. By applying the inverse distance weighting interpolation technique, the spatial distribution of the sediment magnitude was obtained. In total, a stock of 3.4×10^6 m^3 of sediment could be measured in the reservoir. According to [66], the sediment has an average density of 1.12 g/cm^3. Therefore, the total mass of sediment in the reservoir is approximately 3.8×10^6 t.

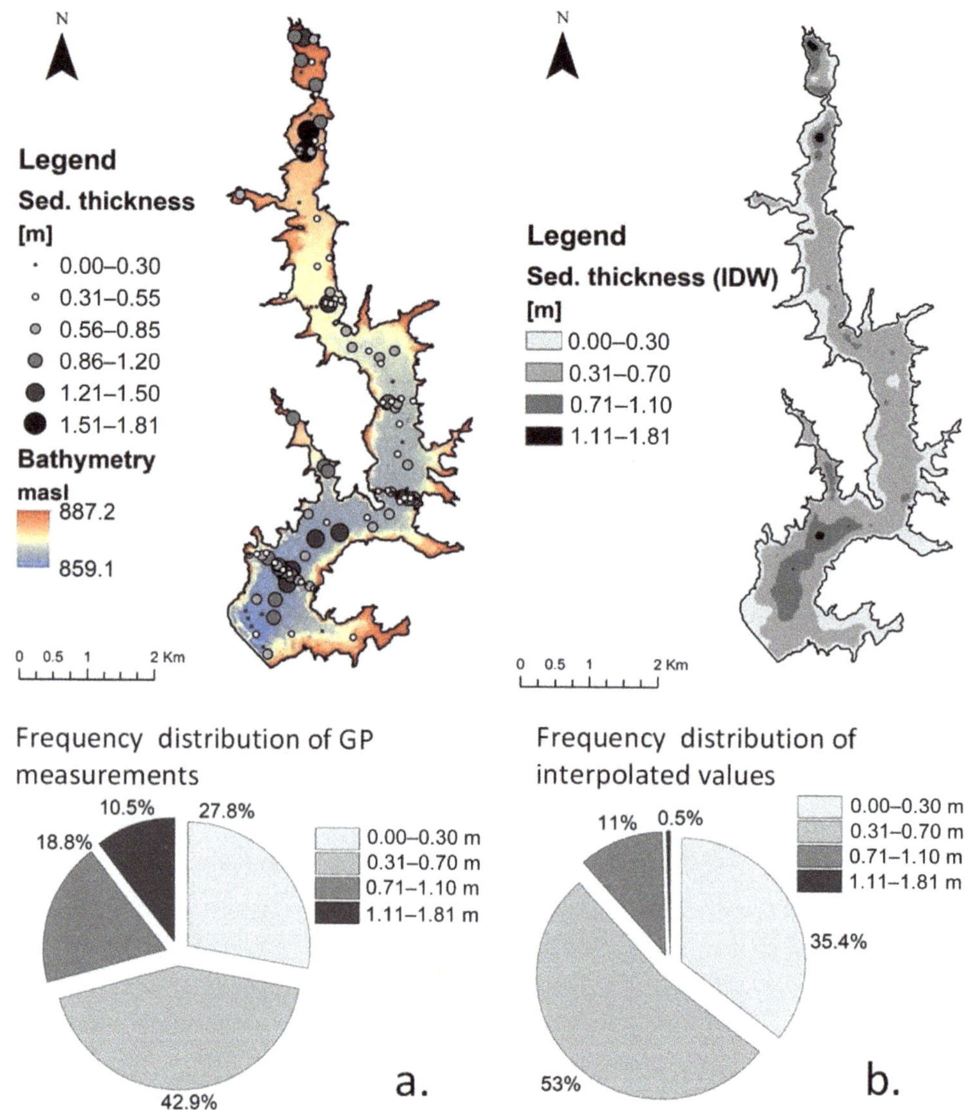

Figure 21. (**a**) Locations of GP measurement and the visualization of the measured value. The pie chart shows the frequency distribution of the measured values of sediment magnitude (**b**). The interpolated map of sediment thickness based on the GP measurements with distribution frequency of the interpolated values. Adapted from [12].

4. Discussion

4.1. Comparison of the Approach to Literature Findings

Several studies were conducted in the Alto Iguacu area in regard to soil erosion [71,72]. Reference [72] conducted a similar study in the Passaúna catchment even though the methodology followed to calculate erosion was different. The soil loss and sediment input were calculated in a yearly time step and the calculation of the C Factor was based only on a LULC map. Despite the similarities in the spatial distribution patterns, the findings from our study indicate that the soil loss is lower than the amount calculated by [72] (Table 4).

Our results show that almost 63% of the catchment had very slight, slight, or moderate soil loss, against 52% found by [72]. Major differences were also observed in areas with very severe and catastrophic soil loss. [72] calculated that 33% of the catchment had more than 100 t ha^{-1} a^{-1} of soil loss, while our results showed that only 12.7% of the catchment had more than 100 t ha^{-1} a^{-1}.

Table 4. Comparison of results by this study with results by [72].

Soil Erosion Classes (Annual Mean)	Present Study (%)	[72] (%)
Very Slight (< 2 t ha^{-1} a^{-1})	55	52.0
Slight (2–5 t ha^{-1} a^{-1})	3.5	
Moderate (5–10 t ha^{-1} a^{-1})	3.7	
High (10–50 t ha^{-1} a^{-1})	15.8	10.0
Severe (50–100 t ha^{-1} a^{-1})	9.0	5.0
Very Severe (100–500 t ha^{-1} a^{-1})	11.3	33.0
Catastrophic (>500 t ha^{-1} a^{-1})	1.4	

Reference [73] conducted another study in the Passaúna watershed, but focused mostly on the continuous monitoring of suspended solids in the Passaúna river before entering the reservoir. In his study, [73] collected 33 large-volume river samples between February 2018 and July 2019. In his study, also measurements from one intensively measured high-flow event of October 2018 were included. The point where the measurements were conducted collects water from 55% of the overall Passaúna Reservoir catchment. For this case, [73] calculated an annual average flux of 10,800 t a^{-1}. This value is approximately 300% lower than the value calculated for sediment input from 55% of the catchment from this study. Reference [73] explains the relatively low flux values with the importance of episodic high-flow events, whose dynamics are not properly described and, in this case, are strongly underestimated by the derived rating curves of suspended solids.

Other regional studies such as that by [71] or the more holistic study by [36] show similar patterns of soil loss in the area of Parana and Alto Iguacu. However, the information presented in these studies is too coarse and cannot be directly compared with our findings.

4.2. *Sediment Input from Catchment vs. Reservoir Sediment Stock*

By comparing the results from the two approaches, it could be observed that the sediment stock is 229% higher than the overall sediment input from the catchment, as calculated from the model (Figure 22). The discrepancies in the results of the modeling are rather high. However, when we refer to the sediment stock, all the material entering the reservoir is included, including here the organic and mineral material that was inside the reservoir before impoundment or which was deposited during the construction phase of the reservoir. In addition, based on the definition by [14], USLE (and subsequently RUSLE) accounts only for the sheet and rill fractions of the soil loss. Therefore, further factors have to be considered to reach a complete estimate of sediment input from the catchment (Table 5).

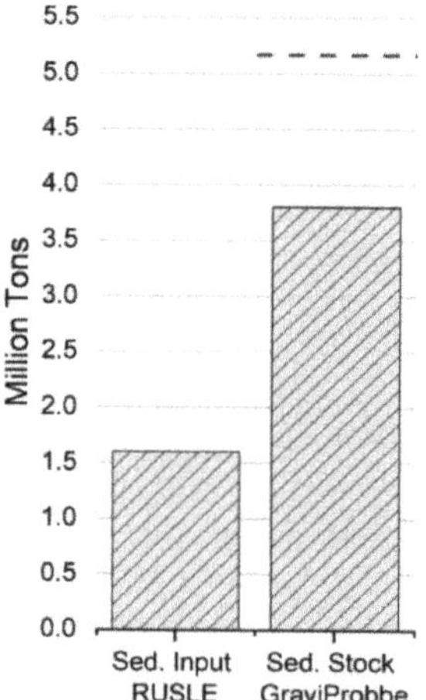

Figure 22. Comparison of the sediment stock in the reservoir with the sediment input from the catchment. The dashed line shows the margin of error as described in [12].

Table 5. Overview of factors creating inconsistencies.

	Factors Creating Errors
In reservoir	Internal production Existing biological stock Errors of the measuring concept Trapping efficiency of reservoir
In catchment	Errors associated with RUSLE calculations Errors associated with SDR calculations Non-inclusion of gully erosion in RUSLE Non-inclusion of channel erosion in RUSLE

One important factor that can create bias in the sediment budget is the error created from the measuring and processing technique in the assessment of reservoir sedimentation. After the interpolation, the frequency distribution of sediment magnitude values changes significantly as shown in the pie charts of Figure 21a,b. This indicates that the interpolation technique has a significant effect on the overall results. The average sediment magnitude of the raster is 40 cm, which is 30% smaller than the average of all measurements (57 cm). An underestimation of the average value from the interpolation technique shows an underestimation of the calculated sediment volume.

In order to properly compare the interpolated map with the measured values, the spatial component should also be taken into consideration. This means that if most of the measurements are located in the thalweg (disproportionally with its surface compared also to the bank slope areas), the average value for the measurements will be higher than the average from the interpolated values, as most of the accumulation is expected to be in the thalweg. Therefore, the reservoir was divided into two parts, thalweg and reservoir bank slope, as shown in Figure 23. For each of the compartments, the average sediment thickness from the GP measurements was calculated. Finally, an overall average value for the whole reservoir was calculated as shown in Equation (9).

$$M = (M_t \cdot A_t + M_b \cdot A_b)/(A_t + A_b), \tag{9}$$

where M_t and M_b are the averages of the measurements in the thalweg and reservoir bank respectively, while A_t and A_b are the areas of the aforementioned compartments.

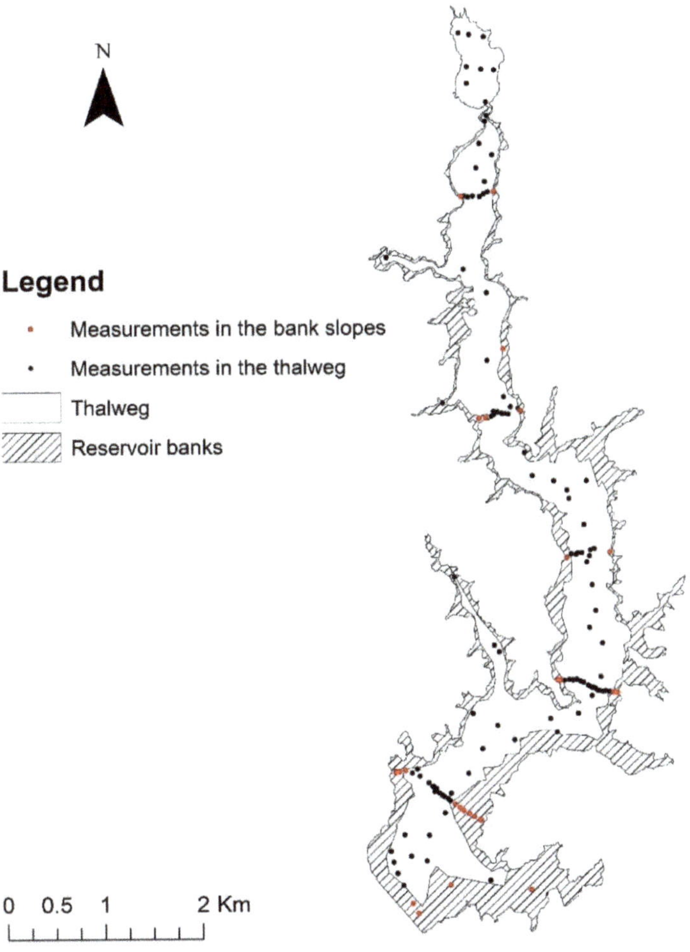

Figure 23. Division of the reservoir into compartments for the calculation of a representative averaged sediment thickness of the reservoir via the GP.

Based on Equation (9), the average sediment magnitude measured in the reservoir is 62 cm, thus 36% higher than the mean raster average (40 cm). Hence, the interpolation can lead to an underestimation of the sediment stock of up to 36%.

Another important factor, which can affect the sediment balance in the Passaúna reservoir, is the contribution of internal production to the sediment stock. Apart from acting as a sink, the reservoir acts also as a source of particles. Due to the climatic conditions and the relatively high nutrient availability (mesotrophic state), the reservoir is productive for plankton communities. Therefore, the autochthonous material created in the reservoir can play an important role in the sediment balance of the system. In other studies, it was observed that the autochthonous material can account for up to 75% of the sediment stock [74]. Even if the exact share of internally produced sediment cannot be defined, the high LOI (>20%) in the main basin of the reservoir in comparison to average values of ~10% at the inflow underlines the importance of autochthonous production. Moreover, before flooding, the reservoir area was not cleaned from the existing biomass. Several trees and former vegetation areas are still visible at the reservoir bottom (Figure 24). This organic material (sometimes degraded) also plays its role in the bias created when comparing both approaches.

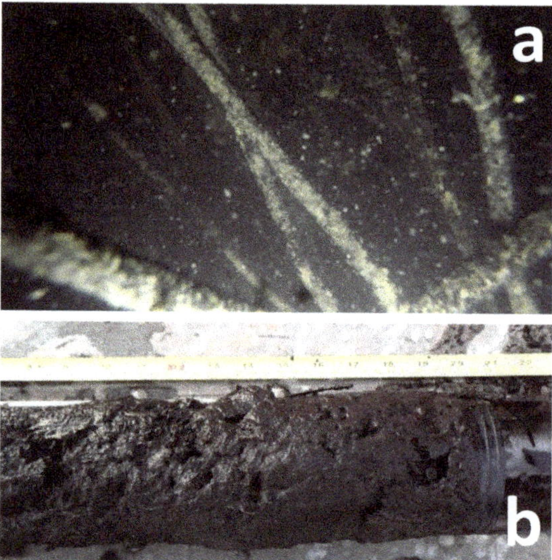

Figure 24. (**a**). Image from vegetation in the Passaúna reservoir bottom; (**b**). Sediment core from Passaúna Reservoir.

One of the most discussed limitations of RUSLE is its inability to represent also gully and stream bank erosion [75–77]. Even with the calculation of the SDR based on the connectivity index, the uncertainties about the prediction of gully and streambank erosion are still present. In comparison to sheet and rill erosion, gully erosion is generally less investigated. However, various studies [78–80] showed that gullies substantially contribute to the sediment budget at a catchment scale. They do not only contribute as a sediment source, but also increase the efficiency of sediment transport from uplands to the valley bottom and river channels, as most of the sediments generated from rill and inter-rill erosion that are not connected to gully structures are deposited at the foot of the hillslopes [81].

Reference [82] estimated that 47–83% of the sediment occurred from gully erosion. In addition, [81] indicated in a review study, that worldwide, gullies can represent 10–94% of the total sediment yield from water erosion. When referring to the sediment input from the catchment, it is still unknown to what extent the gully structures contribute to this budget for the case of Passaúna.

4.3. Limitations of Normalized Difference Vegetation Index (NDVI)-Based Approaches for the Estimation of the C Factor in Forested Areas

When water is the erosive agent, there are three main phases that characterize erosion. The first phase is the detachment of soil particles. In this phase, the potential energy of the raindrops due to its absolute elevation is transformed into kinetic energy. The free fall of the raindrops due to gravity causes remobilization of soil particles when the drops hits the soil surface. The second phase is the transport of the detached material by the accumulated flow, and the final phase of erosion is deposition, which occurs when the transport forces are depleted [14,83,84]. In the C Factor results before correction, a similarity in the values of plant-covered arable land and forest areas was observed. Despite the similarities in the cover canopy between planted arable land and forest (according to the NDVI values), the topsoil's physical properties between these two classes are completely different. While in the erosion component associated with the rain splash, both LULC classes behave similarly due to the similar protection by the plant canopy, in the component of erosion associated with runoff, forest and arable land behave differently. The soil surface below the plant coverage in arable lands is basically bare and facilitates the detachment of soil particles from surface runoff. In contrast to this, the soil in intact forests is normally covered by low vegetation (grass or meadows) and leaf litter, which hinders the creation of runoff and reduces the soil particle detachment. In addition, the soil is more compact in forested areas than in arable land where tillage almost always takes place. In its original form, the C Factor has a direct relation to the soil loss ratio (SLR) [15]. The SLR is a product of five sub-factors, which are prior land use, canopy cover, surface cover, surface roughness, and soil moisture. All of the former factors, except the canopy cover, are associated with the conditions of the soil surface, indicating the importance of the top soil conditions for soil movement initiation. Therefore, the C Factor cannot be quantified only by taking into consideration the vegetation index (canopy cover) but should also include the properties of the soil surface, especially in non-agricultural areas [85–88].

4.4. Management Implications

In terms of management, an analysis was performed on how the soil loss and sediment input from the catchment could be reduced by afforestation in the most problematic areas characterized by high soil loss rates. For this purpose, three scenarios were investigated; more specifically, afforestation of areas with more than 100, 200, and 250 t ha^{-1} a^{-1} of soil loss, which account respectively for 12%, 5%, and 3% of the catchment area (Figure 25). From our calculations, it was found that with full afforestation of these areas, a reduction of 50%, 27%, and 26% of the annual sediment input could be achieved. Such a measure, apart from tackling the soil degradation in the catchment, can also contribute to significantly increasing the reservoir lifetime.

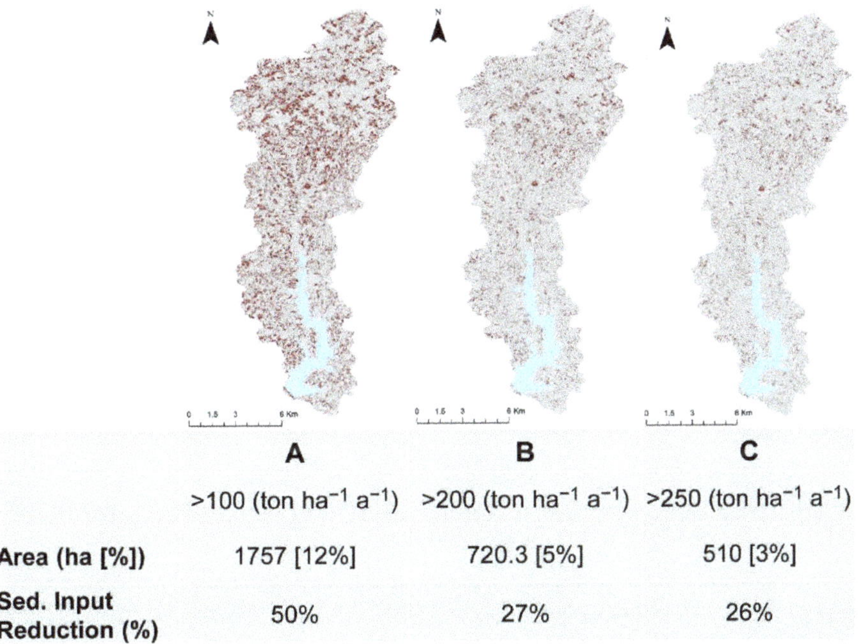

Figure 25. Areal coverage of problematic areas for scenarios (**A–C**) and the respective reduction of sediment input from afforestation. The area in percentage refers to the percentage of the catchment that each management scenario affects.

October and September are the most important months in regard to sediment input and soil loss (Figure 20). For October in particular, the combination of the RUSLE factors is the most effective for producing the highest amount of soil loss. Figure 26 shows the combination of C and R Factors for the three most characteristic months of the year. In the case of the C Factor, October has the same values as July, which is one of the driest and coldest months of the year and has the lowest vegetation cover. As far as the R Factor is concerned, the erosivity is as high as the erosivity in the month of January, which is the month with the highest rainfall. In the case of October, the worst possible combination is present as the rainfall erosivity is maximal, while the vegetation cover is minimal. This combination of factors produces the highest soil loss from a system. In the case of proper land management implementation, like crop rotation, applying crop residues and cover crops in the unprotected soil during the winter and spring months (April–October), a significant reduction of the sediment input could be achieved [89–91].

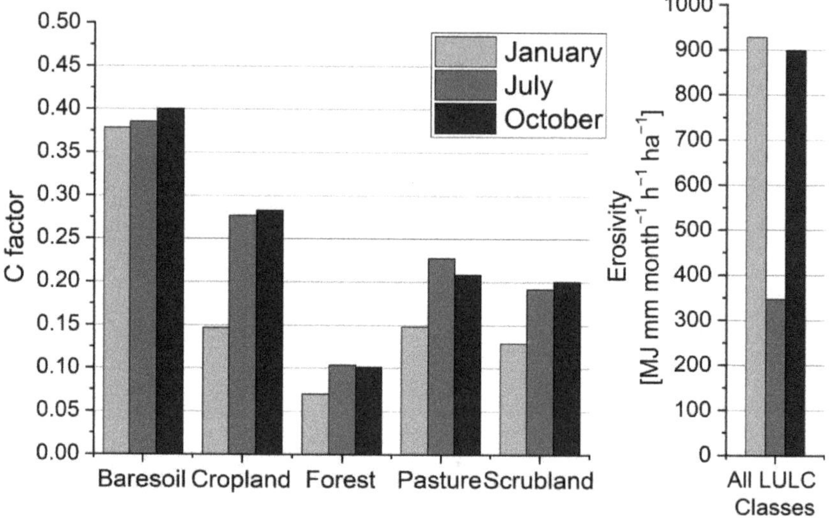

Figure 26. C Factor and R Factor for three months (January, July, October).

4.5. *Uncertainties of the Sediment Yield Model for the Passaúna Catchment*

RUSLE was developed as a tool for long-term soil loss calculation. By calculating the C Factor from a certain scene in 2017 or 2018, it is assumed that the LULC of that specific month has not changed during the last 20 years (rain data available for approx. 20 years). This is to a certain extent not correct. In Parana state, from 1990 until 2019, there has been an increase of almost 45% in arable land and a 5% annual increase in urban areas [92]. Most of this area that was transformed into agricultural land used to be forest, which suggests a gradual increase in erosion in the last 20 years.

This is one of the major drawbacks of the method. However, this drawback can also represent an opportunity. In the case of available precipitation and NDVI data for single months for the entire investigated period, RUSLE could be adapted also for calculation of the actual sediment input and soil loss from that certain month of that specific year. In this way, a calibrated model could be used to derive an accurate balance of sediment input for each month and not only a long-term monthly average of sediment input as in most applications so far.

4.6. Benefits from the Integration of Sentinel-2 Data in Erosion Modeling

The use of vegetation index for calculation of the land cover factor from freely available data is not new. Several studies were conducted based on this principle. However, the spatial and temporal resolution of the images (Landsat or MODIS) in most of the existing literature is relatively low (around 30–250 m) compared to the also freely available Sentinel-2 data [32,93–96]. Improved spatial accuracy and flyover frequency of satellite imagery leads to better erosion modeling results [97,98]. In the tropics and sub-tropics, it is likely that a sequence of satellite scenes show high cloud cover, leading to large data gaps. This emphasizes the importance of short flyover intervals in order to represent fast-changing conditions in the catchment. Furthermore, by the application of more advanced processing steps, more specific information can be derived about the investigated area (for example, the degree of soil sealing). Certain information can be used, as in the case of this study, for a better mapping of erosion and sediment input from urban or semi-urban areas.

Despite the prevailing discrepancy between the sediment yield and accumulated mass inside the reservoir, the model is fully capable of representing the spatial and temporal patterns of soil loss and sediment yield. The application of RUSLE-based models in monthly resolution as decision support systems, due to the increased performance of the model in both spatial and temporal dimension, can lead to an improved river basin management. Specifically, as shown in the previous section, the inclusion of NDVI data can enable the planning of management activities not only in high spatial accuracy but also aimed at the temporal dimension by highlighting the most problematic months of the year. The latter is of high importance as it may lead to reduced catchment management costs.

5. Conclusions

In this study, the modeled sediment yield from a catchment with the sediment stock in a reservoir at the outlet of the catchment is compared. For assessment of the sediment yield from the catchment, a RUSLE-based model was used. The error margins of the RUSLE model results can fluctuate considerably. Therefore, the models have to be coupled with validation measures. This study shows that reservoirs can be used as validation points, despite some limitations. The use of reservoirs as validation points represents a good opportunity, as they collect almost entirely the incoming sediment. Reservoir sediment stock measurements are often easier to achieve than conventional continuous sediment flux monitoring, which produce a high sampling effort and need to deal also with large errors due to the high variability in the river stretches. The assessment of the siltation status of the reservoir creates added value for every operator. In the case of complex systems, however, as shown in this study, several other factors can affect the reservoir sediment balance and, therefore, be misleading regarding the aim of the research. Reservoirs of lower process complexity (e.g., in mountainous areas, low organic material input, or low temperatures) can be more easily used as validation points.

This study showed that the most important factors that create discrepancies for the case of the Passaúna sediment budget are associated mostly with the sediment yield model. On the other hand, when including the errors because of the interpolation technique, the underestimation of sediment yield from the model may become even greater. Even though we fully agree that a RUSLE-based model can reproduce the spatial and temporal patterns of sediment yield from a catchment, the comparison of the approaches in this study shows that there are clear limitations of using modeling approaches for reservoir sediment stock or reservoir lifetime assessment. The two components of the model (RUSLE and SDR model) do not allow the usage of only one calibration point as it is impossible to track the source of error. In order to increase the accuracy of the results, models including the effects of channel, gully, and artificial ditches are needed, and these models need to be calibrated with complementary measures (river-suspended solids and bedload monitoring, calibration of models with erosion plots, or quantification of gully erosion).

Author Contributions: Conceptualization, K.S.; Data curation, K.S., R.A.A., N.W. and K.P.; Formal analysis, M.B.S. and R.K.; Funding acquisition, S.F.; Investigation, M.D., R.A.A. and R.K.; Project administration, S.H. and S.F.; Resources, R.K.; Software, N.W.; Supervision, S.F.; Visualization, K.S. and K.P.; Writing—original draft, K.S.; Writing—review & editing, S.H., M.D., R.A.A., N.W., M.B.S. and S.F. All authors have read and agreed to the published version of the manuscript.

Funding: This research was funded by the German Federal Ministry of Education and Research within Grant 02WGR1431A, in the framework of the research project MuDak-WRM (https://www.mudak-wrm.kit.edu/ Accessed on 9 April 2021).

Institutional Review Board Statement: Not applicable.

Informed Consent Statement: Informed consent was obtained from all subjects involved in the study.

Acknowledgments: Special thanks go to the staff of the laboratory and the technical and scientific staff of PPGERHA (Post-graduate Program on Water Resources and Environmental Engineering) at the Universidade Federal do Parana, who supported our measurement campaigns.

Conflicts of Interest: The authors declare no conflict of interest.

References

1. Montgomery, D.R. *Dirt: The Erosion of Civilizations*; Univ of California Press: Berkeley, CA, USA, 2012; ISBN 0520272900.
2. Dotterweich, M. The history of human-induced soil erosion: Geomorphic legacies, early descriptions and research, and the development of soil conservation—A global synopsis. *Geomorphology* **2013**, *201*, 1–34. [CrossRef]
3. Reusser, L.; Bierman, P.; Rood, D. Quantifying human impacts on rates of erosion and sediment transport at a landscape scale. *Geology* **2015**, *43*, 171–174. [CrossRef]
4. Hooke, R.B. On the history of humans as geomorphic agents. *Geology* **2000**, *28*, 843–846. [CrossRef]
5. Pimentel, D.; Allen, J.; Beers, A.; Guinand, L.; Linder, R.; McLaughlin, P.; Meer, B.; Musonda, D.; Perdue, D.; Poisson, S.; et al. World Agriculture and Soil Erosion. *BioScience* **1987**, *37*, 277–283. [CrossRef]
6. FAO. *Soil Erosion: The Greatest Challenge to Sustainable Soil Management*; FAO: Rome, Italy, 2019.
7. Quinton, J.N.; Govers, G.; van Oost, K.; Bardgett, R.D. The impact of agricultural soil erosion on biogeochemical cycling. *Nat. Geosci.* **2010**, *3*, 311–314. [CrossRef]
8. Merritt, W.S.; Letcher, R.A.; Jakeman, A.J. A review of erosion and sediment transport models. *Environ. Model. Softw.* **2003**, *18*, 761–799. [CrossRef]
9. Verstraeten, G.; Poesen, J. Using sediment deposits in small ponds to quantify sediment yield from small catchments: Possibilities and limitations. *Earth Surf. Process. Landf.* **2002**, *27*, 1425–1439. [CrossRef]
10. Boix-Fayos, C.; de Vente, J.; Martínez-Mena, M.; Barberá, G.G.; Castillo, V. The impact of land use change and check-dams on catchment sediment yield. *Hydrol. Process.* **2008**, *22*, 4922–4935. [CrossRef]
11. Odhiambo, B.K.; Ricker, M.C. Spatial and isotopic analysis of watershed soil loss and reservoir sediment accumulation rates in Lake Anna, Virginia, USA. *Environ. Earth Sci.* **2012**, *65*, 373–384. [CrossRef]
12. Sotiri, K.; Hilgert, S.; Mannich, M.; Bleninger, T.; Fuchs, S. Implementation of comparative detection approaches for the accurate assessment of sediment thickness and sediment volume in the Passaúna Reservoir. *J. Environ. Manag.* **2021**. [CrossRef]
13. Carneiro, C.; Kelderman, P.; Irvine, K. Assessment of phosphorus sediment–water exchange through water and mass budget in Passaúna Reservoir (Paraná State, Brazil). *Environ. Earth Sci.* **2016**, *75*, 564. [CrossRef]
14. Wischmeier, W.H.; Smith, D.D. *Predicting Rainfall Erosion Losses: A Guide to Conservation Planning*; U.S. Department of Agriculture: Washington, DC, USA, 1978.
15. Renard, K.G.; Foster, G.R.; Weesies, D.K.; McCool, D.K.; Yoder, D.C. *Predicting Soil Erosion by Water: A Guide to Conservation Planning with the Revised Universal Soil Loss Equation (RUSLE)*; FAO: Washington, DC, USA, 1997.
16. Desmet, P.J.J.; Govers, G. Modelling topographic potential for erosion and deposition using GIS. *Int. J. Geogr. Inf. Sci.* **1997**, *11*, 603–610. [CrossRef]
17. inVEST- Natural Capital Project. Available online: http://releases.naturalcapitalproject.org/invest-userguide/latest/sdr.html (accessed on 30 January 2020).
18. Abdo, H.; Salloum, J. Spatial assessment of soil erosion in Alqerdaha basin (Syria). *Modeling Earth Syst. Environ.* **2017**, *3*, 26. [CrossRef]
19. Marques, V.; Ceddia, M.; Antunes, M.; Carvalho, D.; Anache, J.; Rodrigues, D.; Oliveira, P.T. USLE K-Factor Method Selection for a Tropical Catchment. *Sustainability* **2019**, *11*, 1840. [CrossRef]
20. Clemente, E.; Oliveira, A.; Fontana, A.; Martins, A.; Schuler, A.; Fidalgo, E.; Monteiro, J. *Erodibilidade dos Solos da Região Serrana do Rio de Janeiro Obtida por Diferentes Equações de Predição Indireta*; Embrapa Solos: Rio de Janeiro, Brazil, 2017.
21. Schick, J.; Bertol, I.; Cogo, N.P.; González, A.P. Erodibilidade de um Cambissolo Húmico sob chuva natural. *Rev. Bras. Ciênc. Solo* **2014**, *38*, 1906–1917. [CrossRef]

22. Silva, M.; Freitas, P.; Blancaneaux, P.; Curi, N.; de Lima, J. Relaçao entre parâmetros da chuva e perdas de solo e determinaçao da erodibilidade de um latossolo vermelho-escuro em Goiânia (GO). *Rev. Bras. Ciência Solo* **1997**, *21*, 131–137.
23. Gee, G.W.; Or, D. 2.4 Particle-size analysis. *Methods Soil Anal. Part 4 Phys. Methods* **2002**, *5*, 255–293.
24. Bouyoucos, G.J. The clay ratio as a criterion of susceptibility of soils to erosion. *J. Am. Soc. Agron.* **1935**, *27*, 738–741. [CrossRef]
25. Rufino, R.L.; Biscaia, R.; Merten, G.H. Determinação do potencial erosivo da chuva do estado do Paraná, através de pluviometria: Terceira aproximação. *Rev. Bras. Ciência Solo* **1993**, *17*, 439–444.
26. Waltrick, P.C.; Machado, M.A.d.M.; Dieckow, J.; Oliveira, D.d. Estimativa da Erosividade de Chuvas no Estado do Paraná Pelo Método da Pluviometria: Atualização Com Dados de 1986 A 2008. *Rev. Bras. Ciência Solo* **2015**, *39*, 256–267. [CrossRef]
27. Lombardi Neto, F.; Moldenhauer, W.C. Rainfall Erosivity: Its Distribution and Relationship with Soil Loss at Campinas, Brasil. *Bragantia* **1992**, *51*, 189–196. [CrossRef]
28. Risse, L.M.; Nearing, M.A.; Laflen, J.M.; Nicks, A.D. Error Assessment in the Universal Soil Loss Equation. *Soil Sci. Soc. Am. J.* **1993**, *57*, 825–833. [CrossRef]
29. Ferreira, V.A.; Weesies, G.A.; Yoder, D.C.; Foster, G.R.; Renard, K.G. The site and condition specific nature of sensitivity analysis. *J. Soil Water Conserv.* **1995**, *50*, 493–497.
30. Estrada-Carmona, N.; Harper, E.B.; DeClerck, F.; Fremier, A.K. Quantifying model uncertainty to improve watershed-level ecosystem service quantification: A global sensitivity analysis of the RUSLE. *Int. J. Biodivers. Sci. Ecosyst. Serv. Manag.* **2017**, *13*, 40–50. [CrossRef]
31. Nearing, M.A.; Romkens, M.J.M.; Norton, L.D.; Stott, D.E.; Rhoton, F.E.; Laflen, J.M.; Flanagan, D.C.; Alonso, C.V.; Binger, R.L.; Dabney, S.M. Measurements and models of soil loss rates. *Science* **2000**, *290*, 1300–1301. [CrossRef]
32. Almagro, A.; Thomé, T.C.; Colman, C.B.; Pereira, R.B.; Marcato Junior, J.; Rodrigues, D.B.B.; Oliveira, P.T.S. Improving cover and management factor (C-factor) estimation using remote sensing approaches for tropical regions. *Int. Soil Water Conserv. Res.* **2019**, *7*, 325–334. [CrossRef]
33. da Silva Santos, L. Sensitivity of Sediment Budget Calculations for an Applicable Reservoir's Lifetime. Master's Thesis, Karlsruhe Institute of Technology, Karlsruhe, Germany, 2019.
34. Durigon, V.L.; Carvalho, D.F.; Antunes, M.A.H.; Oliveira, P.T.S.; Fernandes, M.M. NDVI time series for monitoring RUSLE cover management factor in a tropical watershed. *Int. J. Remote Sens.* **2014**, *35*, 441–453. [CrossRef]
35. Panagos, P.; Borrelli, P.; Meusburger, K.; van der Zanden, E.H.; Poesen, J.; Alewell, C. Modelling the effect of support practices (P-factor) on the reduction of soil erosion by water at European scale. *Environ. Sci. Policy* **2015**, *51*, 23–34. [CrossRef]
36. Borrelli, P.; Robinson, D.A.; Fleischer, L.R.; Lugato, E.; Ballabio, C.; Alewell, C.; Meusburger, K.; Modugno, S.; Schütt, B.; Ferro, V.; et al. An assessment of the global impact of 21st century land use change on soil erosion. *Nat. Commun.* **2017**, *8*, 2013. [CrossRef]
37. Breiman, L. Random Forests. *Mach. Learn.* **2001**, *46*, 5–32. [CrossRef]
38. Tucker, C.J.; Sellers, P.J. Satellite Remote Sensing of Primary Production. *Int. J. Remote Sens.* **1986**, *7*, 1395–1416. [CrossRef]
39. Ridd, M.K. Exploring a V-I-S (vegetation-impervious surface-soil) model for urban ecosystem analysis through remote sensing: Comparative anatomy for cities†. *Int. J. Remote Sens.* **1995**, *16*, 2165–2185. [CrossRef]
40. Kaspersen, P.; Fensholt, R.; Drews, M. Using Landsat Vegetation Indices to Estimate Impervious Surface Fractions for European Cities. *Remote Sens.* **2015**, *7*, 8224–8249. [CrossRef]
41. van der Knijff, J.M.F.; Jones, R.J.A.; Montanarella, L. *Soil Erosion Risk Assessment in Italy;* Citeseer: Princeton, NJ, USA, 1999.
42. Walling, D.E. The sediment delivery problem. *J. Hydrol.* **1983**, *65*, 209–237. [CrossRef]
43. Vigiak, O.; Borselli, L.; Newham, L.T.H.; McInnes, J.; Roberts, A.M. Comparison of conceptual landscape metrics to define hillslope-scale sediment delivery ratio. *Geomorphology* **2012**, *138*, 74–88. [CrossRef]
44. Croke, J.; Mockler, S.; Fogarty, P.; Takken, I. Sediment concentration changes in runoff pathways from a forest road network and the resultant spatial pattern of catchment connectivity. *Geomorphology* **2005**, *68*, 257–268. [CrossRef]
45. Cavalli, M.; Trevisani, S.; Comiti, F.; Marchi, L. Geomorphometric assessment of spatial sediment connectivity in small Alpine catchments. *Geomorphology* **2013**, *188*, 31–41. [CrossRef]
46. Hamel, P.; Chaplin-Kramer, R.; Sim, S.; Mueller, C. A new approach to modeling the sediment retention service (InVEST 3.0): Case study of the Cape Fear catchment, North Carolina, USA. *Sci. Total Environ.* **2015**, *524–525*, 166–177. [CrossRef]
47. de Rosa, P.; Cencetti, C.; Fredduzzi, A. A GRASS Tool for the Sediment Delivery Ratio Mapping. *PeerJ* **2016**. [CrossRef]
48. Grauso, S.; Pasanisi, F.; Tebano, C. Assessment of a Simplified Connectivity Index and Specific Sediment Potential in River Basins by Means of Geomorphometric Tools. *Geosciences* **2018**, *8*, 48. [CrossRef]
49. Borselli, L.; Cassi, P.; Torri, D. Prolegomena to sediment and flow connectivity in the landscape: A GIS and field numerical assessment. *Catena* **2008**, *75*, 268–277. [CrossRef]
50. Jamshidi, R.; Dragovich, D.; Webb, A.A. Distributed empirical algorithms to estimate catchment scale sediment connectivity and yield in a subtropical region. *Hydrol. Process.* **2014**, *28*, 2671–2684. [CrossRef]
51. Saavedra, C. Estimating Spatial Patterns of Soil Erosion and Deposition in the Andean Region Using Geo-Information Techniques. Ph.D. Thesis, Wageningen University, Wageningen, The Netherlands, 2005.
52. Elçi, Ş.; Bor, A.; Çalışkan, A. Using numerical models and acoustic methods to predict reservoir sedimentation. *Lake Reserv. Manag.* **2009**, *25*, 297–306. [CrossRef]
53. Krasa, J.; Dostal, T.; Jachymova, B.; Bauer, M.; Devaty, J. Soil erosion as a source of sediment and phosphorus in rivers and reservoirs—Watershed analyses using WaTEM/SEDEM. *Environ. Res.* **2019**, *171*, 470–483. [CrossRef] [PubMed]

54. True, D.G. *Penetration of Projectiles into Seafloor Soils*; Defense Technical Information Center: Fort Belvoir, VA, USA, 1975.
55. Beard, R.M. A Penetrometer for Deep Seafloor Exploration. In Proceedings of the OCEANS 81, Boston, MA, USA, 16–18 September 1981.
56. Osler, J.; Furlong, A.; Christian, H.; Lamplugh, M. The integration of the free fall cone penetrometer (FFCPT) with the moving vessel profiler (MVP) for the rapid assessment of seabed characteristics. *Int. Hydrogr. Rev.* **2006**, *7*, 45–54.
57. Stoll, R.D. Measuring sea bed properties using static and dynamic penetrometers. In Proceedings of the Sixth International Conference on Civil Engineering in the Oceans, Baltimore, MD, USA, 20–22 October 2004; pp. 386–395.
58. Stark, N.; Kopf, A. Detection and Quantification of Sediment Remobilization Processes Using a Dynamic Penetrometer. In Proceedings of the OCEANS'11 MTS/IEEE KONA, Waikoloa, HI, USA, 19–22 September 2011.
59. Seifert, A.; Kopf, A. Modified dynamic CPTU penetrometer for fluid mud detection. *J. Geotech. Geoenviron. Eng.* **2012**, *138*, 203–206. [CrossRef]
60. Albatal, A.; Stark, N. Rapid sediment mapping and in situ geotechnical characterization in challenging aquatic areas. *Limnol. Oceanogr. Methods* **2017**, *15*, 690–705. [CrossRef]
61. Hilgert, S.; Sotiri, K.; Fuchs, S. Advanced Assessment of Sediment Characteristics Based on Rheological and Hydroacoustic Measurements in a Brazilian Reservoir. In Proceedings of the 38th IAHR World Congress, Panama City, Panama, 1–6 September 2019; 2019.
62. Kirichek, A.; Rutgers, R. *Water Injection Dredging and Fluid Mud Trapping Pilot in the Port of Rotterdam*; CEDA Dredging Days: Rotterdam, The Netherlands, 2019.
63. Kirichek, A.; Shakeel, A.; Chassagne, C. Using in situ density and strength measurements for sediment maintenance in ports and waterways. *J. Soils Sediments* **2020**. [CrossRef]
64. Morris, G.; Fan, J. *Reservoir Sedimentation Handbook*; McGraw-Hill Book, Co.: New York, NY, USA, 2010.
65. Rahmani, V.; Kastens, J.; deNoyelles, F.; Jakubauskas, M.; Martinko, E.; Huggins, D.; Gnau, C.; Liechti, P.; Campbell, S.; Callihan, R.; et al. Examining Storage Capacity Loss and Sedimentation Rate of Large Reservoirs in the Central U.S. Great Plains. *Water* **2018**, *10*, 190. [CrossRef]
66. Sotiri, K.; Hilgert, S.; Fuchs, S. Sediment classification in a Brazilian reservoir: Pros and cons of parametric low frequencies. *Adv. Oceanogr. Limnol.* **2019**, *10*. [CrossRef]
67. Embrapa Solos. *Mapa de Solos de Estado de Parana*; Embrapa Solos: Rio de Janeiro, Brazil, 2007.
68. Mannigel, A.R.; de Passos, M.; Moreti, D.; da Rosa Medeiros, L. Fator erodibilidade e tolerância de perda dos solos do Estado de São Paulo. *Acta Scientiarum. Agron.* **2002**, *24*, 1335–1340. [CrossRef]
69. Silva, A.; Silva, M.; Curi, N.; Avanzi, J.; Ferreira, M. Rainfall erosivity and erodibility of Cambisol (Inceptisol) and Latosol (Oxisol) in the region of Lavras, Southern Minas Gerais State, Brazil. *Rev. Bras. Ciência Solo* **2009**, *33*, 1811–1820. [CrossRef]
70. Duraes, M.F.; de Mello, C.R.; Beskow, S. Sediment yield in Paraopeba River Basin—MG, Brazil. *Int. J. River Basin Manag.* **2016**, *14*, 367–377. [CrossRef]
71. Duraes, M.; Filho, J.; Oliveira, V. Water erosion vulnerability and sediment delivery rate in upper Iguaçu river basin—Paraná. *RBRH* **2016**, *21*. [CrossRef]
72. Saunitti, R.M.; Fernandes, L.A.; Bittencourt, A.V.L. Estudo do assoreamento do reservatório da barragem do rio Passaúna-Curitiba-PR. *Bol. Parana. Geociências* **2004**, *54*, 54. [CrossRef]
73. Wagner, A. Event-Based Measurement and Mean Annual Flux Assessment of Suspended Sediment in Meso Scale Catchments. Ph.D. Thesis, Karlsruhe Institute of Technology, Karlsruhe, Germany, 2019.
74. Koszelnik, P.; Gruca-Rokosz, R.; Bartoszek, L. An isotopic model for the origin of autochthonous organic matter contained in the bottom sediments of a reservoir. *Int. J. Sediment Res.* **2017**. [CrossRef]
75. Quinton, J.N. Erosion and sediment transport. In *Environmental Modelling: Finding Simplicity in Complexity*; John Wiley & Sons Ltd.: London UK, 2004.
76. Belyaev, V.R.; Wallbrink, P.J.; Golosov, V.N.; Murray, A.S.; Sidorchuk, A.Y. A comparison of methods for evaluating soil redistribution in the severely eroded Stavropol region, southern European Russia. *Geomorphology* **2005**, *65*, 173–193. [CrossRef]
77. Alewell, C.; Borrelli, P.; Meusburger, K.; Panagos, P. Using the USLE: Chances, challenges and limitations of soil erosion modelling. *Int. Soil Water Conserv. Res.* **2019**, *7*, 203–225. [CrossRef]
78. Wallbrink, P.J.; Murray, A.S.; Olley, J.M.; Olive, L.J. Determining sources and transit times of suspended sediment in the Murrumbidgee River, New South Wales, Australia, using fallout 137Cs and 210Pb. *Water Resour. Res.* **1998**, *34*, 879–887. [CrossRef]
79. Walling, D.E. Tracing suspended sediment sources in catchments and river systems. *Sci. Total Environ.* **2005**, *344*, 159–184. [CrossRef]
80. Wilkinson, S.N.; Prosser, I.P.; Rustomji, P.; Read, A.M. Modelling and testing spatially distributed sediment budgets to relate erosion processes to sediment yields. *Environ. Model. Softw.* **2009**, *24*, 489–501. [CrossRef]
81. Poesen, J.; Nachtergaele, J.; Verstraeten, G.; Valentin, C. Gully erosion and environmental change: Importance and research needs. *CATENA* **2003**, *50*, 91–133. [CrossRef]
82. Poesen, J.; Vanwalleghem, T.; de Vente, J.; Knapen, A.; Verstraeten, G.; Martínez-Casasnovas, J.A. *Gully Erosion in Europe*; Wiley-Interscience: Hoboken, NJ, USA, 2006; ISBN 9780470859209.
83. Morgan, R.P.C. *Soil Erosion*; Blackwell Publishing: Hoboken, NJ, USA, 1979; ISBN 0582486920.

84. Werner, C.G. *Soil Conservation in Kenia*; Springer: Berlin/Heidelberg, Germany, 1980.
85. Wang, G.; Wente, S.; Gertner, G.Z.; Anderson, A. Improvement in mapping vegetation cover factor for the universal soil loss equation by geostatistical methods with Landsat Thematic Mapper images. *Int. J. Remote Sens.* **2002**, *23*, 3649–3667. [CrossRef]
86. Zhang, Y.; Yuan, J.; Liu, B. Advance in researches on vegetation cover and management factor in the soil erosion prediction model. *Ying Yong Sheng Tai Xue Bao J. Appl. Ecol.* **2002**, *13*, 1033–1036.
87. Zhang, W.; Zhang, Z.; Liu, F.; Qiao, Z.; Hu, S. Estimation of the USLE Cover and Management Factor C Using Satellite Remote Sensing: A Review. In Proceedings of the 2011 19th International Conference on Geoinformatics, Shanghai, China, 24–26 June 2011.
88. Panagos, P.; Borrelli, P.; Meusburger, K.; Alewell, C.; Lugato, E.; Montanarella, L. Estimating the soil erosion cover-management factor at the European scale. *Land Use Policy* **2015**, *48*, 38–50. [CrossRef]
89. Sullivan, P. Overview of Cover Crops and Green Manures. 2003. Available online: https://cpb-us-e1.wpmucdn.com/blogs.cornell.edu/dist/e/4211/files/2014/04/Overview-of-Cover-Crops-and-Green-Manures-19wvmad.pdf (accessed on 30 January 2020).
90. Sullivan, P. Applying the Principles of Sustainable Farming. 2003. Available online: https://ipm.ifas.ufl.edu/pdfs/Applying_the_Principles_of_Sustainable_Farming.pdf?pub=295%5D (accessed on 30 January 2020).
91. SoCo Project Team. Adressing Soil Degradation in EU Agriculture: Relevant Processes, Practices and Policies. 2009. Available online: https://esdac.jrc.ec.europa.eu/ESDB_Archive/eusoils_docs/other/EUR23767_Final.pdf (accessed on 30 January 2020).
92. Zalles, V.; Hansen, M.C.; Potapov, P.V.; Stehman, S.V.; Tyukavina, A.; Pickens, A.; Song, X.-P.; Adusei, B.; Okpa, C.; Aguilar, R.; et al. Near doubling of Brazil's intensive row crop area since 2000. *Proc. Natl. Acad. Sci. USA* **2019**, *116*, 428–435. [CrossRef]
93. Zdruli, P.; Karydas, C.G.; Dedaj, K.; Salillari, I.; Cela, F.; Lushaj, S.; Panagos, P. High resolution spatiotemporal analysis of erosion risk per land cover category in Korçe region, Albania. *Earth Sci. Inform.* **2016**, *9*, 481–495. [CrossRef]
94. Pham, T.G.; Degener, J.; Kappas, M. Integrated universal soil loss equation (USLE) and Geographical Information System (GIS) for soil erosion estimation in A Sap basin: Central Vietnam. *Int. Soil Water Conserv. Res.* **2018**, *6*, 99–110. [CrossRef]
95. Grauso, S.; Verrubbi, V.; Peloso, A.; Zini, A.; Sciortino, M. *Estimating the C-Factor of USLE/RUSLE by Means of NDVI Time-Series in Southern Latium. An Improved Correlation Model*; ENEA: Rome, Italy, 2018.
96. Chuenchum, P.; Xu, M.; Tang, W. Estimation of Soil Erosion and Sediment Yield in the Lancang-Mekong River Using the Modified Revised Universal Soil Loss Equation and GIS Techniques. *Water* **2020**, *12*, 135. [CrossRef]
97. Gianinetto, M.; Aiello, M.; Polinelli, F.; Frassy, F.; Rulli, M.C.; Ravazzani, G.; Bocchiola, D.; Chiarelli, D.D.; Soncini, A.; Vezzoli, R. D-RUSLE: A dynamic model to estimate potential soil erosion with satellite time series in the Italian Alps. *Eur. J. Remote Sens.* **2019**, *52*, 34–53. [CrossRef]
98. Karydas, C.; Bouarour, O.; Zdruli, P. Mapping Spatio-Temporal Soil Erosion Patterns in the Candelaro River Basin, Italy, Using the G2 Model with Sentinel2 Imagery. *Geosciences* **2020**, *10*, 89. [CrossRef]

MDPI

Article

Impact of Forest Conversion to Agriculture on Hydrologic Regime in the Large Basin in Vietnam

Nguyen Cung Que Truong [1,*], Dao Nguyen Khoi [2], Hong Quan Nguyen [3,4] and Akihiko Kondoh [1]

1 Center for Environmental Remote Sensing, Chiba University, Chiba 263-8522, Japan; kondoh@faculty.chiba-u.jp
2 Faculty of Environment, University of Science, Vietnam National University—Ho Chi Minh City, Ho Chi Minh 700000, Vietnam; dnkhoi@hcmus.edu.vn
3 Institute for Circular Economy Development, Vietnam National University—Ho Chi Minh City, Ho Chi Minh 700000, Vietnam; nh.quan@iced.org.vn
4 Center of Water Management and Climate Change, Institute for Environment and Resources, Vietnam National University—Ho Chi Minh City, Ho Chi Minh 700000, Vietnam
* Correspondence: cungque@gmail.com; Tel.: +81-43-251-1111

Abstract: Deforestation due to agricultural land expansion occurred greatly during 1994 to 2005 with a high proportion of forests being converted into agriculture in the upstream Dong Nai river basin in Vietnam. Most of these conversions included expansions of coffee plantations in Dak Lak and Lam Dong provinces, which are in the world's Robusta coffee production area. The aim of this study is to quantify the impact on the water cycle due to the conversion of forest to coffee plantations in a tropical humid climate region by the application of a hydrological model: soil and water assessment tool (SWAT). The model was calibrated with climate data from 1980–1994, validated with climate data from 1995–2010, and verified with statistical indicators such as Nash–Sutcliffe efficiency (NSE), percent bias (PBIAS), and ratio of the root mean square error (RSR). The simulations indicated that forest conversions into agriculture (expansion of coffee plantations) had significantly increased surface runoff (SUR) while actual evapotranspiration (ET), soil water content (SW), and groundwater discharge (GW) decreased. These changes are mainly related to the decrease in infiltration and leaf area index (LAI) post land cover changes. However, the soil was not thoroughly destroyed after deforestation due to the replacement of the lost forest with crops and vegetation. Therefore, changes in infiltration were marginal and not sufficient to bring large changes in the annual flow. Higher reductions in ET and SW were proposed, resulting in reduced streamflow in the dry season at the basin where the proportion of agricultural land was higher than the forest cover. Besides the plantation expansion, which resulted in streamflow reductions in the dry season, an existing problem was over-irrigation of coffee plantations that could likely deplete groundwater resources. Hence, balancing economic benefits by coffee production and mitigating groundwater depletion issues should be prioritized for land use management in the study area.

Keywords: Dong Nai river basin; LUCC; flow regime; SWAT; coffee plantation

Citation: Truong, N.C.Q.; Khoi, D.N.; Nguyen, H.Q.; Kondoh, A. Impact of Forest Conversion to Agriculture on Hydrologic Regime in the Large Basin in Vietnam. *Water* **2022**, *14*, 854. https://doi.org/10.3390/w14060854

Academic Editor: Theodore Endreny

Received: 8 February 2022
Accepted: 4 March 2022
Published: 9 March 2022

Publisher's Note: MDPI stays neutral with regard to jurisdictional claims in published maps and institutional affiliations.

1. Introduction

The population of Vietnam has increased rapidly since 1960 and reached nearly 96 million by 2017. The Vietnamese government initiated a series of economic reforms in the mid-1980s, which were aimed at stimulating economic growth, the most remarkable policy being the New Economic Zones program. This policy resulted in large-scale displacement of residents to uninhabited areas due to expansion of the agricultural land. These factors led to the conversion of land use and land cover in Vietnam after mid-1980s [1]. Furthermore, since 1998, the Vietnamese government implemented multiple national No programs for poverty reduction and programs for socio-economic development in mountainous areas; for example, Decision No.133/1998/QD-TTg dated 23 July 1998, Decision No.135/1998/QD-TTg dated 31 July 1998, and Decision No.143/2001/QD-TTg dated 27 September 2001. These decisions highlighted migration and land reclamation as important projects of the program, which in turn led to a major disturbance in land use.

As described in [2], forested area declined sharply between 1994 and 2005 throughout the study area, which was the upstream Dong Nai river basin (UDNB) (Figure 1), due to the conversion to agricultural land caused by exponential population growth. The proportions of forest and agricultural land in 1994 were 73% and 23%, respectively, which subsequently changed to 51% and 40%, respectively, in 2005. It led to an immediate increase in total discharge, streamflow, and abundant, low, and scanty runoffs. This study was initiated to complement the results of Truong et al., (2018) [2], and analyze the elements, which alter the streamflow, through the evaluation of the changes in the hydrological components by using the soil and water assessment tool (SWAT) model [3].

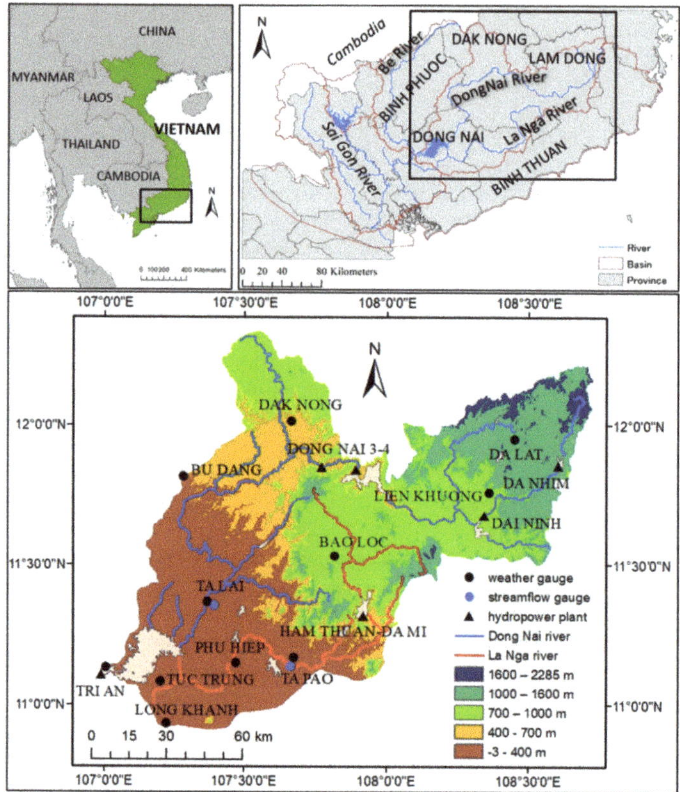

Figure 1. The UDNB with weather, stream gauges and hydropower-plant locations. (Modified from [2]).

Research on the impact of land use/land cover change (LUCC) on hydrology has increased since the 1960s. Several reviews [4–9] have focused on the changes in the annual streamflow and neglected the changes in flow regime. These studies revealed that increased forest cover reduced the annual streamflow and vice versa; additionally, the total annual water yield increased with the increase in lost forest percentage [10]. However, the studies on the effects of LUCC in large tropical river basins could not estimate similar relationships [6], whereas water resource management usually requires information on regional (>1000 km^2) or large-scale catchments [11]. Furthermore, the effect of forest cover change on flooding, particularly in developing countries, is an ongoing discussion [12,13].

Bradshaw et al., (2007) [12] shows that flood frequency is positively correlated with natural forest area loss by using generalized linear modeling. Yet, Van Dijk et al., (2009) [13] reanalyzed the data used in [12] and suggested that the removal of trees does not affect large flood events. It is also known that forests present finite capabilities to retain large amounts of precipitation, especially during extreme rainfall events, even if the forest cover percentage is significantly high [14,15]. Defining a threshold above which forest cover is no longer effective in reducing a flood is a challenge in forest hydrology [16].

The following summaries of the study in a large tropical rainforest river basin present different perspectives on the overall impact of vegetation on streamflow. Few catchment studies indicated that the annual discharge increased with forest-to-crop expansion due to low evapotranspiration and infiltration rates.

Thanapakpawin et al., (2006) [17] assessed hydrological regimes with land use change at the Mae Chaem river basin (3853 km^2, elevation 255–2565 m) in northwestern Thailand; this was achieved by three forest-to-crop expansion scenarios and a crop-to-forest reversal scenario. The results showed that unregulated runoff increased with the conversion from forest to crops, owing to decreasing evapotranspiration and irrigation diversion directly influencing discharge magnitude and significantly varying water yields. Similarly, Costa et al., (2003) [18] evaluated the effect of land use changes on discharge in the upper Tocantins basin (175,360 km^2). The authors estimated the annual and seasonal mean discharges with two datasets: (1) agricultural land use of 30.2% (1960) and climate data during 1949–1968; and (2) agricultural land use of 49.2% (1995) and climate data during 1979–1998. Although precipitation did not significantly differ between the two periods, the annual discharge increased by 24%, rainy season discharge increased by 28%, and seasonal peaks occurred about one month earlier. Post land cover changes, reduced infiltration, increased the rainy season surface flow and actual evapotranspiration reduction increased the discharge throughout the year.

Studies conducted in Vietnam also concluded that forest gain can decrease annual runoff while forest loss affects in the opposite way. Nguyen et al., (2014) [19] simulated water discharge in the Srepok watershed in the central highlands by using the SWAT hydrological model. The increase in forest cover area from 50.45% in 2000 to 79.59% in 2010 resulted in the reduction in surface runoff percentage by half. The impact of land use changes on hydrological processes in the Be River catchment was investigated with SWAT by Khoi and Suetsugi (2014) [20]. The results indicated that deforestation increased the surface runoff and soil water content (over 10%) while actual evapotranspiration, water yield, and annual flow increased marginally (approx. 1%).

In contrast, other studies showed that vegetation has no impact on streamflow. Wilk et al., (2001) [21] used the Hydrologiska Byråns Vattenbalansavdelnin (HBV) hydrology model, which is a conceptual model that simulates daily discharge with input data of daily rainfall and temperature, and monthly estimates of potential evaporation to determine the rainfall or change in runoff regime after a decrease in forest cover from 80% (1965) to 27% (1992) at the Upper Nam Pong Basin (area 12,100 km^2, elevation 300–1400 m) in northeastern Thailand. However, no detectable trends in river discharge were observed possibly due to a significant number of remaining trees and secondary growth on agricultural land. Beck et al., (2013) [22] examined the effects of afforestation on the streamflow for 12 mesoscale catchments (area 23–346 km^2) in Puerto Rico. However, the correlation between changes in the forest area and changes in streamflow was insignificant. The three possible reasons were data errors, heterogeneity in catchment response, and streamflow generation in the headwater areas, whereas changes in forest area mainly occurred in the drier lowlands. Similarly, the spatial variations of LUCC on streamflow were estimated in the study by Liu et al., (2020) [23].

In addition, the effects of logging methods on water yield and streamflow in the tropical forest watershed were conducted by Malmer (1992) [24]. Malmer observed that a combination of burning and no soil disturbance substantially increased the water yield as compared to cases that experienced soil disturbance and loss of infiltrability. Additionally, it led to high-speed runoff during storms in Mendolong, Malaysia, which included six catchments with areas varying from 3.4 to 18.2 ha.

The purpose of this study is to quantify the impacts on the water cycle due to the conversion of forest to coffee plantations in the UDNB, a large tropical rainforest basin, by assessing the changes in the water balance components, such as the actual evapotranspiration (ET—actual evapotranspiration during the time step; measured in mm), surface runoff (SUR—surface runoff contribution to streamflow during the time step; measured in mm H_2O), groundwater discharge (GW—groundwater contribution to streamflow during the time step; measured in mm), and soil water content (SW—amount of water in soil profile at the end of the time period; measured in mm). Effective water resource management and strategic planning needs to consider the effect of LUCC and water availability, particularly in the agricultural sector. The findings from this study will be significant to decision-makers working for integrated river basin management for the development of land use adaptation and mitigation strategies.

2. Materials and Methods

2.1. Study Area

UDNB is located in Vietnam in the central highlands, which is a tropical humid zone receiving southwest monsoons; 90% of the annual rainfall (the average rainfall from 1993 to 2012 was 2415 mm/year) occurs during the rainy season (May–October) while the dry season (November–April) receives the remaining rainfall. Since the basin extends from the mountains to the lower plains, the temperature varies significantly throughout. The average temperature ranges between 18 and 26 °C [2] (Figure 2).

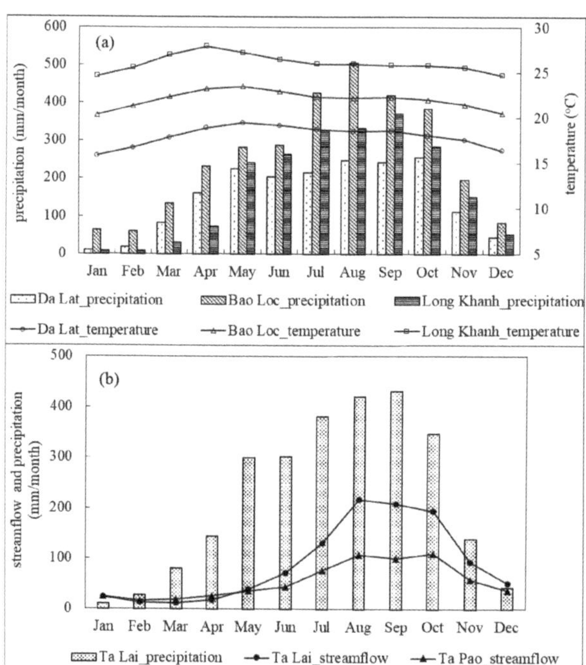

Figure 2. Mean monthly temperature and precipitation at different elevations in three stations Da Lat (>1000 m), Bao Loc (400~1000 m), and Long Khanh (<400 m) (**a**). Mean monthly streamflow compared to precipitation at Ta Lai and Ta Pao stations (**b**). (Precipitation data are not available at Ta Pao station).

The main stem of the Dong Nai River originates from the high hill (elevation 1000 m to 2000 m) north of Lam Dong province, where it is called Da Nhim River. Its flow course initially follows the southwestward direction and later turns to the west forming the border line between Lam Dong and Dak Nong provinces. Thereafter, Dong Nai River heads to the southeast and crosses Dong Nai province in the southwest. The La Nga River originating from Lam Dong province lies to the south of the basin and merges into the Dong Nai River before it flows into the Tri An reservoir [25] (Figures 1 and 3). The high-gradient streams indicated a steep slope and rapid flow of water.

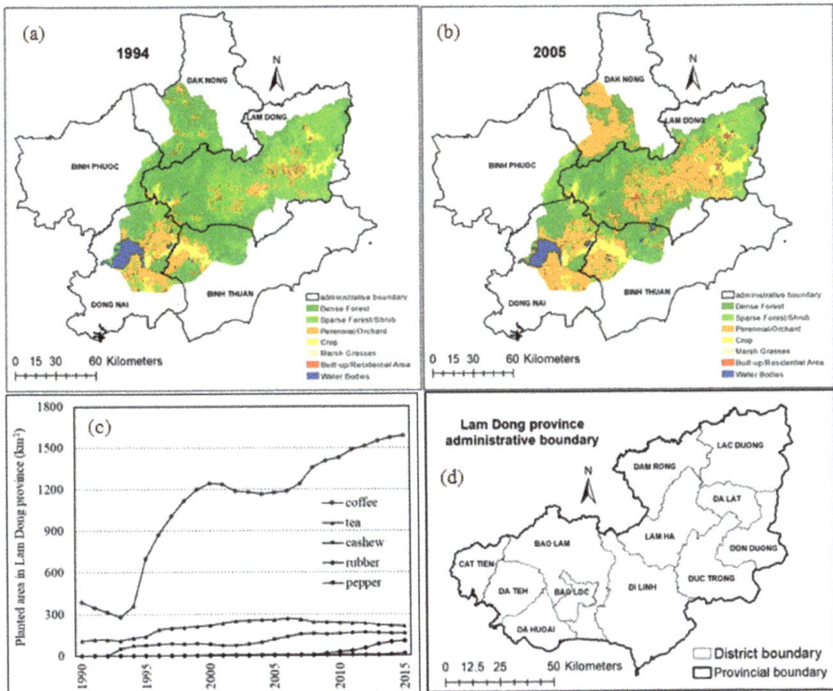

Figure 3. Land-use changes of UDNB in 1994 and 2005 (**a**,**b**); Trends in the planted areas of multi-year industrial crops (**c**) and administrative boundaries of Lam Dong province (**d**) (Data source: Lam Dong Statistical Office).

2.2. Land Use/Land Cover Change in the Study Area

According to [2], land use changed intensely between1994 to 2005; a high proportion (area approximately 3343 km^2, which accounts for 31.12% forest cover) of forest land was converted to agricultural land. Dense forest and sparse forest/shrub of the forest area constituted 73% whereas agricultural land, which included perennial orchard and crops constituted 24% of the total area (14,706 km^2) in 1994. The forest cover decreased to 51% in 2005 while agriculture land increased to 45%. The majority of the land use changes were in Lam Dong and Dak Nong provinces. Figure 2 shows land cover maps of the UDNB for 1994 and 2005 classified from Landsat 5 TM images (path 124 row 52, 7 January 1994; and 22 February 2005) (reference from [2]).

The coffee sector in Vietnam grew exponentially throughout the 1990s due to government mandates and incentives in the form of favorable credit, subsidized inputs, and low-cost land for exporting crops. Since the early 2000s, Vietnam became the world's second-largest coffee producer with 30% contribution to the GDP of the central highlands, the largest Robusta coffee production area worldwide [26]. In this region, Lam Dong and Dak Lak (Dak Lak was divided into two separate provinces: Dak Lak and Dak Nong since

2004) are the largest coffee producing provinces, which cover nearly 3442 km^2 (68%) of the total 5065 km^2 area of coffee plantations of Vietnam in 2003 [27]. The coffee plantations in Lam Dong province increased by 819.43 km^2 (from 355.95 km^2 to 1175.38 km^2) (Figure 2). Five districts including Bao Lam, Bao Loc, Duc Trong, Di Linh, and Lam Ha observed major land use changes in Lam Dong province and comprised the largest coffee plantations (over 90% of Lam Dong's total coffee plantations) (Table 1). However, the area under coffee plantations increased from approximately 1500 km2 in 1994 to 2400 km2 in 1999 and subsequently to 2800 km^2 in 2001 (accounting for 51.25% of agricultural land and 14.28% of the total area) [28]. Thus, the majority of land use conversions in the study area accounted for expansion of coffee plantation.

Table 1. Coffee Planted Areas by Districts in Lam Dong Province (km^2).

Year	1995	1999	2000	2001	2002	2003	2004	2005
Lac Duong	2.01	14.02	16.55	16.55	16.70	16.52	13.77	10.40
Dam Rong	-	-	-	-	-	-	-	30.94
Da Lat	8.03	34.29	38.41	36.76	34.4	34.6	33.12	33.45
Don Duong	4.48	14.60	17.02	13.56	10.11	8.56	8.56	7.96
Lam Ha	217.50	333.53	344.36	344.94	346.30	343.22	340.17	320.61
Duc Trong	18.07	92.26	101.61	109.54	93.79	86.03	80.91	78.79
Di Linh	229.63	373.32	382.92	378.34	362.96	361.92	361.62	361.63
Bao Lam	183.63	262.36	263.64	262.94	257.94	257.08	257.69	259.47
Bao Loc	30.55	61.44	64.41	72.74	68.86	69.82	68.27	69.39
Cat Tien	2.00	2.09	1.70	0.57	0.55	0.30	0.30	0.00
Da Huoai	3.31	10.11	9.34	6.3	3.58	1.26	0.87	0.62
Da Teh	1.20	2.13	3.63	3.48	2.52	2.37	2.12	2.12
Total	701.04	1200.15	1243.59	1237.39	1190.01	1181.68	1167.4	1175.38

Data source: Lam Dong Statistical Office.

2.3. The Soil and Water Assessment Tool (SWAT) Model

Hydrological modeling was selected to quantify the impacts of LUCC and climate variability on the hydrological parameters. Globally, many researchers have confirmed the high performance and algorithms of SWAT [1,29–36]. SWAT model [36,37] is a continuous, long term, distributed parameter model that was developed to predict the long-term impacts of land management practices on water, sediment and agricultural yields in large complex watersheds with varying soils, and land use and management conditions. SWAT requires specific information about weather, soil properties, topography, vegetation, and land management to model the physical processes associated with water movement, sediment movement, etc.

A watershed is divided into multiple subwatersheds, which are further subdivided into hydrological response units (HRUs) that comprise homogeneous land use, land management, and topographical and soil characteristics [3]. The hydrologic cycle simulated by SWAT is based on the water balance equation:

$$SW_t = SW_0 + \sum_{i=1}^{t}\left(R_{day} - Q_{surf} - E_a - w_{seep} - Q_{gw}\right) \quad (1)$$

where SW_t is the final soil water content (mm H_2O), SW_0 is the initial soil water content on day i, t is the time, R_{day} is the amount of precipitation on day i, Q_{surf} is the amount of surface runoff on day i, E_a is the amount of evapotranspiration on day i, w_{seep} is the amount of water entering the vadose zone from the soil profile on day i, and Q_{gw} is the amount water return flow on day i (mm H_2O). A detailed description of the model is given in the SWAT theoretical documentation [38].

Input Data

SWAT model requires a digital elevation model (DEM), land use/land cover, soil properties, meteorological, and observed streamflow data.

- DEM. The DEM of the basin was derived from SRTM30 data that have been published by the United States National Aeronautics and Space Administration (NASA) (Figure 1).
- Soil properties. We used the soil map of the world developed by the Food and Agriculture Organization (FAO) of the United Nations.
- Land use and land cover data. The land use maps of 1994 and 2005 with seven land cover classes, dense forest, sparse forest/Shrub, perennial/orchard, crop land, built-up/residential, marsh/grasses, and water body, which were classified from the Landsat imagery, were used (Figure 2) [2]. To assess the performance of SWAT under LUCC, the daily hydrographs for the two land use maps 1994 and 2005 with the same climatic conditions and model parameters were simulated.
- Meteorological data. Data for daily precipitation (mm), maximum and minimum temperatures (°C), solar radiation (Wm^{-2}), wind speed (ms^{-1}), and relative humidity (%) were provided by the provincial department of natural resources and environment (DONRE). The study area was located in a tropical humid zone receiving southwest monsoon; 80% of the annual rainfall occurred during the rainy season (May–October) while the dry season (November–April) received the remaining rainfall. As UDNB extends from the high hills to the low plain area, temperature and precipitation vary significantly (Figure 3).
- Streamflow data. Daily streamflow data (m^3/s) at Ta Lai station from 1 January 1987 to 31 December 2010 were collected and data at Ta Pao station from 1 January 1980 to 31 December 2010 (Figure 1) were compared with the modeled surface flow.

2.4. Model Setup and Performance Evaluation

In this study, we used the geographic information system interface ArcSWAT to parameterize the model. Basin delineation was implemented by delineating the stream network from SRTM30 gridded DEM data, and assigning Tri An dam as the outlet of the study basin. The basin was divided into 111 sub-basins (SB) (Figure 3). Later, 979 HRUs were created based on the map combinations of land use, soil, and slope map. The observed meteorological data and streamflow data were used for calibration (1980/1987–1994) and performance validation (1995–2010) in the daily time step of the flow simulation. The uncertainty analysis was conducted by sequential uncertainty fitting (SUFI-2) method, which was implemented in the SWAT-CUP [39]. The final calibrated parameters for the basins of Dong Nai and La Nga rivers are presented in Table 2.

Table 2. Highly Sensitive Final Calibrated Parameters.

Parameter	Description of Parameter	Range	Best Simulation	
			Dong Nai	La Nga
CN2.mgt	Initial SCS CN II value	−0.5–0.5	−0.312	−0.098
CH_K2.rte	Channel effective hydraulic conductivity	−0.01–500	219.299	35.765
ALPHA_BF.gw	Baseflow alpha factor	0–1	1.203	1.003
SOL_AWC.sol	Available water capacity	−0.5–0.5	0.942	0.559
GWQMN.gw	Threshold water depth in the shallow aquifer for flow	0–5000	1654	2021
REVAPMN.gw	Threshold water depth in the shallow aquifer for "revap"	0–1000	925	925
GW_DELAY.gw	Groundwater delay	0–500	8.219	33.853
SOL_Z.sol	Soil depth	−0.2–0.2	0.000	0.000
GW_REVAP.gw	Groundwater 'revap' coefficient	0.02–0.2	0.288	0.072
SOL_K.sol	Saturated hydraulic conductivity	−0.5–0.5	0.220	0.220

Ten land surface response parameters significantly affected the streamflow simulation, and therefore, they were the most sensitive parameters of the model. Among these, CN2 and SOL_AWC directly govern surface response by controlling the SUR that directly contributes to the streamflow. Curve number II (CN2), which is a function of watershed properties that includes soil type, land use and treatment, ground surface condition, and antecedent moisture conditions [40], adjusts the soil humidity for different land uses to estimate the surface runoff. Low values of CN2 reflect decreased SUR and increased baseflow. Yet, the climate in the study area has distinct wet and dry seasons. The curve number method is able to account for this by using the empirical rainfall-runoff relationships for dry, average, and wet antecedent wetness conditions (CNI, CNII, and CNIII). Dile et al. [41] indicated that the curve number method works better in the wet season (high rainfall conditions) that it is able to be useful for hydrological simulation in tropical regions. The average streamflow in the dry season (low rainfall conditions) is small, resulting in only small errors in estimating streamflow. SOL_AWC is the volume of water available for plant uptake when the soil is at field capacity and can be estimated by determining the quantity of water released between the field capacity of soil and the point of permanent wilting [40]. Low values of SOL_AWC indicate a low soil capability to maintain its humidity, which subsequently increases the amount of water available for surface runoff and percolation. One of the parameters governing subsurface response is GW_REVAP, which controls the amount of water that will move from the shallow aquifer to the root zone as a result of soil moisture depletion, and the amount of direct groundwater uptake from deep-rooted trees and shrubs [42]. A low GW_REVAP value reflects restricted movement of water from the superficial aquifer to the root zone, while a high value indicates that the transfer rate is close to the rate of evapotranspiration.

According to Moriasi et al. [43] three quantitative statistics, Nash–Sutcliffe efficiency (NSE), percent bias (PBIAS), and the root mean square error (RSME)—standard deviation (STDEV) of measured data ratio (RSR) were used for model evaluation. Model performance can be judged based on the general performance ratings (Table 3) obtained by these values. RMSE values less than half the STDEV of measured data indicate an acceptable error range. According to that, less than 0.5 RSR value as the most stringent "very good" rating, and two less stringent ratings of 10% and 20% greater than this value for the "good" and "satisfactory" ratings, respectively [14,44]. The formulae for these statistics are given below.

$$\mathrm{NSE} = 1 - \left[\frac{\sum_{i=1}^{n} \left(Y_i^{obs} - Y_i^{sim} \right)^2}{\sum_{i=1}^{n} \left(Y_i^{obs} - Y^{mean} \right)^2} \right] \tag{2}$$

$$\mathrm{PBIAS} = \left[\frac{\sum_{i=1}^{n} \left(Y_i^{sobs} - Y_i^{sim} \right) * 100}{\sum_{i=1}^{n} Y_i^{obs}} \right] \tag{3}$$

$$\mathrm{RSR} = \frac{RSME}{STDEV_{obs}} = \frac{\sqrt{\sum_{i=1}^{n} \left(Y_i^{obs} - Y_i^{sim} \right)^2}}{\sqrt{\sum_{i=1}^{n} \left(Y_i^{obs} - Y^{mean} \right)^2}} \tag{4}$$

where Y_i^{obs} corresponds to the ith observation for the constituent being evaluated, Y_i^{sim} is the ith simulated value for the constituent being evaluated, Y^{mean} is the mean of observed data for the constituent being evaluated, and n is the total number of observations.

SWAT model for the Dong Nai upstream river basin was calibrated by comparing the simulated and observed streamflow data at two gauge stations, Ta Lai (main stream of the Dong Nai river) and Ta Pao (main stream of the LaNga river) (Figure 1). The comparison of calibration and validation data (red line) of the model with the observed data (blue line) at the Ta Lai and Ta Pao stations is shown in Figure 4. It indicates that the model closely replicated the observational data during the calibration period. The statistical evaluations are shown in Table 4 also suggest a good agreement between the measured and simulated streamflow. PBIAS was approximately 12.92% and 0.35% for the calibration period and −7.47% and 1.20% for the validation period at the Ta Lai and Ta Pao stations, respectively. The correlation coefficients NSE were 0.88 and 0.85 at Ta Lai station, and 0.67 and 0.51 at Ta Pao station for the monthly streamflow in the calibration and the validation periods, respectively. Although the RSR for the validation period at the Ta Pao station was 0.70 (within the range of "satisfactory" benchmarks), which was comparatively less accurate, the results of the performance were still considered satisfactory, which indicates that the fundamental rainfall-runoff relationship is well documented. Thus, we affirm that these results indicate "good performance".

Table 3. General Performance Ratings of Statistical Indices for Monthly Streamflow Simulation.

Rating	NSE	PBIAS (%)	RSR
Very good	0.75 < NSE ≤ 1.00	PBIAS < ±10.00	0.00 ≤ RSR ≤ 0.50
Good	0.65 < NSE ≤ 0.75	10.00 ≤ PBIAS < ±15.00	0.50 < RSR ≤ 0.60
Satisfactory	0.50 < NSE ≤ 0.65	15.00 ≤ PBIAS < ±25.00	0.60 < RSR ≤ 0.70
Unstatisfactory	NSE ≤ 0.50	PBIAS ≥ ±25.00	RSR > 0.70

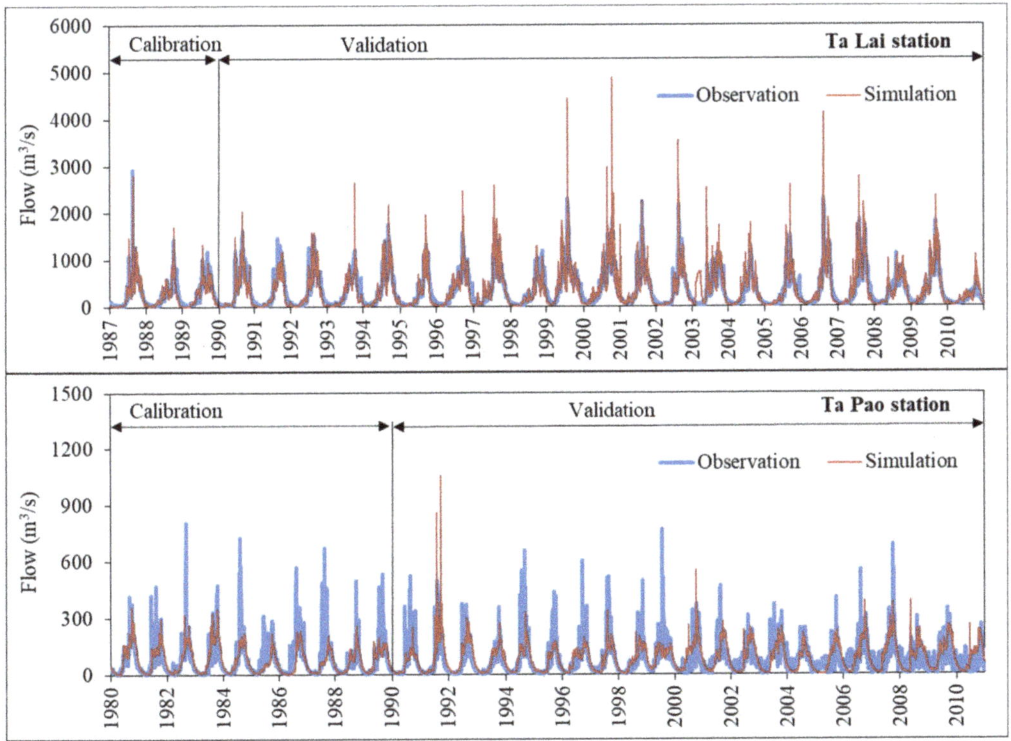

Figure 4. Comparison of simulation results and observated data at Ta Lai and Ta Pao stations.

Table 4. Model performance for calibration and validation.

Station	Calibration (1980/1987–1994)			Validation (1995–2010)		
	NSE	PBIAS	RSR	NSE	PBIAS	RSR
Ta Lai	0.88	12.92	0.34	0.85	−7.47	0.38
Ta Pao	0.67	0.35	0.58	0.51	1.20	0.70

3. Results

The results of model simulation using land use data 1994 (Landuse 1994) and 2005 (Landuse 2005) of ET, SUR, SW, and GWQ are given in Figure 5. Under the impact of LUCC, deforestation and expanding coffee plantations increased the SUR drastically by 35% at both Ta Lai and Ta Pao stations. Accordingly, the streamflow in the rainy season in both Ta Lai and Ta Pao stations increased by 5% and 81% (6% and 1% of the annual streamflow), respectively. Contrastingly, the streamflow in the dry season at Ta Pao station reduced 23% (11.06 m^3/s). (Figure 6).

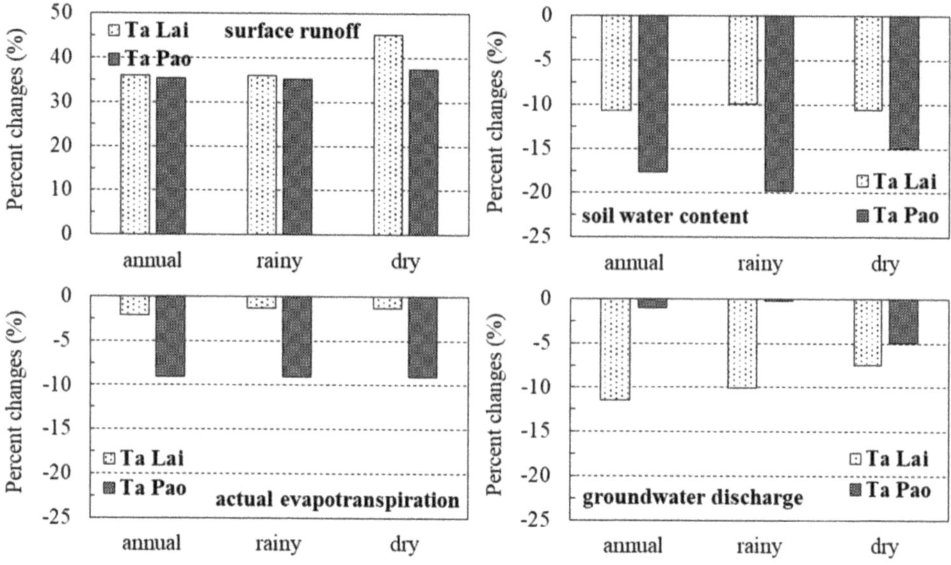

Figure 5. Annual and seasonal changes of hydrological components under LUCC (rainy season: May to October; dry season: November–April).

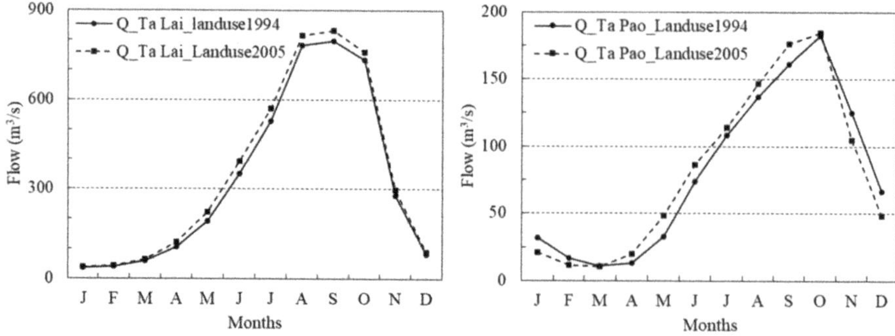

Figure 6. Comparison of monthly streamflow (Q–m^3/s) between Landuse 1994 and 2005 at Ta Lai and Ta Pao stations.

In contrast, the ET, SW, and GWQ showed a downward trend. Annually, ET insignificantly decreased with a reduction of 2%, and SW decreased by 10% at Ta Lai station. These reductions were greater by 9% of ET and 18% of SW at Ta Pao station. GWQ was reduced significantly by 11% at Ta Lai and insignificantly by 1% at Ta Pao. ET did not differ between the dry season and rainy season, whereas SW and GWQ exhibited lower reductions in the dry season than the rainy season, and SUR in the dry season increased considerably in the rainy season.

Total ET in the SWAT model comprises the evaporation from the canopy surface, transpiration, evaporation from the soil, and groundwater evapotranspiration (Revap) [37]. Revap is the movement of water from the underlying shallow aquifer to the unsaturated zone in response to water demand for evapotranspiration. Revap can be calculated separately in regions where the saturated zone is within the root zone [45]. In this study, ET results did not include revap. Temporarily, ET was mainly controlled by soil water availability in the hot-dry season (March and April), leaf area index (LAI) in the early rainy season (May and June), and atmospheric conditions in the mid- to late- rainy season (July to October) and cool-dry season [46]. However, SWAT simulated ET for two land use maps (1994 and 2005) with the same climatic condition, soil data, and model parameters. Reductions of LAI due to forest conversion to plantations was the main cause of decreased ET.

There were large discrepancies in the SW, ET, and GWQ comparison results between the two stations, Ta Lai and Ta Pao. ET and SW were highly affected by seasonal variations at Ta Lai than Ta Pao station. On the other hand, GWQ remained unaffected in the rainy season and only decreased by approximately 5% in the dry season. To investigate the effects of LUCC in Dong Nai river subbasin (DN) (Ta Lai station) and La Nga river subbasin (LN) (Ta Pao station), we compared a fraction of main land use in 1994 and 2005. Subbasin areas of DN and LN were 10,639 km^2 (72.35% of the study area) and 4067 km^2 (27.65% of the study area), respectively. Figure 7 shows the fraction of land use in DN and LN. Forest (dense and sparse) and agriculture (orchard and crop) were the dominant land use patterns, occupying about 78.2% and 17.8%, respectively, of total DN area, and 59.7% and 38.5% of the total LN area, respectively, in 1994. The ratio of two land use types in 2005 changed in the range of 38.7–57.6% at DN and 36.0–61.0% at TL. In other words, forest area decreased by 26.4% and 39.7% from 1994 to 2005. Conversely, agricultural land increased by 117.5% and 58.3% at DN and LN, respectively. Despite similarities in land use change trends between the two subbasins, there is a difference in the catchment scale, the proportion of forest and agricultural land, and ratio of changed areas of each subbasin. The area of lost forest is less in DN than LN; contrastingly, the area of plantation-replaced-forest is much larger in DN than LN. This suggests that the amount of decrease in ET and SW is proportional to the ratio of lost forest area, and as the areas of plantation-replaced-forest increased, reductions in the infiltration rate lowered. In addition, the conversion from forest to agriculture in DN mainly occurred in the headwater area in Lam Dong and Dak Nong provinces. Meanwhile, this conversion also appeared in LN in the headwater area, Lam Dong province, and was scattered in Binh Thuan and Dong Nai provinces at the downstream (Figure 2). The spatial distribution of LUCC also affected the hydrology.

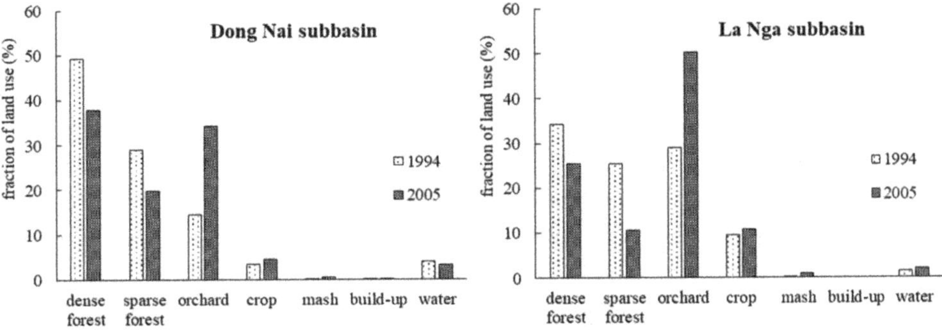

Figure 7. Fraction of land uses in Dong Nai and Ta Lai river subbasins in 1994 and 2005 (calculated from [2]).

4. Discussions

The results of this study validate the previously reported results, which state that land use changes involving conversion of forests to agricultural expansions increase the surface runoff due to reduced evapotranspiration and infiltration. While studying the dam functions in forests, Kuraji [47] collected data from experimental basins around the world and reported that forests have a positive effect on flood mitigation and a negative effect on drought mitigation. Deforestation and conversion to arable land/grassland are usually accompanied by an increase in surface runoff or total discharge [5,48–50]; particularly, forest replacement by annual crops in tropical rainforest regions tends to increase annual streamflow quantities and storm flow events with higher peak discharges [51]. Surface runoff increases because of reduced evapotranspiration of the replaced vegetation due to the forest vegetation intercepts, which loses more water than other land use types [52,53]. Additionally, reductions in soil water content and groundwater discharge may be because the infiltration rate of forest land is higher than other land-use types [6].

Tan et al. [30] concluded that interception and infiltration rates were higher in forests than other land cover types; hence deforestation caused an increase in surface runoff and reduced water movement within the soil layers. Additionally, after deforestation or forest fires, the infiltration rates decreased due to the loss of tree cover. However, in the study area the lost forest was replaced immediately with crops and vegetation. Moreover, as discussed in [20], agricultural area with more shade trees were replaced quickly with secondary forests at any abandoned areas. Trees in these areas have higher transpiration rates outside the original forest. As a matter of fact, significant numbers of trees on agricultural land as well as secondary growth invading abandoned plots exist. This suggests that the soil was not thoroughly destroyed after deforestation and changes in infiltration were marginal. Infiltration reductions increase the streamflow during the rainy season; however, it is insufficient to reduce the dry season streamflow at Ta Lai station.

In the study area, irrigation of the coffee plantations directly affected the depletion of groundwater and fundamentally disrupted the regional hydrological system. Since coffee smallholders draw groundwater for dry season irrigation, the amount of water presently used exceeds the crop water requirement, as rightly stated, "smallholders irrigate more than twice the recommended level, with the belief that yield increases linearly with irrigation amount" [26,54,55]. As mentioned above, the results show SW and GWQ in the dry season reduced by 14.89% and 4.96%, respectively, at Ta Pao station (Figure 5) with the increase in the coffee plantation area. According to the previous studies [26,54–58], irrigation during the dry season is crucial to achieve high coffee yields as it assists in breaking flower bud dormancy and inducing fruit setting. Groundwater is the major source for coffee irrigation in the Central Highlands. The water supply for the coffee tree consists of a micro-basin irrigation system; every tree stands in a planting hole with dimensions of 2.6 m $\times$ 2.6 m $\times$ 0.2 m. The water is pumped up to the plantations and watered through a hose with 100 mm application depth. D'haezea et al. [53] concluded that the present groundwater abstraction at the Ea Tul watershed in the Dak Lak province is not sustainable in the dry year as it exceeds the safe aquifer yield. Controlled agricultural (coffee) expansion is necessary to solve the issue of water depletion and to ensure a sustainable environment.

5. Conclusions

Human activities, such as LUCC, especially conversion of forests to agricultural lands are the main factors that affect the streamflow regime. Quantifying the contribution of LUCC to the flow regime is important for water resources planning and management. In the present study area, forest coverage decreased from 73% in 1994 to 51% in 2005; in contrast, agricultural land increased from 24% in 1994 to 45% in 2005. The majority of land use conversions in the study area were due to expansion of coffee plantations. The impacts of LUCC on streamflow regime by assessing the changes of each hydrological component, such as actual evapotranspiration (ET), surface runoff (SUR), soil water content (SW), and ground water discharge (GW) were estimated based on the SWAT model. The significant findings from the analysis in this study were: (1) Deforestation and conversion to agriculture increases SUR drastically due to reductions in ET, (2) in contrast, SW and GWQ tend to decrease due to reduced infiltration after land cover change, (3) these reductions eventually increase in proportion; subsequently, decreasing the streamflow in the dry season when the proportion of agricultural land is higher than forest cover, (4) uncontrolled expansion of coffee plantations in conjunction with the existing coffee irrigation system will lead to severe water imbalances in the future.

Author Contributions: Conceptualization and methodology, N.C.Q.T. and A.K.; analysis, N.C.Q.T. and D.N.K.; data curation, H.Q.N.; writing—original draft preparation, N.C.Q.T.; writing—review and editing, D.N.K. and H.Q.N. All authors have read and agreed to the published version of the manuscript.

Funding: This research received no external funding.

Institutional Review Board Statement: Not applicable.

Informed Consent Statement: Not applicable.

Data Availability Statement: Not applicable.

Acknowledgments: We would like to thank the Center for Environmental Remote Sensing, Chiba University for supporting me as a cooperative researcher.

Conflicts of Interest: The authors declare no conflict of interest.

References

1. Pham, T.T.H.; Moeliono, M.; Nguyen, T.H.; Nguyen, H.T.; Vu, T.H. *The Context of REDD+ in Vietnam: Drivers, Agents and Institutions*; CIFOR Occasional Paper No. 75; Center for International Forestry Research (CIFOR): Bogor, Indonesia, 2012. [CrossRef]
2. Truong, N.C.Q.; Nguyen, H.Q.; Kondoh, A. Land Use and Land Cover Changes and Their Effect on the Flow Regime in the Upstream Dong Nai River Basin, Vietnam. *Water* **2018**, *10*, 1206. [CrossRef]
3. Arnold, J.G.; Moriasi, D.N.; Gassman, P.W.; Abbaspour, K.C.; White, M.J.; Srinivasan, R.; Santhi, C.; Harmel, R.D.; Griensven, A.V.; Van Liew, M.W.; et al. SWAT: Model Use, Calibration, and Validation. *Trans. ASABE* **2012**, *55*, 1491–1508. [CrossRef]
4. Hibert, A.R. Forest treatment effects on water yield. In *Forest Hydrology, Proceedings of the International Symposium on Forest Hydrology, State College, PA, USA, 29 August–10 September 1965*; Sopper, W.E., Lull, H.W., Eds.; Pergamon: Oxford, UK, 1967; pp. 527–543.
5. Bosch, J.M.; Hewlett, J.D. A review of catchment experiments to determine the effect of vegetation changes on water yield and evapotranspiration. *J. Hydrol.* **1982**, *55*, 3–23. [CrossRef]
6. Bruijnzeel, L.A. *Hydrology of Moist Tropical Forests and Effects of Conversion: A State of Knowledge Review*; UNESCO Humid Tropics Programme Publication, Free University: Amsterdam, The Netherlands, 1990; 224p.
7. Stednick, J.D. Monitoring the effects of timber harvest on annual water yield. *J. Hydrol.* **1996**, *176*, 79–95. [CrossRef]
8. Brown, A.E.; Zhang, L.; McMahon, T.A.; Western, A.W.; Vertessy, R.A. A review of paired catchment studies for determining changes in water yield resulting from alterations in vegetation. *J. Hydrol.* **2005**, *310*, 28–61. [CrossRef]
9. Brown, A.E.; Western, A.W.; McMahon, T.A.; Zhang, L. Impact of forest cover changes on annual streamflow and flow duration curves. *J. Hydrol.* **2013**, *483*, 39–50. [CrossRef]
10. Cou, L. Land use/cover change impacts on hydrology in large river basins: A review. In *Terrestrial Water Cycle and Climate Change: Natural and Human-Induced Impacts*; Tang, Q., Oki, T., Eds.; John Wiley & Sons, Inc.: Hoboken, NJ, USA, 2016; pp. 103–134.
11. Rientjes, T.H.M.; Hailel, A.T.; Kebede, E.; Mannaerts, C.M.M.; Habib, E.; Steenhuis, T.S. Changes in land cover, rainfall and stream flow in Upper Gilgel Abbay catchment, Blue Nile basin—Ethiopia. *Hydrol. Earth Syst. Sci.* **2011**, *15*, 1979–1989. [CrossRef]
12. Bradshaw, C.J.A.; Sodi, N.S.; Peh, K.S.H.; Brook, B.W. Global evidence that deforestation amplifies flood risk and severity in the developing world. *Glob. Change Biol.* **2007**, *13*, 2379–2395. [CrossRef]
13. Van Dijk, A.I.J.M.; Van Noordwijk, M.; Calder, I.R.; Bruijnzeel, L.A.; Schellekens, J.; Chappell, N.A. Forest-flood relation still tenuous—comment on "Global evidence that deforestation amplifies flood risk and severity in the developing world" by C.J.A. Bradshaw, N.S. Sodi, K.S.-H. Peh and B.W. Brook. *Glob. Change Biol.* **2009**, *15*, 110–115. [CrossRef]
14. Kastridis, A.; Theodosiou, G.; Fotiadis, G. Investigation of flood management and mitigation measures in ungauged NATURA protected watersheds. *Hydrology* **2021**, *8*, 170. [CrossRef]
15. Sapountzis, M.; Kastridis, A.; Kazamias, A.-P.; Karagiannidis, A.; Nikopoulos, P.; Lagouvardos, K. Utilization and uncertainties of satellite precipitation data in flash flood hydrological analysis in ungauged watersheds. *Glob. NEST J.* **2021**, *23*, 388–399. [CrossRef]
16. De Jong, C. European perspectives on forest hydrology. In *Forest Hydrology: Processes, Management and Assessment*; Amatya, D., Williams, T., Bren, L., De Jong, C., Eds.; CABI: Wallingford, UK, 2016; pp. 69–87.

17. Thanapakpawin, P.; Richey, J.; Thomas, D.; Rodda, S.; Campbell, B.; Logsdon, M. Effects of landuse change on the hydrologic regime of the Mae Chaem river basin, NW Thailand. *J. Hydrol.* **2006**, *334*, 215–230. [CrossRef]
18. Costa, M.H.; Aurélie, B.; Jeffrey, A.C. Effects of large-scale changes in land cover on the discharge of the Tocantins River, Southeastern Amazonia. *J. Hydrol.* **2003**, *283*, 206–217. [CrossRef]
19. Nguyen, T.N.Q.; Nguyen, D.L.; Nguyen, K.L. Effect of land use change on water discharge in Srepok watershed, Central Highland, Viet Nam. *Int. Soil Water Conserv. Res.* **2014**, *2*, 74–86. [CrossRef]
20. Khoi, D.N.; Suetsugi, T. Impact of climate and land-use changes on hydrological processes and sediment yield-a case study of the Be River catchment, Vietnam. *Hydrol. Sci. J.* **2014**, *59*, 1095–1108. [CrossRef]
21. Wilk, J.; Andersson, L.; Plermkamon, V. Hydrological impacts of forest conversion to agriculture in a large river basin in northeast Thailand. *Hydrol. Processes* **2001**, *15*, 2729–2748. [CrossRef]
22. Beck, H.E.; Bruijnzeel, L.A.; Van Dijk, A.I.J.M.; McVicar, T.R.; Scatena, F.N.; Schellekens, J. The impact of forest regeneration on streamflow in 12 meso-scale humid tropical catchments. *Hydrol. Earth Syst. Sci.* **2013**, *10*, 3045–3102. [CrossRef]
23. Liu, Z.; Cuo, L.; Li, Q.; Liu, X.; Ma, X.; Liang, L.; Ding, J. Impacts of Climate Change and Land Use/Cover Change on Streamflow in Beichuan River Basin in Qinghai Province, China. *Water* **2020**, *12*, 1198. [CrossRef]
24. Malmer, A. Water-yield changes after clear-felling tropical rainforest and establishment of forest plantation in Sabah, Malaysia. *J. Hydrol.* **1992**, *34*, 77–94. [CrossRef]
25. Japan International Cooperation Agency (JICA): Nippon Koei Co., Ltd. The Master Plan Study on Dong Nai River and Surrounding Basins Water Resources Development: Final Report; Vol. 4. Appendix II: Topography and Geology, Appendix III: Meteorology and Hydrology. 1996. Available online: http://open_jicareport.jica.go.jp/617/617/617_123_11309523.html (accessed on 20 September 2017).
26. ICO (International Coffee Organization). Country Coffee Profile: Vietnam. 2019, ICC-124-9. Available online: http://www.ico.org/documents/cy2018-19/icc-124-9e-profile-vietnam.pdf (accessed on 20 July 2020).
27. The World Bank—Agriculture and Rural Development Department. Coffee Sector Report. Report No. 29358-VN. June 2004. Available online: http://documents1.worldbank.org/curated/en/458011468172449689/pdf/293580VN0Coffe1ver0P0826290 1Public1.pdf (accessed on 20 July 2020).
28. Cheesman, J.; Bennett, J. Managing Groundwater Access in the Central Highlands (Tay Nguyen), Viet Nam. ISSN 1832-7435. Australian Centre for International Agricultural Research (ACIAR), Research Report No. 1. May 2005. Available online: https://crawford.anu.edu.au/people/academic/jeff-bennett/managing-groundwater-access-central-highlands-tay-nguyen-viet-nam (accessed on 7 February 2022).
29. Mango, L.M.; Melesse, A.M.; McClain, M.E.; Gann, D.; Setegn, S.G. Land use and climate change impacts on the hydrology of the upper Mara River Basin, Kenya: Results of a modeling study to support better resource management. *Hydrol. Earth Syst. Sci.* **2011**, *15*, 2245–2358. [CrossRef]
30. Pervez, M.S.; Henebry, G.M. Assessing the impacts of climate and land use and land cover change on the freshwater availability in the Brahmaputra River basin. *J. Hydrol. Reg. Stud.* **2015**, *3*, 285–311. [CrossRef]
31. Tan, M.L.; Ibrahim, A.L.; Yusop, Z.; Duan, Z.; Ling, L. Impacts of land-use and climate variability on hydrological components in the Johor River basin, Malaysia. *Hydrol. Sci. J.* **2015**, *60*, 873–889. [CrossRef]
32. Li, D.; Tian, Y.; Liu, C.; Hao, F. Impact of land-cover and climate changes on runoff of the source regions of the Yellow River. *J. Geogr. Sci.* **2004**, *14*, 330–338. [CrossRef]
33. Jun, T. Combined impact of climate and land use changes on streamflow and water quality in eastern Massachusetts, USA. *J. Hydrol.* **2009**, *379*, 268–283. [CrossRef]
34. Karlsson, I.B.; Sonnenborg, T.O.; Refsgaard, J.C.; Trolle, D.; Børgesen, C.D.; Olesen, J.E.; Jeppesen, E.; Jensen, K.H. Combined effects of climate models, hydrological model structures and land use scenarios on hydrological impacts of climate change. *J. Hydrol.* **2016**, *535*, 301–317. [CrossRef]
35. Hua, G.; Qi, H.; Tong, J. Annual land seasonal streamflow responses to climate and land-cover changes in the Poyang Lake basin, China. *J. Hydrol.* **2008**, *355*, 106–122. [CrossRef]
36. Arnold, J.G.; Srinivasan, R.; Muttiah, R.S.; Williams, J.R. Large area hydrologic modeling and assessment part I: Model development. *J. Am. Water Resour. Assoc.* **1998**, *34*, 73–89. [CrossRef]
37. Arnold, J.G.; Fohrer, N. SWAT2000: Current capabilities and research opportunities in applied watershed modelling. *Hydrol. Processes* **2005**, *19*, 563–572. [CrossRef]
38. Neitsch, S.L.; Arnold, J.G.; Kiniry, J.R.; Williams, J.R. Soil and Water Assessment Tool: Theoretical Documentation Version. 2009. Available online: http://swat.tamu.edu/documentation/ (accessed on 20 October 2015).
39. SWAT-CUP: SWAT Callibration and Unertainty Programs—A User Manual. Available online: http://swat.tamu.edu/software/swat-cup/ (accessed on 20 October 2015).
40. Veith, T.L.; Van Liew, M.W.; Bosch, D.D.; Arnold, J.G. Parameter Sensitivity and Uncertainty in SWAT: A Comparison across Five USDA-ARS Watersheds. *Am. Soc. Agric. Biol. Eng.* **2010**, *53*, 1477–1486.
41. Dile, Y.T.; Karlberg, L.; Srinivasan, R.; Rockstrom, J. Investigation of the curve number method for surface runoff estimation in tropical regions. *JAWRA J. Am. Water Resour. Assoc.* **2016**, *52*, 1155–1169. [CrossRef]
42. Liew, M.W.; Veith, T.L.; Bosch, D.D.; Arnold, J.G. Suitability of SWAT for the Conservation Effects Assessment Project: Comparison on USDA Agricultural Research Service Watershed. *J. Hydrol. Eng.* **2007**, *12*, 173–189. [CrossRef]

43. Moriasi, D.N.; Arnold, J.G.; Van Liew, M.W.; Bingner, R.L.; Harmel, R.D.; Veith, T.L. Model evaluation guidelines for systematic quantification of accuracy in watershed simulations. *Trans. ASABE* **2007**, *50*, 885–900. [CrossRef]
44. Singh, J.; Knapp, H.V.; Arnold, J.G.; Demissie, M. Hydrologic modeling of the Iroquois River watershed using HSPF and SWAT. *J. Am. Water Resour. Assoc.* **2005**, *41*, 361–375. [CrossRef]
45. Olanrewaju, O.A.; Huade, G.; Vincent, E.A.P.; Okke, B. Comparison of MODIS and SWAT evapotranspiration over a complex terrain at different spatial scales. *Hydrol. Earth Syst. Sci.* **2018**, *22*, 2775–2794. [CrossRef]
46. Li, Z.; Zhang, Y.; Wang, S.; Yuan, G.; Yang, Y.; Cao, M. Evapotranspiration of a tropical rain forest in Xishuangbanna, southwest China. *Hydrol. Processes* **2010**, *24*, 2405–2416. [CrossRef]
47. Kuraji, K. *Green-Dam Function (Water Source Protection Function) of Forest and Its Enhancement*; Japan Afforestation Flood Control Association: Tokyo, Japan, 2003; 76p, Available online: http://www.uf.a.u-tokyo.ac.jp/~{}kuraji/Midorinodam.pdf (accessed on 14 September 2016). (In Japanese)
48. Nakano, H. Effect on streamflow of forest cutting and change in regrowth on cut-over area [in Japanese with Enghlish abstract]. *Bull. Gov. For. Exp. Stn.* **1971**, *240*, 1–251.
49. Chang, J.H. Hydrology in Humid Tropical Asia. In *Hydrology and Water Management in the Humid Tropics*; Bonell, M., Hufschmidt, M.M., Gladwell, J.S., Eds.; Cambridge University Press: Cambridge, UK, 1993; pp. 55–66. ISBN 0-521-02002-6.
50. Lal, R. Challenges in agriculture and forest hydrology in the humid tropics. In *Hydrology and Water Management in the Humid Tropics*; Bonell, M., Hufschmidt, M.M., Gladwell, J.S., Eds.; Cambridge University Press: Cambridge, UK, 1993; pp. 395–404. ISBN 0-521-02002-6.
51. Gerold, G.; Leemhuis, C. Effects of "ENSO-events" and rainforest conversion on river discharge in Central Sulawesi (Indonesia). In *Tropical Rainforests and Agroforests under Global Change*; Tscharntke, T., Leuschner, C., Veldkamp, E., Faust, H., Guhardja, E., Bidin, A., Eds.; Springer: Berlin, German, 2010; pp. 327–350. ISBN 978-3-642-26259-3.
52. Munoz-Villers, L.E.; McDonnell, J.J. Land use change effects on runoff generation in a humid tropical montane cloud forest region. *Hydrol. Earth Syst. Sci.* **2013**, *17*, 3543–3560. [CrossRef]
53. Ma, X.; Xu, J.; Luo, Y.; Aggarwa, S.P.; Li, J. Response of hydrological processes to land- cover and climate changes in Kejie watershed, South-West China. *Hydrol. Processes* **2009**, *23*, 1179–1191. [CrossRef]
54. D'haeze, D.; Deckers, J.; Raes, D.; Tran, A.P.; Nguyen, D.M.C. Over-irrigation of *Coffea canephora* in the Central Highlands of Vietnam revisited: Simulation of soil moisture dynamics in Rhodic Ferralsols. *Agric. Water Manag.* **2003**, *63*, 185–202. [CrossRef]
55. Ahmad, A. An institutional analysis of changes in land use pattern and water scarcity in dak lak province. In Proceedings of the at the Nordic Conference on "Institutions, Livelihoods and the Environment: Change and Response in Mainland Southeast Asia", Copenhagen, Denmark, 27–29 September 2000; University of Lund: Lund, Sweden, 2000.
56. Amarasinghe, U.A.; Chu, T.H.; D'haeze, D.; Tran, Q.H. Toward sustainable coffee production in Vietnam: More coffee with less water. *Agric. Syst.* **2015**, *136*, 96–105. [CrossRef]
57. D'haezea, D.; Raesa, D.; Deckersa, J.; Phong, T.A.; Loi, H.V. Groundwater extraction for irrigation of Coffea canephora in Ea Tul watershed, Vietnam—a risk evaluation. *Agric. Water Manag.* **2005**, *73*, 1–19. [CrossRef]
58. Cheesman, J.; Tran, V.H.S.; Bennett, J. Managing Groundwater Access in the Central Highlands (Tay Nguyen), Viet Nam. ISSN 1832-7435. Australian Centre for International Agricultural Research (ACIAR), Research Report No. 6. November 2007. Available online: https://crawford.anu.edu.au/people/academic/jeff-bennett/managing-groundwater-access-central-highlands-tay-nguyen-viet-nam (accessed on 7 February 2022).

 *water*

Article

River Buffer Effectiveness in Controlling Surface Runoff Based on Saturated Soil Hydraulic Conductivity

Hatma Suryatmojo [1,*] and Ken'ichirou Kosugi [2]

[1] Laboratory of Watershed Management, Faculty of Forestry, Universitas Gadjah Mada, Bulaksumur, Yogyakarta 55281, Indonesia
[2] Laboratory of Erosion Control, Graduate School of Agriculture, Kyoto University, Sakyo-ku, Kyoto 606-8501, Japan; kos@kais.kyoto-u.ac.jp
* Correspondence: hsuryatmojo@ugm.ac.id

Abstract: In tropical Indonesia, rainforests are managed by an intensive forest management system (IFMS). The IFMS has promoted selective logging for timber harvesting and intensive line planting to enrich the standing stock. The implementation of the IFMS has reduced the forest canopy cover, disturbed the surface soil, changed the soil hydraulic properties, and increased direct runoff and soil erosion. Investigation of the IFMS impact on soil hydraulic properties and the generation of surface runoff using a saturated hydraulic conductivity model is needed. Soil hydraulic properties were investigated on 11 plots, including one virgin forest plot and 10 plots at different operational periods of the IFMS. A two-dimensional saturated soil water flow simulation was applied to generate surface runoff from different periods of the IFMS. The main parameters of canopy cover, net rainfall, and saturated hydraulic conductivity were used in the simulations. A simulation scenario of a surface runoff hydrograph in different forest operations was used to analyze the river buffer effectiveness. The results showed that fundamental IFMS activities associated with mechanized selective logging and intensive line planting have reduced the soil hydraulic conductivity within the near-surface profile. The recovery time for near-surface K_s on non-skidder tracks was between 10 and 15 years, whereas on the skidder tracks it was more than 20 years. Forest disturbances have altered the typical surface hydrological pathways, thereby creating the conditions for more surface runoff on disturbed surfaces than on undisturbed surfaces. Maintaining the buffer area is an effective means to reduce the peak discharge and surface runoff in the stream channel.

Keywords: tropical forest; river buffer; surface runoff; peak discharge; saturated hydraulic conductivity

Citation: Suryatmojo, H.; Kosugi, K. River Buffer Effectiveness in Controlling Surface Runoff Based on Saturated Soil Hydraulic Conductivity. *Water* **2021**, *13*, 2383. https://doi.org/10.3390/w13172383

Academic Editor: Brigitta Schütt

Received: 21 July 2021
Accepted: 26 August 2021
Published: 30 August 2021

1. Introduction

In forested areas, the movement of water between the atmosphere and soil plays an important role in the storage capacity of the land. Forest canopies serve as a barrier against precipitation reaching the ground. Global evidence suggests that changes in interception loss, evapotranspiration, infiltration, and stormflow pathways caused by various degrees of forest conversion can alter the timing and magnitude of direct runoff and baseflow for an unpredictable period of time [1–5]. Some studies have investigated changes in hydrological variables within the soil profile that may have implications for the partitioning and movement of subsurface stormflow [6–11]. An understanding of the spatial variability of soil hydraulic properties is important to accurately determine the subsurface flux of water [12] and its variation following disturbances of surface soil [4,7–9,13–16]. In forested hillslopes, soil water has been observed to increase rapidly and gently in the region downslope from tree stems, particularly at points close to the tree stems [17].

Land conversion and timber extraction are likely to alter the biodiversity and hydrologic responses of forested areas. In tropical Indonesia, rainforests are managed by an intensive forest management system (IFMS). The IFMS has promoted selective logging for timber harvesting and intensive line planting to enrich the standing stock. Timber extraction using heavy machines destroys the soil structure, which affects water and nutrient cycling, and accelerates runoff and soil erosion rates [5,6,18]. Heavy machines in timber collection areas and on skidder roads can increase soil compaction by up to 40% of natural conditions [19], and 10–30% of the soil surface may be denuded due to logging roads, skidder tracks, and log landings [6,20]. The use of heavy equipment tends to compact the topsoil, setting in motion a negative spiral of reduced infiltrability and an increased frequency of surface runoff flow and sheet erosion, thereby hindering the establishment of a new protective layer of vegetation and litter [6].

The implementation of the IFMS has reduced the forest canopy cover, destroyed the surface soil, changed the hydraulic conductivity, and increased direct runoff and soil erosion. Soil compaction has been considered the principal form of damage associated with logging, restricting root growth, and reducing productivity [9]. Timber harvest has been shown to have a significant impact on the watershed hydrology and serve as sediment sources and transport pathways on cleared land [4,5,21–29]. Data from rainfall simulation experiments and saturated soil hydraulic conductivity (K_s) measurements in a disturbed upland watershed have convincingly shown that the various land cover types within the fragmented landscape differ in their ability to infiltrate rainwater [13,22,30].

Different forest treatments and disturbances affect soil hydraulic properties in different ways [4,13,25,28]. When the forest soil is subjected to numerous disturbances due to surface opening, logging road development, timber cutting, falling tree hits, and heavy machines, there is an accompanying change in the intrinsic properties of the soil, which alters the soil's hydrological balance. The effects of forest regrowth are believed to be able to restore soil hydraulic properties. We investigated the impact of the IFMS on soil hydraulic conductivity and the generation of surface runoff in different river buffer scenarios.

2. Materials and Methods

2.1. Study Area

The study site is located in the headwater region of the Katingan watershed, one of the largest in Central Kalimantan, Indonesia. The site is in the Sari Bumi Kusuma (SBK) concession area, a private forest company (00°36′–01°10′ S, 111°39′–112°25′ E) located in the lowland part of Bukit Baka hills. This location is part of a high-biodiversity area known as the "Heart of Borneo."

The mean annual rainfall from 2001 to 2012 was 3631 mm, with the highest average monthly precipitation (353 mm) occurring in November. The lowest average monthly precipitation (209 mm) was recorded in August [31]. According to the forest climate classification system of Schmidt and Ferguson [32], the area is a type A (very wet) tropical rainforest (monthly average rainfall > 100 mm). Because the location is between 5° N and 5° S, the study site is also classified as having an equatorial climate [33], which is generally characterized by high precipitation and high temperatures throughout the year. The soil in the region is classified as Ultisol (USDA soil taxonomy classification). This group of soils was previously called red–yellow podzolic soils. These red–yellow podzols appear to be more widely distributed in Indonesia and are a major soil in the lowland area of Kalimantan.

The soil structure is dominated by angular blocky material and the soil texture is dominated by silty clays, which contain 40% or more clay and 40% or more silt. The soil bulk density in the virgin forest is 0.636 g cm. Manual land clearing in the line-planted and cleared areas has increased soil bulk density by 23–32%; this has increased by 11% in the logged area and by 73% in the skidder tracks of mechanized logging. Ten years after plantation, the values in the line-planted, cleared, and logged areas recovered to levels similar to that of virgin forest, but that for skidder-track areas was still 52% higher than that in virgin forest [34].

The Intensive Forest Management System (IFMS) is a new forest management system for tropical Indonesian forests. The main activities are timber harvesting using a selective logging method and forest rehabilitation with intensive line planting. A stricter cutting regime coupled with several additional selective logging controls is imposed to minimize the impact of logging on skid trails. This includes the alignment of skid trails along the contour, construction of cross-drains at 45–60° along the skid trails, and skid trail deactivation at the end of the skid trail line. Furthermore, no logging is allowed within a buffer area of at least 20 m on both sides of perennial streams. The IFMS phase after selective logging is forest rehabilitation with intensive line planting. Intensive line planting involves line clearing and line rehabilitation. About 15–20% of the forest area is a clear-cut line to enrich the standing stock using an intensive strip-line planting system.

2.2. Experimental Plots

Field observations of soil hydraulic properties were conducted on 11 plots, including 1 virgin forest plot and 10 plots at different operational blocks of the IFMS, as shown in Figure 1. Field measurements of infiltration were made using a portable double-ring infiltrometer. The inner ring was 13 cm in diameter and 24 cm high, and the outer ring was 30 cm in diameter and 15 cm high. A total of 123 infiltrometer tests were performed in the 10 plots of forest operation blocks from the first year until 10 years after forest treatments began, and in the undisturbed forest (virgin forest) plot. In each plot of the IFMS, the infiltrometer test was performed at four locations with three repetitions based on differences in topography: a line-planted area, cleared area, logged area, and an area with skidder tracks, as shown in Figure 2. In the virgin forest plot, the infiltrometer test was performed at upper, middle, and lower slopes of virgin forest blocks. These test sites were selected using a random sampling method. The tests were conducted after a minimum of 2 days without rainfall [34].

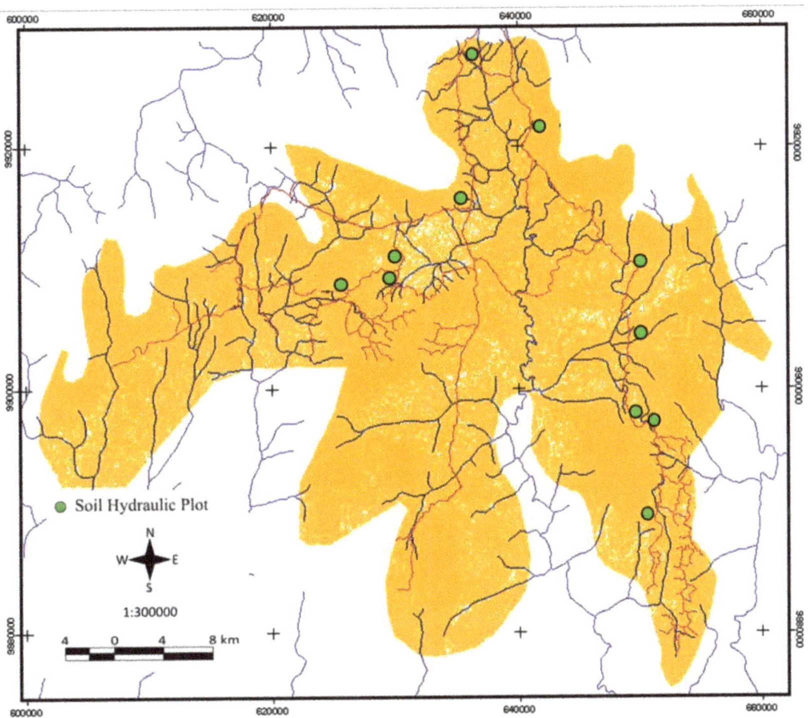

Figure 1. Location of the experimental plots for soil hydraulic properties.

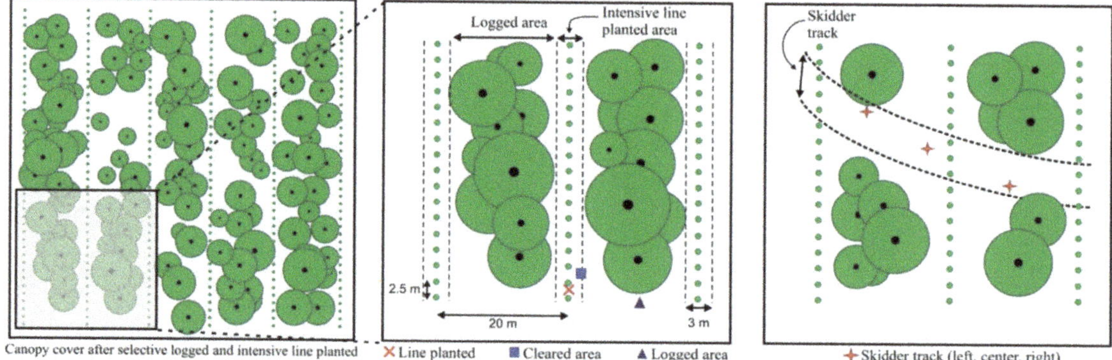

Figure 2. Infiltrometer test sites in line-planted, cleared, logged, and skidder-track areas.

Each infiltrometer was driven 10 cm into the ground by hammering a wooden platform placed on top of the device. Great care was taken to minimize soil disturbance within the inner ring. Water was added to both rings, but measurements were made only in the inner ring. The outer ring provided a buffer that reduced boundary effects caused by the cylinder and by lateral flow at the bottom of the ring. A plastic sheet was placed in the inner ring. The outer ring was filled with water to 5 cm and the water level was maintained. The inner ring was filled with water to create a ponding depth of 5 cm for the constant head of water at the skidder track, and of 10 cm for all other sites. The plastic sheet was removed and the level of water was recorded continuously each 10 min at the skidder track and each 15 min at all other sites. Cumulative infiltration and elapsed time were recorded over a 3–4 h period to ensure that steady-state infiltrability had been attained. The infiltrability (IR) was calculated from the cumulative infiltration as a function of time [34].

2.3. Saturated Hydraulic Conductivity

Saturated hydraulic conductivity (K_s) is one of the most important hydraulic properties affecting water flow in soils. Soil type, spatial, and seasonal variability, in addition to scale dependency, are key factors that make it more difficult to accurately measure the saturated hydraulic conductivity [35,36]. Saturated hydraulic conductivity (K_s) is one of the properties of the flowpaths in almost all modes of streamflow generation [10,37]. This parameter was estimated from infiltration measurements taken in each plot, using the falling-head method [38]:

$$K_s = \frac{L}{t_1} \ln\left(\frac{b_0 + L}{b_1 + L}\right) \tag{1}$$

where b_0 is ponded water in the soil column at $t = 0$, b_1 is the height of ponded water at t, and L is the length of the soil column.

An additional 32 undisturbed soil samples were collected from the plots. Sampling points at the skidder track plots were selected at locations aged 1–12 years after forest treatments began and eight soil core samples were taken from the virgin forest plot at soil depths of 0–10, 10–20, 20–30, 30–40, 40–50, 90–100, 120–130, and 150 cm.

In the laboratory, a water retention test using a pressure plate was used to measure the volumetric water content, θ. Undisturbed soil core samples were placed on an aluminum tray and slowly saturated by adding water from the bottom over a 24 h period. Soil water retention curves were measured using the pressure plate method [38] for matric pressure heads (ψ) of −5, −10, −30, −60, −100, −150, −200, and −500 cm. After measuring the water content at ψ = −500 cm, the soil samples were resaturated from the bottom over a 24 h period.

Then, the K_s of each sample was measured using the falling head method [39]. The observed hydraulic data sets were analyzed using physical-based models for soil water retention and hydraulic conductivity functions using the lognormal distribution (LN) model [38]. The LN model is effective for analyzing hydraulic properties and soil water movements in connection with the soil pore-size distribution [40–43]. Based on this model, water retention and hydraulic conductivity can be expressed as:

$$Se = \frac{(\theta - \theta_r)}{(\theta_s - \theta_r)} = Q\left(\frac{\ln(\psi / \psi_m)}{\sigma}\right) \quad (2)$$

where S_e is the effective saturation; θ_s and θ_r are the saturated water content and residual water content, respectively; K_s is the saturated hydraulic conductivity; ψ_m is the matric potential head at an S_e of 0.5, which is related to the median pore radius by the capillary pressure function; σ (dimensionless) is the standard deviation of the log-transformed pore radius and represents the width of the pore radius distribution; and Q denotes the complementary normal distribution function [40].

The relationship between hydraulic conductivity K and ψ [40] was derived by substituting Equation (2) into the pore structure model proposed by [44]:

$$K(\psi) = K_s\left\{Q\left[\frac{\ln(\psi / \psi_m)}{\sigma}\right]\right\}^{0.5}\left\{Q\left[\frac{\ln(\psi / \psi_m)}{\sigma} + \sigma\right]\right\}^2 \quad (3)$$

Equations (2) and (3) produce adequate descriptions of the measured hydraulic properties of various field soils [41–43,45]. The LN model was used to obtain the best fit to the observed data sets. The best fit was achieved by minimizing the residual sum of squares (RSS) between the computed and measured θ and K values. To fit the water retention model to the observed water retention $\theta(\psi)$ curves, ψ_m and σ of the LN model were optimized by minimizing the RSS. θ_s was fixed at a maximum measured θ value for each soil.

2.4. Surface Runoff Generation

A two-dimensional saturated soil water flow simulation using the Fortran program was applied to generate surface runoff from different periods of the IFMS. We used a two-dimensional calculation domain divided by the finite element mesh. Figure 3 shows a side view of the element as the calculated area and the node as the point output of seepage discharge. Within it, unequal triangular elements were computed using topography and soil depth field data and the θ and ψ measurements. The surface interface in the region's slope was assigned the same gradients as the average values measured in the analysis region. On the surface, downslope, and bottom boundaries, the seepage face boundary condition was imposed with different interval depths, following the soil layers.

Based on the soil profile investigation, surface soil disturbance from IFMS operations was found to have an effect at a soil depth of 0–15 cm. The scenario setting for the two-dimensional saturated soil water flow simulation model is shown in Figure 3. The soil layer depths were based on the soil profile investigation. We found no difference in the soil layers between −60 to −100 cm and −100 to −150 cm, therefore we used the soil layer of −60 to −100 cm to represent the parent material horizon. The slope gradient was given the same gradient, of 30°, as the average measured in the operational blocks. The element of the calculated area was designed with a maximum length of 20 cm for the detailed calculation at the point output (node).

Water discharge in a forested catchment includes vertical drainage in surface soil and downslope drainage in subsurface soil (Figure 4). In vertical drainage, rainfall is supplied to the soil surface and infiltrates the soil profile unless the rainfall intensity is greater than the soil permeability. The water moves vertically in surface soil (unsaturated zone) and discharges at the bottom of the soil profile. Vertical drainage is considered to be an input into subsurface soil (saturated zone). Runoff generation was created at the edge of river channel after vertical drainage in surface soil and downslope drainage in surface soil reached their maxima.

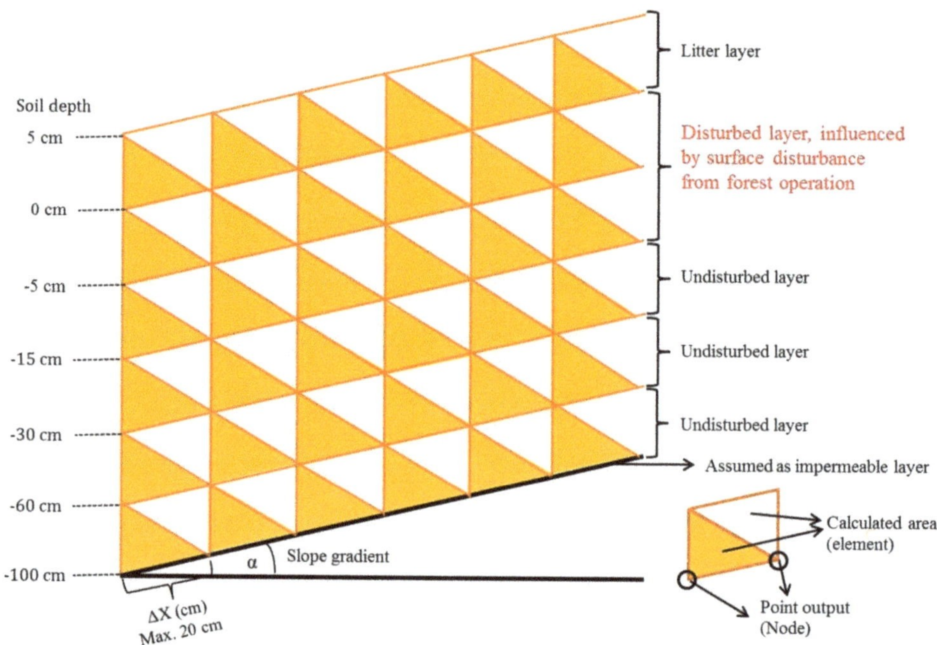

Figure 3. Scenario setting for two-dimensional saturated soil water flow simulation model.

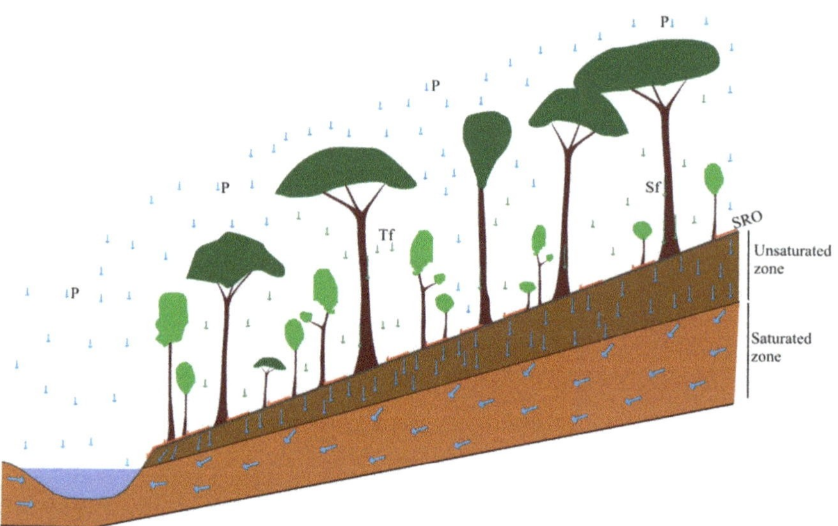

Figure 4. The process of water discharge in a forested catchment. (P = precipitation; Tf = throughfall; Sf = stemflow; SRO = surface runoff).

Several parameters were established for different periods of forest operation and different types of forest treatment. The main parameters of canopy cover, net rainfall, and K_s were used in the simulations. In the present study, canopy cover density between plots was measured from a 1 ha permanent sample plot (PSP). PSP is a long-term observation site of forest growth, for measuring diameter increment, volume increment, and stand structure dynamics. In the PSP, forest vegetation is measured using a nested cover quadrats method.

Each type of vegetation was measured at 25 subplots (total in 1 hectare). The subplot was classified into 20 × 20 m for trees with a diameter >20 cm (at 1.3 m above the land surface). The tree canopy cover was calculated using a conversion equation from tree diameter to canopy area for each species. Each forest operational plot and virgin forest was established with three sets of PSPs located randomly on the upper, middle, and lower slope.

3. Results and Discussion

3.1. Soil Hydraulic Conductivity

Volumetric water content and hydraulic conductivity were analyzed to assess recovery across different forest management periods. Figure 5 shows water retention $\theta(\psi)$ curves for a skidder track area during 0–12 years following treatment.

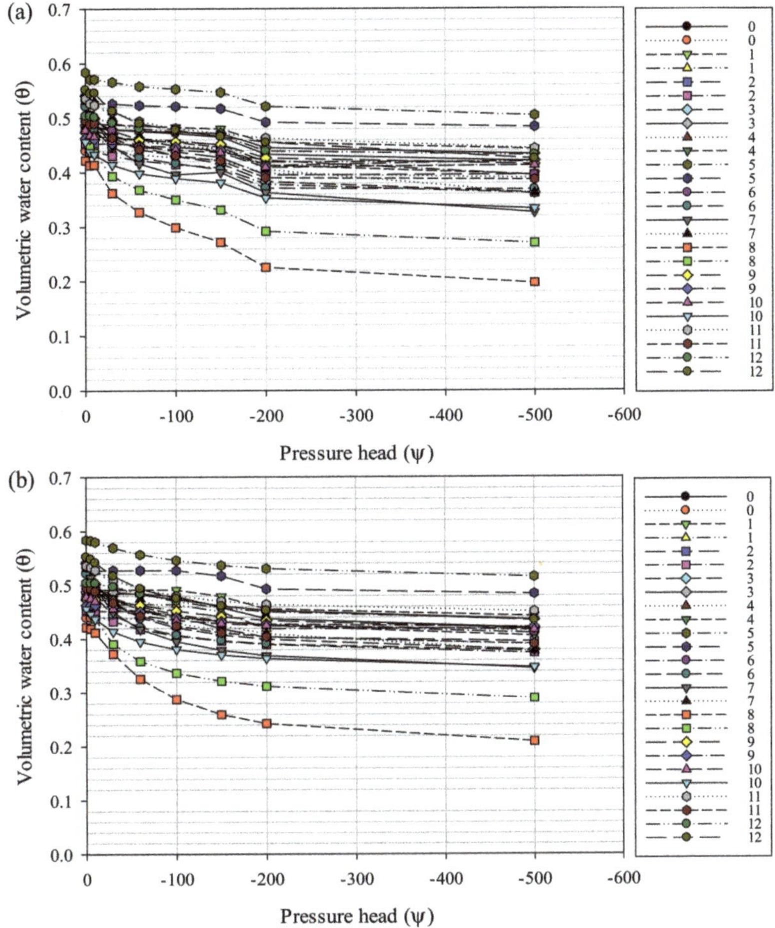

Figure 5. Water retention $\theta(\psi)$ curves for a skidder track area 0–12 years after IFMS treatment. (**a**) Observed $\theta(\psi)$. (**b**) Estimated $\theta(\psi)$.

No large differences were found between observed $\theta(\psi)$ and estimated $\theta(\psi)$. Water retention curves showed large changes in the range of $0 > \psi > -60$ cm. The greatest changes tended to occur in the 8 year old site, indicating a lower soil moisture content and higher soil compaction at that site. The different θ values in the skidder track may be associated with different levels of soil compaction caused by different volumes of tractor traffic.

Water retention curves in the virgin forest (Figure 6) showed large changes in the range of $0 > \psi > -30$ cm, indicating the existence of soil macropores. The curves for surface soils (0–50 cm deep) tended to show greater changes than subsurface soils (50–150 cm deep). In surface soil, there was intensive root growth that increased the number of macropores and the porosity.

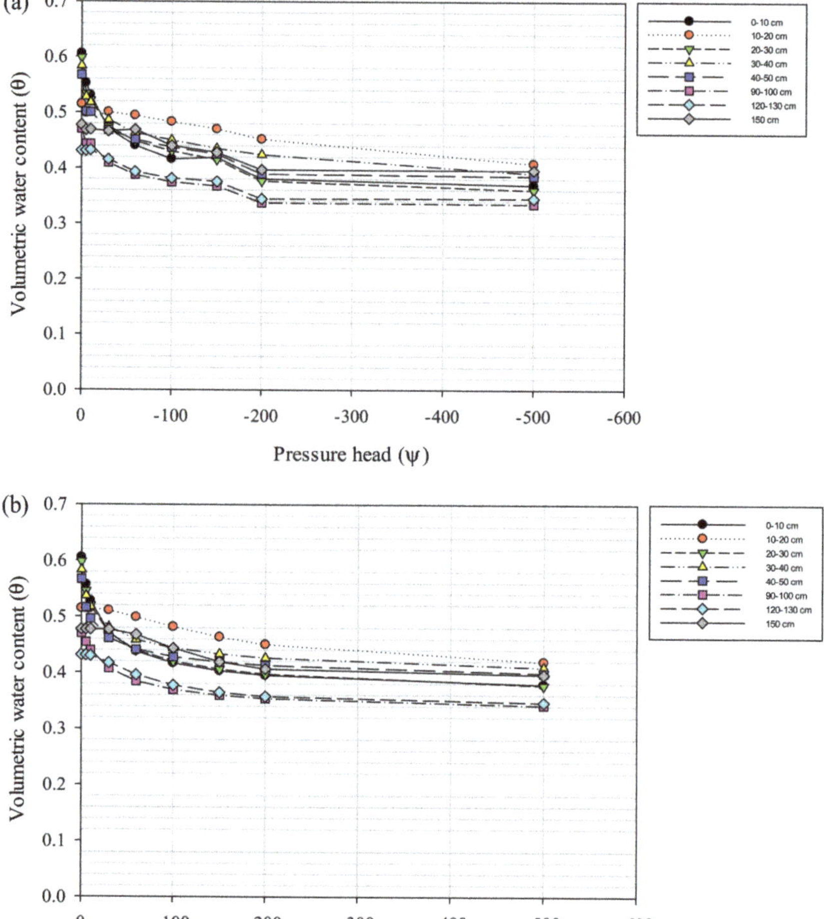

Figure 6. Water retention $\theta(\psi)$ curves in the virgin forest at different soil depths. (**a**) Observed $\theta(\psi)$. (**b**) Estimated $\theta(\psi)$.

The saturated hydraulic conductivity (K_s) at different slope locations tended to increase according to site age, i.e., from the 1 year old site to the 10 year old site (Figure 7). Values were higher at upper slopes than in the middle and at lower slopes. Those in skidder tracks were lower than at other test sites, particularly in 7 to 10 year old sites. The recovery of K_s tended to take longer in skidder tracks than in other areas. In 6 to 10 year old sites, no large differences in K_s values were found between test plots. These results indicate that soil hydraulic conductivity needs at least an estimated 15 years to recover and to reach the K_s levels of a virgin forest.

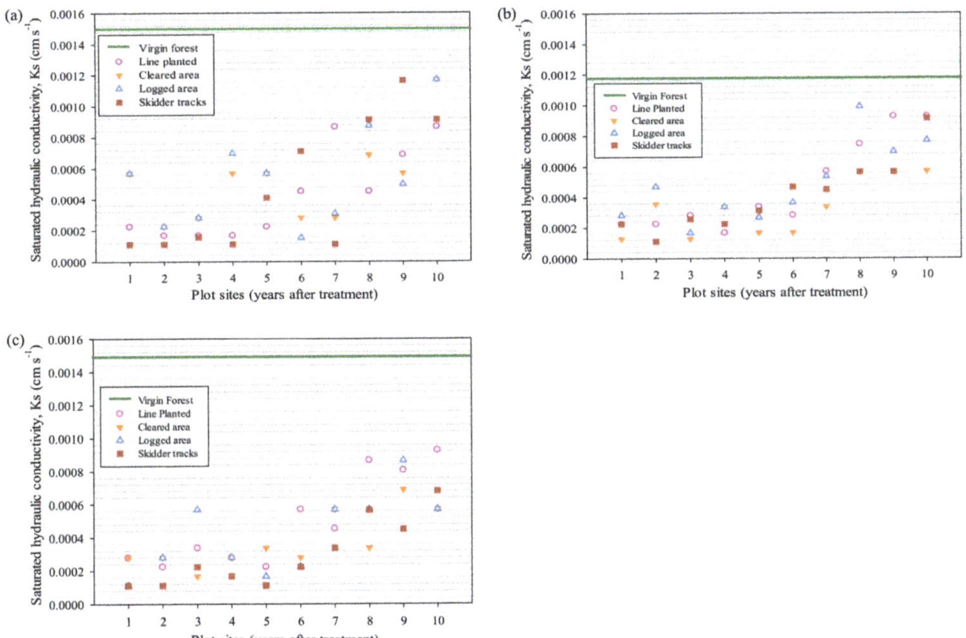

Figure 7. Saturated hydraulic conductivity (K_s) at all 11 plots, calculated from infiltration measurements taken in each area. (**a**) Upper slope. (**b**) Middle Slope. (**c**) Lower slope.

Figure 8a shows the relationship between observed and estimated θ using the LN model. There was good correspondence between observed and estimated values (data fall closely around 1:1 line), indicating that the retention model can adequately express the observed water retention curves. Figure 8b shows the relationship between K_s measured from the ring infiltrometer test ($K_{s\text{-}ring}$) and soil core samples ($K_{s\text{-}core}$). The data are widely scattered, which indicates that the results from the infiltrometer tests did not adequately correspond to the K_s values from soil core samples. This may be due to several factors, such as the different locations, sample size, time period, and the different number of samples. Direct measurements of K_s (either in the laboratory using soil core samples previously taken from the field, or directly in the field without removing a soil sample) are preferred to indirect methods (derived from soil textural characteristics). Field methods provide data that better represent the reality of water flow in natural conditions [12,35,36].

3.2. Saturated Hydraulic Conductivity to Generate Surface Runoff

The soil hydraulic properties were used to generate surface runoff (SRO) values. Measurements of saturated hydraulic conductivity (K_s) following the various surface disturbances were used to assess the influence of forest fragmentation on the near-surface hydrologic response. Forest fragmentation results in a mosaic of surfaces with distinct infiltration characteristics. Table 1 shows all of the observed K_s values at different periods and test sites of IFMS treatment. The values were lower for all test sites than for the virgin forest, further confirming that surface soil disturbances reduce the K_s value. The soil hydraulic properties shown in Table 1 were used to estimate surface runoff flow as designed in Figure 3.

Different degrees of canopy cover lead to different levels of canopy interception. The average canopy cover density between plots is shown in Figure 9.

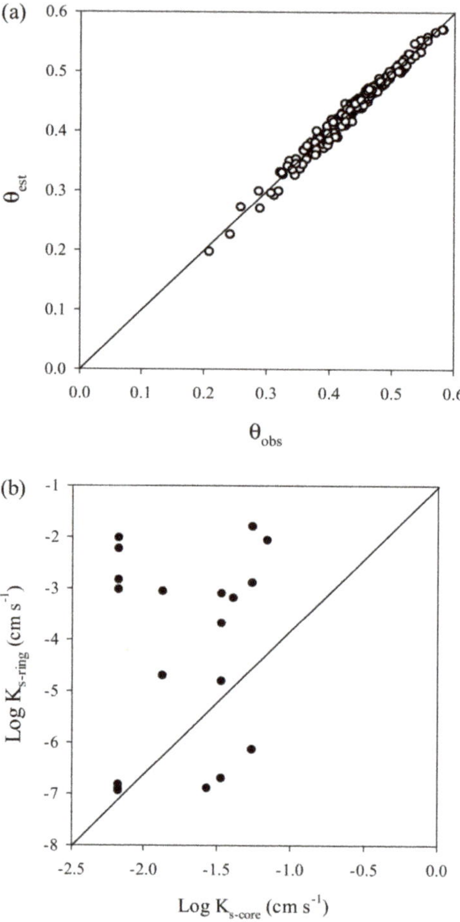

Figure 8. (**a**) Relationship between observed (θ_{obs}) and estimated (θ_{est}) water content. (**b**) Relationship between K_s measured using ring infiltrometer tests ($K_{s\text{-}ring}$) and soil core samples ($K_{s\text{-}core}$).

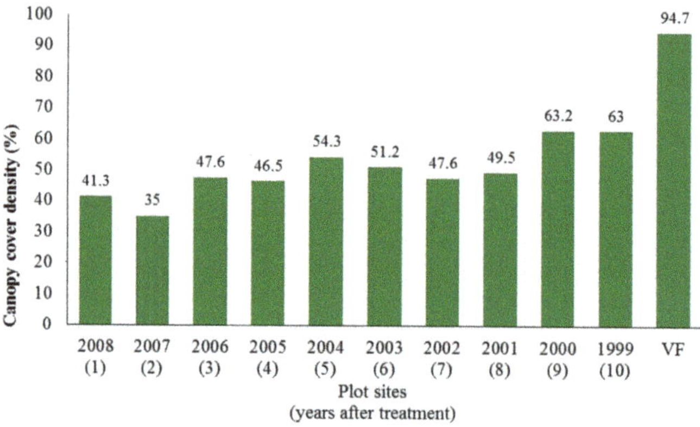

Figure 9. Average canopy cover density measured in permanent sample plots at different IFMS sites and the virgin forest site.

Table 1. Observed soil hydraulic properties in different periods and test sites under IFMS treatment.

Test Sites	Periods	K_s	Ψ_m	$\log(-\Psi_m)$	θ_r	σ	θ_e
Line Planted	1999	0.0009	−41.0198	1.6130	0.3254	1.8264	0.1743
	2000	0.0008	−41.9226	1.6224	0.3275	1.8044	0.1722
	2001	0.0007	−43.3268	1.6368	0.3308	1.7711	0.1689
	2002	0.0006	−44.0213	1.6437	0.3324	1.7551	0.1673
	2003	0.0004	−47.0335	1.6724	0.3389	1.6882	0.1608
	2004	0.0003	−51.2237	1.7095	0.3474	1.6020	0.1523
	2005	0.0002	−53.5833	1.7290	0.3519	1.5565	0.1479
	2006	0.0003	−51.4185	1.7111	0.3478	1.5982	0.1519
	2007	0.0002	−53.3802	1.7274	0.3515	1.5603	0.1482
	2008	0.0002	−51.7840	1.7142	0.3485	1.5910	0.1512
Cleared Area	1999	0.0007	−42.6708	1.6301	0.3293	1.7865	0.1704
	2000	0.0006	−44.0285	1.6437	0.3324	1.7549	0.1673
	2001	0.0005	−45.3997	1.6571	0.3354	1.7239	0.1643
	2002	0.0004	−47.8510	1.6799	0.3406	1.6708	0.1591
	2003	0.0002	−51.9810	1.7158	0.3488	1.5872	0.1509
	2004	0.0003	−49.3260	1.6931	0.3437	1.6401	0.1561
	2005	0.0004	−47.8510	1.6799	0.3406	1.6708	0.1591
	2006	0.0002	−53.5833	1.7290	0.3519	1.5565	0.1479
	2007	0.0003	−49.0211	1.6904	0.3430	1.6464	0.1567
	2008	0.0004	−48.3744	1.6846	0.3417	1.6598	0.1580
Logged Area	1999	0.0007	−43.4338	1.6378	0.3311	1.7686	0.1687
	2000	0.0007	−42.7485	1.6309	0.3295	1.7847	0.1702
	2001	0.0005	−45.1193	1.6544	0.3348	1.7302	0.1649
	2002	0.0005	−45.9036	1.6618	0.3365	1.7128	0.1632
	2003	0.0004	−47.6623	1.6782	0.3403	1.6748	0.1595
	2004	0.0003	−50.5322	1.7036	0.3460	1.6157	0.1537
	2005	0.0004	−47.0335	1.6724	0.3389	1.6882	0.1608
	2006	0.0003	−48.8732	1.6891	0.3427	1.6494	0.1570
	2007	0.0003	−49.8810	1.6979	0.3448	1.6288	0.1549
	2008	0.0002	−52.6517	1.7214	0.3501	1.5742	0.1496
Skidder Tracks	1999	0.0008	−41.6762	1.6199	0.3270	1.8104	0.1727
	2000	0.0007	−43.3331	1.6368	0.3308	1.7710	0.1689
	2001	0.0007	−43.3423	1.6369	0.3308	1.7708	0.1689
	2002	0.0003	−51.2491	1.7097	0.3474	1.6015	0.1523
	2003	0.0003	−51.2387	1.7096	0.3474	1.6017	0.1523
	2004	0.0001	−59.3777	1.7736	0.3620	1.4528	0.1377
	2005	0.0003	−51.2387	1.7096	0.3474	1.6017	0.1523
	2006	0.0004	−47.0833	1.6729	0.3390	1.6871	0.1607
	2007	0.0001	−59.3777	1.7736	0.3620	1.4528	0.1377
	2008	0.0001	−56.9932	1.7558	0.3580	1.4942	0.1417
Virgin Forest	top soil	0.0012	−39.2047	1.5933	0.3209	1.8721	0.1788
	20–30 cm	0.0006	−44.0910	1.6443	0.3325	1.7535	0.1672
	30–40 cm	0.0005	−44.9986	1.6532	0.3346	1.7329	0.1652
	40–50 cm	0.0006	−44.3201	1.6466	0.3331	1.7482	0.1667
	90–100 cm	0.0008	−41.8066	1.6212	0.3273	1.8072	0.1724

Forest operation in the IFMS sites reduced the canopy cover density, which increased with time. Canopy cover density influences the net rainfall that reaches the forest floor. To determine the variation in net rainfall between plots under the different treatments (line-planted, cleared, logged, and skidder-tracked areas), an assumption was developed, as shown in Figure 10.

Figure 10 was developed based on the PSP measurements and field observations. The canopy covers for line-planted, cleared, and skidder-tracked areas were assumed to be similar to those of logged areas in the 8 to 10 year old sites. In the 5 to 7 year old sites, the canopy cover in cleared areas was assumed to be similar to that of line-planted areas. The data on the canopy cover of line-planted areas, when plotted on a graph, followed a curved line that was assumed to represent a vegetation growth curve. In 1 to 4 year-old sites, there were different canopy covers between line-planted and cleared areas, with line-planted areas having more canopy cover than cleared areas that had 1–4 years of succession. No vegetation was planted on skidder tracks; thus, the canopy cover on skidder tracks was affected by the surrounding vegetation. Therefore, the canopy cover on skidder tracks in 1 to 10 year old sites was assumed to occur in straight line.

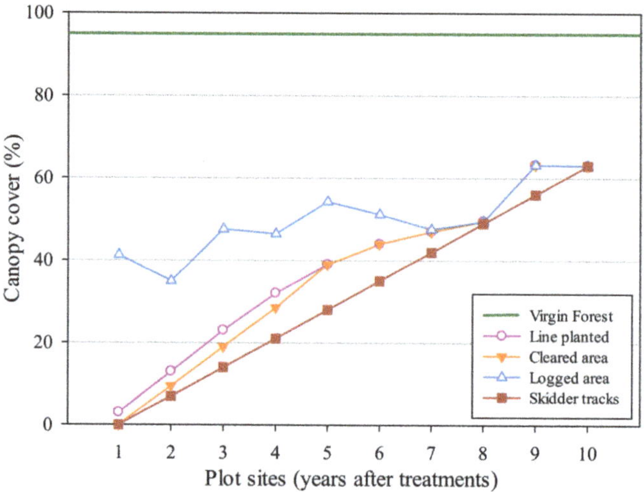

Figure 10. Canopy cover in the plot sites at different forestry treatments.

3.3. *Effectiveness of a River Buffer on Surface Runoff Flow*

A river buffer is used to reduce the impact of logging on runoff and erosion. This simulation combined a buffer area with each treatment area, as shown in Figure 11. The slope gradient was assumed to be 30°, based on the average slope gradient in the study sites.

Several rainfall events were used to generate the surface runoff hydrograph in the scenario without a river buffer (left) and with a river buffer (right) shown in Figure 12.

The SRO hydrograph indicates that there was a significant difference between plots during rainfall events (Figure 12). A dramatic difference was apparent between the virgin forest (VF) site and 1 year old site. Ten years after logging operations were ceased and vegetation was allowed to recover, differences were still found in the hydrograph between the VF site and 10 year old site. This difference indicates that surface disturbances at the 10 year old site continued to affect the hydraulic properties of the forest soil at the time of this study. Figure 13 shows the relationship between rainfall and total SRO.

Figure 13 show the data on SRO at all sites during different rainfall events. IFMS treatment changed the vertical drainage in surface soil. This indicates that forest interception (include canopy interception and forest floor interception) in the VF site contributed to reducing the peak discharge (Qp) in Figure 12. In the VF site, the average canopy interception was 23.8% of rainfall and the average forest interception was 91.7% [46].

Figure 13 shows clear differences in the surface runoff hydrograph between scenarios without and with the use of a buffer. The difference between the VF site and the 10 year old site were slightly smaller than that in the scenario without a buffer area. This indicates that the buffer area has significant influence in reducing the SRO hydrograph in the disturbed sites. To clarify the differences between the scenario with and without a buffer area, the coefficients of Qp and SRO are shown in Figure 14.

Soil hydrologic properties at the disturbed sites were changed and produced higher Qp and SRO coefficients, compared to the VF site (Figure 14). However, the Qp and SRO coefficients declined at disturbed sites over time, indicating that soil hydraulic properties slowly recovered. A smaller Qp coefficient indicates a greater potential for a forested catchment to reduce Qp (Figure 14a). In the early years after IFMS treatment, large gaps in Qp and SRO were found between the scenarios with and without buffer areas. The buffer area can effectively reduce the Qp and SRO coefficients, which declined at each site over time. The buffer area canopy serves as a barrier against precipitation reaching the ground. The high canopy cover density in the buffer area controlled the net precipitation via canopy interception. The treated area had less canopy cover, compacted soils, and low infiltration capacities. Consequently, these conditions reduced the forest interception, evapotranspiration, and infiltration volumes, creating a quick surface runoff response and increasing the percentage of rainfall to surface runoff in the model. Undisturbed soil hydraulic properties played a major role in the high infiltration capacity. The combination between high canopy interception and high infiltration capacity in the buffer area had a significant role in the SRO reduction that occurred in the early period following treatment (Figure 14b). Ten years following the initiation of the IFMS, the SRO coefficient of the disturbed sites was reduced and was similar to that of the VF site. These results suggest that the disturbed sites need at least 15 years to recover their soil hydraulic properties to levels similar to those at the VF site.

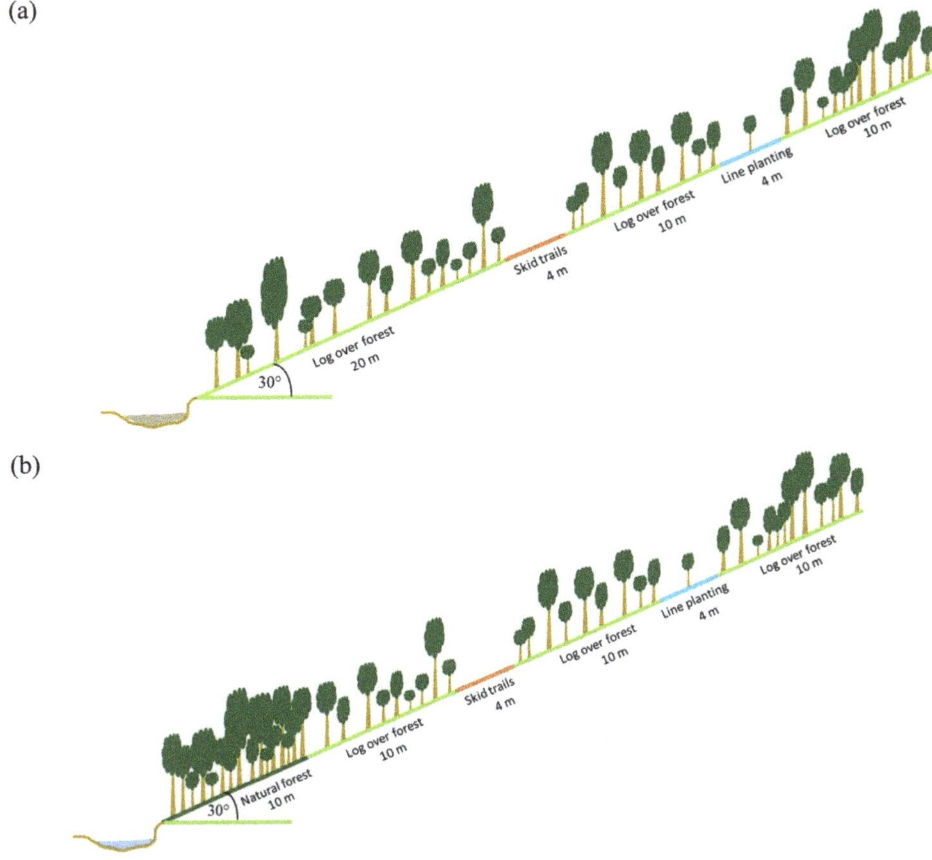

Figure 11. Simulation scenario of a surface runoff hydrograph in different forest operations: (**a**) without river buffer; (**b**) with river buffer.

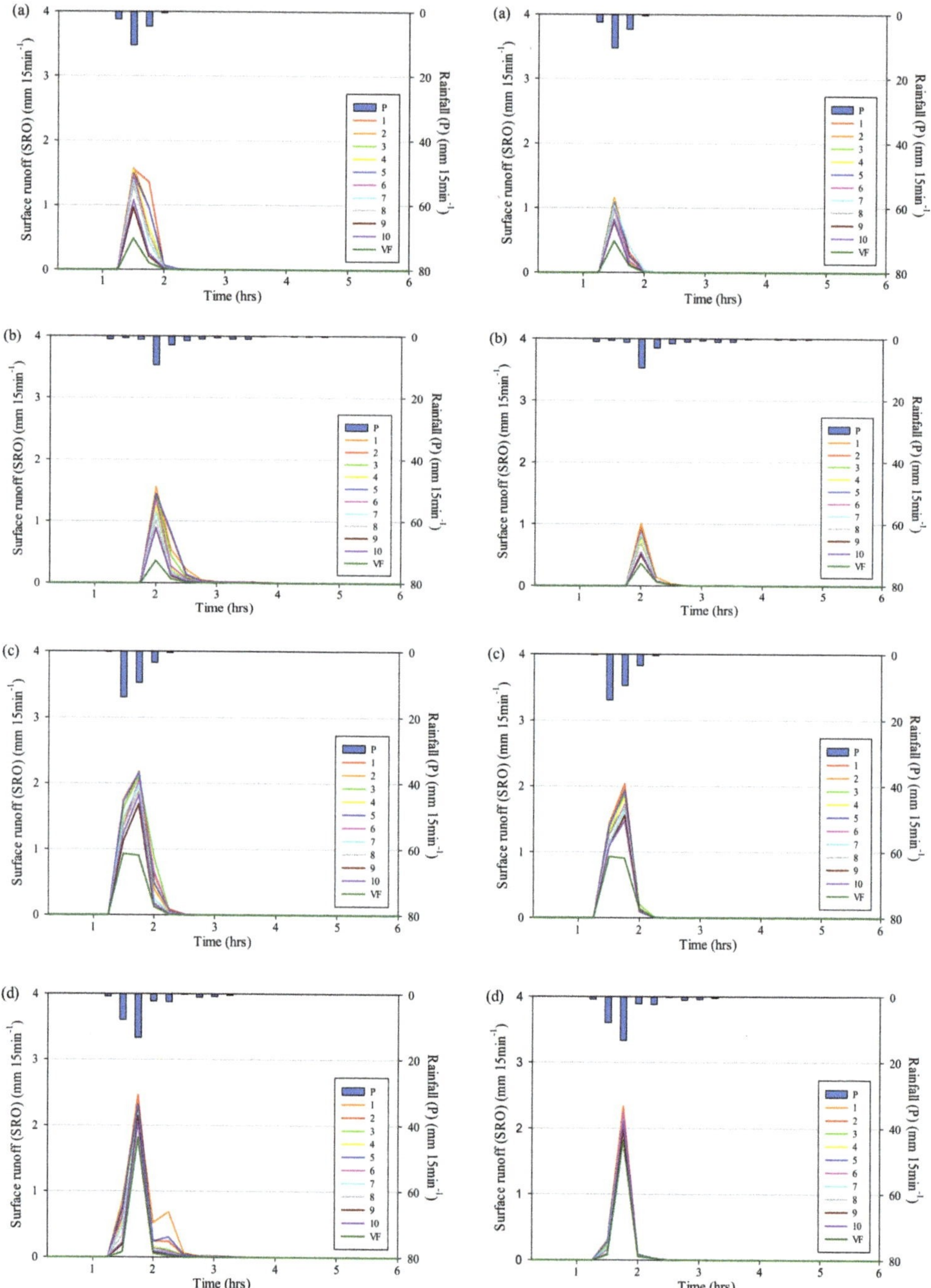

Figure 12. *Cont.*

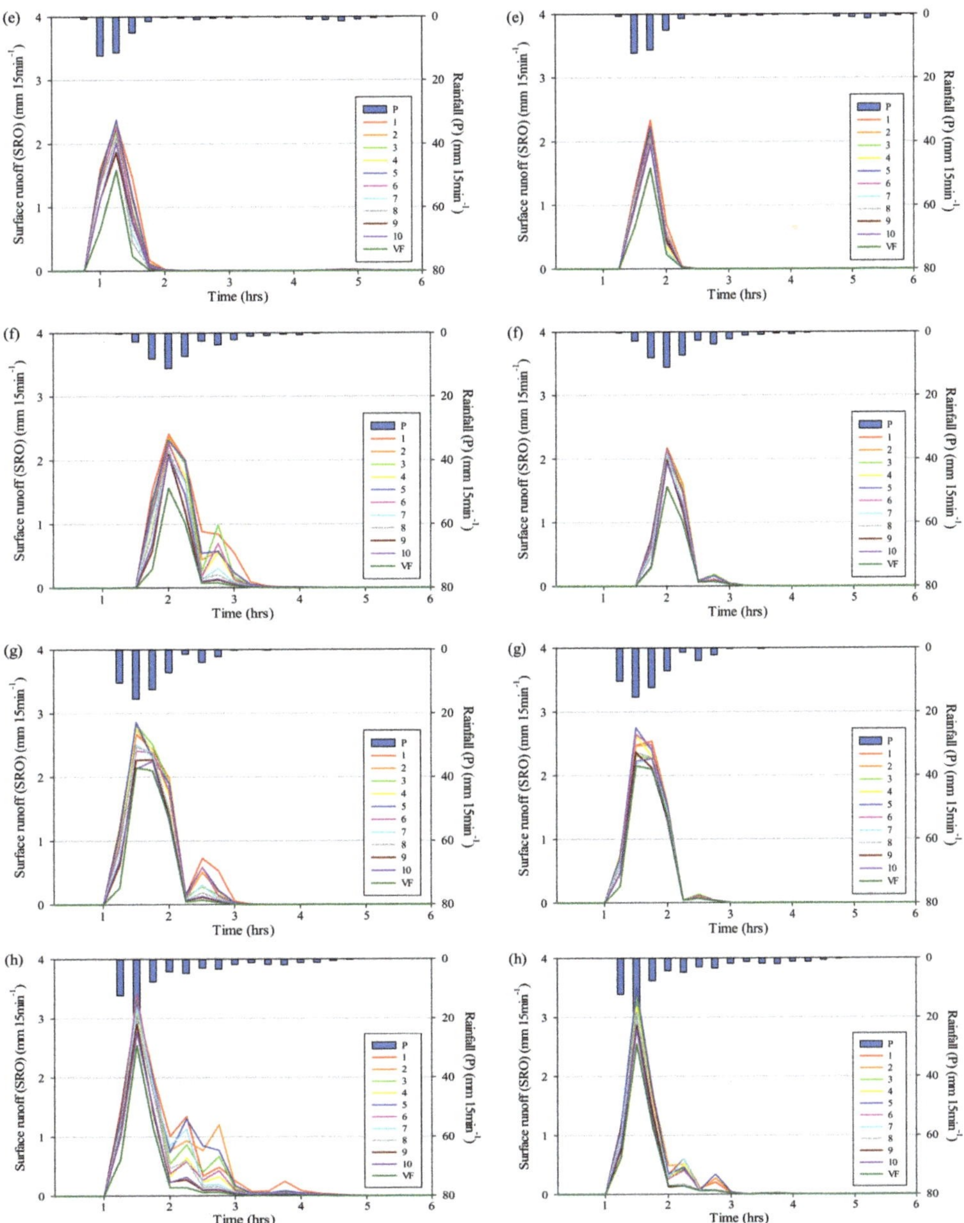

Figure 12. Surface runoff hydrograph generated from rainfall events of: (**a**) 17.8 mm, (**b**) 21.2 mm, (**c**) 27.2 mm, (**d**) 29.2 mm, (**e**) 37.6 mm, (**f**) 41.4 mm, (**g**) 53.4, and (**h**) 66.6 mm.

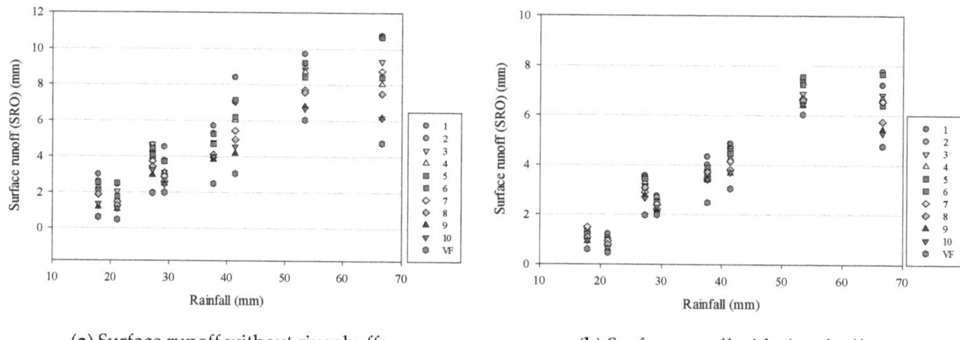

(a) Surface runoff without river buffer

(b) Surface runoff with river buffer

Figure 13. Relationship between rainfall and surface runoff in the different scenarios.

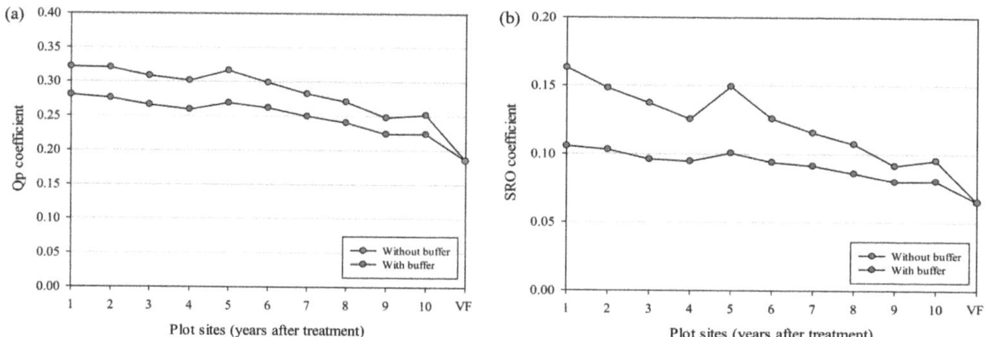

Figure 14. (**a**) Coefficient of Qp to peak rainfall intensity. (**b**) Coefficient of total SRO to the amount of rainfall.

4. Conclusions

Forest disturbance affects soil hydraulic properties, and therefore results in differences in the propensity to generate surface runoff flow in different types of forested sites. The current study showed that soil compaction changed the soil pore distribution and affected soil moisture characteristics. In natural conditions, water retention curves for surface soils (0–50 cm deep) tended to display greater changes in θ than subsurface soils (50–150 cm deep). Intensive root growth at depths of 0–50 cm was effective for increasing the number of macro-pores and the porosity. Low surface K_s values were found on skidder tracks compared to other surface disturbance types. Consolidated surfaces, such as skidder tracks, line-planted areas, cleared areas, and logged areas, contribute disproportionately to the stormflow response because low K_s values (compared to natural conditions) lead to surface runoff for low rainfall intensities. Forest disturbances have altered the typical surface hydrological pathways, thereby creating the conditions for more surface runoff on disturbed surfaces than on undisturbed surfaces. The recovery time for near-surface K_s on non-skidder tracks was estimated to be between 10 and 15 years, whereas in the skidder tracks it was estimated to be more than 20 years. Maintaining the buffer area can effectively reduce the Qp and SRO values in the stream channel.

Author Contributions: Conceptualization, H.S. and K.K.; methodology, H.S. and K.K.; Modelling software, K.K.; Validation, H.S., K.K.; formal analysis, H.S.; investigation, H.S. Both authors have read and agreed to the published version of the manuscript.

Funding: This research received no external funding.

Institutional Review Board Statement: Not applicable.

Informed Consent Statement: Not applicable.

Data Availability Statement: This study did not report any data.

Conflicts of Interest: The authors declare no conflict of interest.

References

1. Bruijnzeel, L.A. Hydrology of moist tropical forest and effects of conversion: A state knowledge review. In *Hydrology of Moist Tropical Forests and Effects of Conversion: A State of Knowledge Review*; UNESCO: Paris, France; Vrije Universiteit: Amsterdam, The Netherlands, 1990.
2. Beshta, R.L.; Pyles, M.R.; Skaugset, A.E.; Surfleet, C.G. Peakflow responses to forest practices in the western cascades of Oregon, USA. *J. Hydrol.* **2000**, *233*, 102–120. [CrossRef]
3. Bruijnzeel, L.A. Hydrological functions of tropical forests: Not seeing the soil for the tree? *Agric. Ecosyst. Environ.* **2004**, *104*, 185–228. [CrossRef]
4. Ziegler, A.D.; Negishi, J.N.; Sidle, R.C.; Noguchi, S.; Abdul Rahim, N. Impact of logging disturbance on hillslope saturated hydraulic conductivity in a tropical forest in Peninsular Malaysia. *Catena* **2006**, *67*, 89–104. [CrossRef]
5. Suryatmojo, H.; Fujimoto, M.; Yamakawa, Y.; Kosugi, K.; Mizuyama, T. Runoff and soil erosion characteristics in different periods of an intensive forest management system in a tropical Indonesian rainforest. *Int. J. Sustain. Dev. Plan.* **2014**, *9*, 830–846. [CrossRef]
6. Van Der Plas, M.C.; Bruijnzeel, L.A. Impact of Mechanized Selective Logging of Rainforest on Topsoil Infiltrability in the Upper Segama Area, Sabah, Malaysia. *Hydrol. Warm Humid Regions IAHS Publ.* **1993**, *216*, 203–211.
7. Malmer, A. Hydrological effects and nutrient losses of forest plantation establishment on tropical rainforest land in Sabah, Malaysia. *J. Hydrol.* **1996**, *174*, 129–148. [CrossRef]
8. Noguchi, S.; Abdul Rahim, N.; Yusop, Z.; Tani, M.; Toshiaki, S. Rainfall-runoff responses and roles of soil moisture variations to the response in Tropical Rain Forst, Bukit Tarek, Peninsular Malaysia. *J. For. Res.* **1997**, *2*, 125–132. [CrossRef]
9. Williamson, J.R.; Neilsen, W.A. The influence of forest site on rate and extent of soil compaction and profile disturbance of skid trails during ground-based harvesting. *Can. J. For. Res.* **2000**, *30*, 1196–1205. [CrossRef]
10. Ziegler, A.D.; Giambelluca, T.W.; Tran, L.T.; Vana, T.T.; Nullet, M.A.; Fox, J.; Pinthong, J.; Maxwell, J.F.; Evett, S. Hydrological consequences of landscape fragmentation in mountainous northern Vietnam: Evidence of accelerated surface runoff flow generation. *J. Hydrol.* **2004**, *287*, 124–146. [CrossRef]
11. Zimmermann, B.; Elsenbeer, H.; De Moraes, J.M. The influence of land-use changes on soil hydraulic properties: Implications for runoff generation. *For. Ecol. Manag.* **2005**, *222*, 29–38. [CrossRef]
12. Kosugi, K.I.; Uchida, T.; Matsuda, S.; Mizuyama, T. Spatial variability of soil hydraulic properties in forested hillslope. *J. For. Res.* **1999**, *4*, 107–114.
13. Ziegler, A.D.; Giambelluca, T.W.; Plondke, D.; Leisz, S.; Tran, L.T.; Fox, J.; Nullet, M.A.; Vogler, J.B.; Troung, D.M.; Vien, T.D. Hydrological consequences of landscape fragmentation in mountainous nortern Vietnam: Buffering of Hortonian surface runoff flow. *J. Hydrol.* **2007**, *337*, 52–67. [CrossRef]
14. Osuji, G.E.; Okon, M.A.; Chukwuma, M.C.; Nwarie, I.I. Infiltration Characteristics of Soils under Selected Land Use Practices in Owerri, Southeastern Nigeria. *World J. Agric. Sci.* **2010**, *6*, 322–326.
15. Hassler, S.K.; Zimmermann, B.; Breugel, M.; Hall, J.S.; Elsenbeer, H. Recovery of saturated hydraulic conductivity under secondary succession on former pasture in the humid tropics. *For. Ecol. Manag.* **2011**, *261*, 1634–1642. [CrossRef]
16. Zimmermann, A.; Schinn, D.S.; Francke, T.; Elsenbeer, H.; Zimmermann, B. Uncovering patterns of near surface saturated hydraulic conductivity in an overland flow-controlled landscape. *Geoderma* **2013**, *195–196*, 1–11. [CrossRef]
17. Liang, W.-L.; Kosugi, K.; Mizuyama, T. Heterogeneous soil water dynamics around a tree growing on a steep hillslope. *Vadose Zone J.* **2007**, *6*, 879–889. [CrossRef]
18. Nussbaum, R.; Hoe, A.L. Rehabilitation of Degraded Sites in Logged-Over Forest Using Dipterocarps. In *Dipterocarps Forest Ecosystem. Towards Sustainable Management*; Schulte, A., Schone, D., Eds.; World Publishing: Singapore, 1996; pp. 446–463.
19. Nussbaum, R.; Anderson, J.; Spencer, T. Planting Dipterocarps for Rehabilitation of Log Landings and Skid Trails in Sabah, Malaysia. In Proceedings of the Fifth Round-Table Conference on Dipterocarps, Chiang Mai, Thailand, 7–10 November 1994; Appanah, S., Khoo, K.C., Eds.; Forest Research Institute Malaysia: Kuala Lumpur, Malaysia, 1994; pp. 299–315.
20. Adam, K.L.; Bruijnzeel, L.A. Managing Tropical Forest Watersheds for Production: Where contra-dictionary theory and practice co-exist. In *Wise Management of Tropical Forests*; Oxford Forestry Institute: Oxford, UK, 1992; pp. 37–75.
21. Bruijnzeel, L.A.; Chritchley, W.R. Environmental impacts of logging moist tropical forests. In *International Hydrology Programme, Series 7*; UNESCO: Paris, France, 1994.
22. Ziegler, A.D.; Giambelluca, T.W. Importance of rural roads as source areas for runoff in mountainous areas of northern Thailand. *J. Hydrol.* **1997**, *196*, 204–229. [CrossRef]
23. Wemple, B.C.; Swanson, F.J.; Jones, J.A. Forest roads and geomorphic process interactions, Cascade Range, Oregon. *Earth Surf. Proc. Landf.* **2001**, *26*, 191–204. [CrossRef]
24. Chappell, N.A.; Douglas, I.; Hanapi, J.M.; Tych, W. Sources of suspended sediment within tropical catchment recovering from selective logging. *Hydrol. Process.* **2004**, *18*, 685–701. [CrossRef]

25. Gholzom, E.H.; Gholami, V. A Comparison between Natural Forests and Reforested Lands in Terms of Runoff Generation Potential and Hydrologic Response (Case Study: Kasilian Watershed). *Soil Water Res.* **2012**, *4*, 166–173. [CrossRef]
26. Schnorbus, M.; Alila, Y. Peak flow regime changes following forest harvesting in a snow-dominated basin: Effects of harvest area, elevation, and channel connectivity. *Water Resour. Res.* **2013**, *49*, 517–535. [CrossRef]
27. Chandler, K.R.; Stevens, C.J.; Binley, A.; Keith, A.M. Influence of tree species and forest land use on soil hydraulic conductivity and implications for surface runoff generation. *Geoderma* **2018**, *310*, 120–127. [CrossRef]
28. Hao, M.; Zhang, J.; Meng, M.; Chen, H.Y.; Guo, X.; Liu, S.; Ye, L. Impacts of changes in vegetation on saturated hydraulic conductivity of soil in subtropical forests. *Sci. Rep.* **2019**, *9*, 1–9.
29. Zhao, M.; Boll, J.; Brooks, E.S. Evaluating the effects of timber harvest on hydrologically sensitive areas and hydrologic response. *J. Hydrol.* **2020**, *593*, 125805. [CrossRef]
30. Fashi, F.H.; Gorji, M.; Shorafa, M. Estimation of soil hydraulic parameters for different land-uses. *Model. Earth Syst. Environ.* **2016**, *2*, 1–7. [CrossRef]
31. Suryatmojo, H.; Fujimoto, M.; Yamakawa, Y.; Kosugi, K.; Mizuyama, T. Water balance changes in the tropical rainforest with intensive forest management system. *Int. J. Sustain. Future Hum. Secur. J.-SustaiN* **2013**, *1*, 56–62. [CrossRef]
32. Schmidt, F.H.; Ferguson, J.H.A. Rainfall types based on wet and dry period ratios for Indonesia with Western New Guinea. 1951. Available online: https://agris.fao.org/agris-search/search.do?recordID=US201300720509 (accessed on 21 November 2020).
33. Tan, K.H. *Soils in the Humid Tropics and Monsoon Region of Indonesia;* CRC Press, Taylor and Francis Group: Boca Raton, FL, USA, 2008; pp. 77–90.
34. Suryatmojo, H. Recovery of Forest Soil Disturbance in the Intensive Forest Management System. *Procedia Environ. Sci.* **2014**, *20*, 832–840. [CrossRef]
35. Fodor, N.; Sandor, R.; Orfanus, T.; Lichner, L.; Rajkai, K. Evaluation method dependency of measured saturated hydraulic conductivity. *Geoderma* **2011**, *165*, 60–68. [CrossRef]
36. Reynolds, W.D.; Bowman, B.T.; Brunke, R.R.; Drudy, C.F.; Tan, C.S. Comparison of Tensio Infiltrometer, Pressure Infiltrometer, and Soil Core Estimates of Saturated Hydraulic Conductivity. *J. Soil Sci. Soc.* **2000**, *64*, 478–484. [CrossRef]
37. Elsenbeer, H. Hydrologic flowpaths in tropical rainforest soilscapes: A review. *Hydrol. Proces.* **2001**, *15*, 1751–1759. [CrossRef]
38. Jury, W.A.; Gardner, W.R.; Gardner, W.H. *Soil Physics*, 5th ed.; John Wiley & Sons, Inc.: Hoboken, NJ, USA, 1991; pp. 77–87.
39. Kluten, A.; Dirksen, C. Hydraulic conductivity and diffusivity: Laboratory methods. In *Methods of Soil Analysis, Part. 1: Physical and Meneralogical Methods, Monograph No. 9;* Klute, A., Ed.; American Society of Agronomy: Madison, WI, USA, 1986.
40. Kosugi, K. Lognormal distribution model for unsaturated soil hydraulic properties. *Water Resour. Res.* **1996**, *32*, 2697–2703. [CrossRef]
41. Kosugi, K. A new model to analyse water retention characteristics of forest soils based on soil pore radius distribution. *J. For. Res.* **1997**, *2*, 1–8. [CrossRef]
42. Kosugi, K. New diagrams to evaluate soil pore radius distribution and saturated hydraulic conductivity of forest soil. *J. For. Res.* **1997**, *2*, 95–101. [CrossRef]
43. Kosugi, K. Effect of pore radius distribution of forest soils on vertical water movement in soil profile. *J. Jpn. Soc. Hydrol. Water Resour.* **1997**, *10*, 226–237. [CrossRef]
44. Mualem, Y. A new model for predicting the hydraulic conductivity of unsaturated porous media. *Water Resour. Res.* **1976**, *12*, 513–522. [CrossRef]
45. Kosugi, K.; Mori, K.; Yasuda, H. An inverse modelling approach for characterization of unsaturated water flow in an organic forest floor. *J. Hydrol.* **2001**, *246*, 96–108. [CrossRef]
46. Suryatmojo, H.; Fujimoto, M.; Kosugi, K.; Mizuyama, T. Effects of Selective Logging Methods on Runoff Characteristics in Paired Small Headwater Catchment. *Procedia Environ. Sci.* **2013**, *17*, 221–229. [CrossRef]

 water

Article

Monetary Valuation of Flood Protection Ecosystem Service Based on Hydrological Modelling and Avoided Damage Costs. An Example from the Čierny Hron River Basin, Slovakia

Igor Gallay, Branislav Olah *, Zuzana Gallayová and Tomáš Lepeška

Department of Applied Ecology, Faculty of Ecology and Environmental Sciences, Technical University in Zvolen, T.G. Masaryka 24, SK-960 01 Zvolen, Slovakia; gallay@tuzvo.sk (I.G.); gallayova@tuzvo.sk (Z.G.); lepeska@tuzvo.sk (T.L.)

* Correspondence: olah@tuzvo.sk; Tel.: +421-45-5206-506

Abstract: Flood protection is considered one of the crucial regulating ecosystem services due to climate change and extreme weather events. As an ecosystem service, it combines the results of hydrological and ecosystem research and their implementation into land management and/or planning processes including several formally separated economic sectors. As managerial and economic interests often diverge, successful decision-making requires a common denominator in form of monetary valuation of competing trade-offs. In this paper, a methodical approach based on the monetary value of the ecosystem service provided by the ecosystem corresponding to its actual share in flood regulating processes and the value of the property protected by this service was developed and demonstrated based on an example of a medium size mountain basin (290 ha). Hydrological modelling methods (SWAT, HEC-RAS) were applied for assessing the extent of floods with different rainfalls and land uses. The rainfall threshold value that would cause flooding with the current land use but that would be safely drained if the basin was covered completely by forest was estimated. The cost of the flood protection ecosystem service was assessed by the method of non-market monetary value for estimating avoided damage costs of endangered infrastructure and calculated both for the current and hypothetical land use. The results identify areas that are crucial for water retention and that deserve greater attention in management. In addition, the monetary valuation of flood protection provided by the current but also by hypothetical land uses enables competent and well-formulated decision-making processes.

Keywords: flood protection; river basin management; ecosystem service; hydrological modelling; monetary valuation; land use change

Citation: Gallay, I.; Olah, B.; Gallayová, Z.; Lepeška, T. Monetary Valuation of Flood Protection Ecosystem Service Based on Hydrological Modelling and Avoided Damage Costs. An Example from the Čierny Hron River Basin, Slovakia. *Water* **2021**, *13*, 198. https://doi.org/10.3390/w13020198

Received: 19 December 2020
Accepted: 12 January 2021
Published: 15 January 2021

1. Introduction

Ecosystem services (ES), as the benefits people obtain from ecosystems [1], are outcomes of ecosystem properties—biophysical conditions, structures, and processes [2], which constitute ecosystem functions [3,4]. In addition, ecosystem services feature highly distinctive spatial and temporal patterns of distribution, quantity, and flows [5]. The importance of spatial relations between the ecosystems providing the services and the areas where those services are utilized is stressed by many authors [6–8]. Ecosystems do not exist in isolation but they compose whole complexes of ecosystems interacting with each other in the landscape [9]. Ecosystem services are usually provided within units that define processes such as watersheds, specific habitats, or other natural units [7,10]. Spatial relations are for certain ES so significant that several authors [11,12] call them landscape services. Spatial relations and the location of the ecosystem are important also for a monetary valuation of the ecosystem service [13–15]. In theory, two "identical" ecosystems protecting the nearby area from natural hazards such as landslides, erosion, or floods provide an identical supply of ecosystem service, but the value of each ES reflects the actual value of the respective protected infrastructure (human lives, property, city size etc.). Flood protection (regulation or mitigation) belongs among those ecosystem services, for which the spatial accent is of crucial importance. Flood protection ES supply addresses the ecosystem's capacity to lower flood hazards caused by heavy precipitation events by reducing the runoff fraction [5]. This reduction corresponds to the damage-limiting efficiency of recent ecosystems compared to the maximum possible runoff.

The service capacity is an ecosystem's potential to deliver services based on biophysical properties, social conditions, and ecological functions. Capacity responds to natural or anthropogenic changes over time and space [2,8,16,17]. River floods are the costliest and most frequent natural hazards both in Europe [18,19] and in the world [20–22]. Direct and indirect economic losses and threat to human lives are continuously growing due to increasing population density, the growth of infrastructure and the increasing frequency of heavy precipitation events due to climate change [23–25]. According to [26], flood protection ES consists of: supply—flood volume regulated by vegetation and soils; service—area of avoided flood damage due to regulation by vegetation or soil; and benefit—avoided costs due to loss of property or infrastructure.

The value of ecosystems and their services can be expressed in different ways as it has complex meaning with several dimensions [11,27,28]. There are three value domains: ecological (biophysical), socio-cultural and economic [1,29,30]. The assessment of an ecosystem's capacity to provide flood protection ecosystem services is often based on biophysical methods [31,32], i.e., by employing hydrological models based on water retention functions of the vegetation and soil cover [15] or assessing storage capacity of floodplains [33]. There are several methods for the monetary valuation of ecosystem services [34–37]. Most methods for the monetary valuation of regulating services apply calculations of replacement costs and avoided damage costs [38,39]. Other methods valuate the benefits (in terms of increased economic welfare) resulting from reduced flooding as a consequence of reduced deforestation [40] or an economic valuation of the storage capacity of natural floodplain wetlands, as well as riparian forest, protecting nearby areas [41,42]. Similarly, a valuation of avoided flood damages for a floodplain's conservation in comparison to a scenario of residential construction is presented by [22]. The expressed willingness to pay method was applied by [43,44] to demonstrate how much people were willing to pay for flood risk reduction.

Flood hazard is defined as the probability that a critical peak discharge is exceeded and which in a combination with the consequent economic damage generates flood risk [45]. As noted by [30], it is useful to apply risk assessment methods in the assessment of flood protection ES. Risk is a combination of hazard (potential source of harm) and vulnerability (magnitude of damage or danger to life) [46,47]. To estimate hazards, it is necessary to know: (a) what is the probability and frequency of the occurrence of each hazardous phenomenon at a given place; (b) the intensity required for them to happen; (c) to what extent their total effect would influence the landscape and/or vulnerability of the social-economic system. Disasters are better viewed as a result of the complex interaction between a potentially damaging physical event (i.e., floods, droughts, fire, earthquakes and storms) and the vulnerability of a society, its infrastructure, economy and environment, which are determined by human behavior [48,49]. Modelling software and geographical information systems (GIS) are highly recommended for assessing risk assessment, natural hazards, and ecosystem services [14,50]. The use of catchment-based hydrologic models provides the basis to reveal the varying importance of factors and processes responsible for the formation of river swellings as well as the capability of different land use types to retain part of the incoming water, which reveals their regulation capacity [31,51–53].

The aim of this article is to develop a method for the monetary valuation of flood protection ecosystem services. The approach is based on a hypothesis that a monetary value of the ecosystem service provided by the ecosystem corresponds to its actual share in flood regulating processes and the value of the property protected by this service. The value of the ES corresponds to the size of the benefit to the society, and it is expressed in the value of property (infrastructure) that is protected by an ecosystem. Since all ecosystems in the river basin (above the protected infrastructure) contribute to the protection of property, the contribution of each ecosystem to this protection was assessed and valuated according to the share of each ecosystem in reducing the total runoff.

2. Materials and Methods

The methodical approach adopted is based on a combination of different bio-geophysical GIS data with terrain mapping and the results of hydrological modelling. The extent of the flooded area (flood hazard) was modelled according to different precipitation volumes and different land uses. The economic valuation of the flood protection ecosystem service was based on the estimated value of the endangered residential infrastructure.

Clarification note: the term "land use" applied in the hydrological analyses is considered, in a broad sense, equal to "ecosystem" in the parts dealing with ecosystem services assessment. The authors are fully aware of terminological difference but for the sake of simplification, these terms were used as synonyms.

2.1. Study Area

Flood regulation ecosystem service was estimated for the Čierny Hron River basin situated in central Slovakia in a mountainous region (Figure 1). This basin is rather specific, as in the years 1960–2017, 13 floods occurred in various locations within the study area [54]. Six of them were flash floods from intense summer precipitations. The total area of the basin is 291.7 ha, the minimum elevation of the basin is 461 m a.s.l., the maximum elevation is 1338 m a.s.l., and the mean elevation is 785 m a.s.l. The mean annual precipitation increases approximately from 780 to 1150 mm as the elevation increases [55]. The mean year one-day maximum precipitation is around 45 mm/day, but the absolute maximum recorded in the period 1981–2010 reached 70–80 mm/day. The mean temperature varies from −3.5 to −5.0 °C in the winter season and from 12 to 16 °C in summer. The dominant soils are loam and sandy loam Cambisols on Paleozoic granitic and metamorphic rocks from highlands to mountainous landscape. Regarding land use, 79% of the basin area is covered by forest, 9% by logging and regeneration logging area, 7% by grasslands, 3% by agricultural land and 2% by rural settlements. Most of the forests are mixed forests. Coniferous forests occupy the highest parts of the area and deciduous forests cover only the lowest part of the basin.

2.2. Modelling Surface Water Runoff

The volume of runoff water was modelled applying the SWAT model [56]. This model requires input parameters to describe soil properties, topography, land use, land management, and climate. It simulates watershed processes by first dividing the basin into sub-watersheds, and then further dividing the sub-watersheds into unique land use, soil, and slope combinations called hydrologic response units (HRUs). Modelled runoff (result of SWAT) served as an input into the HEC-RAS model [57] that showed whether the river channel safely discharges the modelled flow or if it is spilled from the banks. In the HEC-RAS modelling, steady flow analysis with mixed flow regime and critical deep boundary condition was applied. Subsequently, the extent of the flood was modelled using the HEC-GeoRAS extension of the ArcGIS software. The necessary inputs for modelling water runoff in the river channel and the subsequent extent of floods were (in addition to the above-mentioned): shape (channel geometry), capacity of the river channel (cross-sectional cut lines), and Manning's roughness coefficients of surface for the channel and for the land use. A database and polygon layer were created in ArcGIS 10.3 software. Relief inputs into the applied models were derived from 10 m digital elevation mode (DEM 3.5) provided by the Geodesy, Cartography and Cadastre Authority of the Slovak Republic [58], and all analyses used 10 m resolution. Soil characteristics were derived from the soil databases provided by the National Forest Centre [59] for forest soil and the Soil Science and Conservation Research Institute [60] for agricultural soil. National soil databases were transformed accordingly to the FAO classes required by the SWAT model [61]. Climate data input into the SWAT model came from the SWAT Weather Database [62] and from the Slovak Hydrometeorological Institute [63]. River channel transverse profiles were derived from DEM and verified and refined by field measurements. Land use was derived from actual (2018) high-resolution (25 cm/px) orthophoto maps [58] and refined by field mapping. Spatial information about the location, size, and type of buildings and other infrastructure in the area was derived from several sources: Geodesy, Cartography and Cadastre Authority of the Slovak Republic [58], Slovak Road Administration [64], Slovak Environment Agency [65], and OpenStreetMap [66]. Precipitation and river discharge data were provided by the Slovak Hydrometeorological Institute [63] and the Slovak Water Management Enterprise [67].

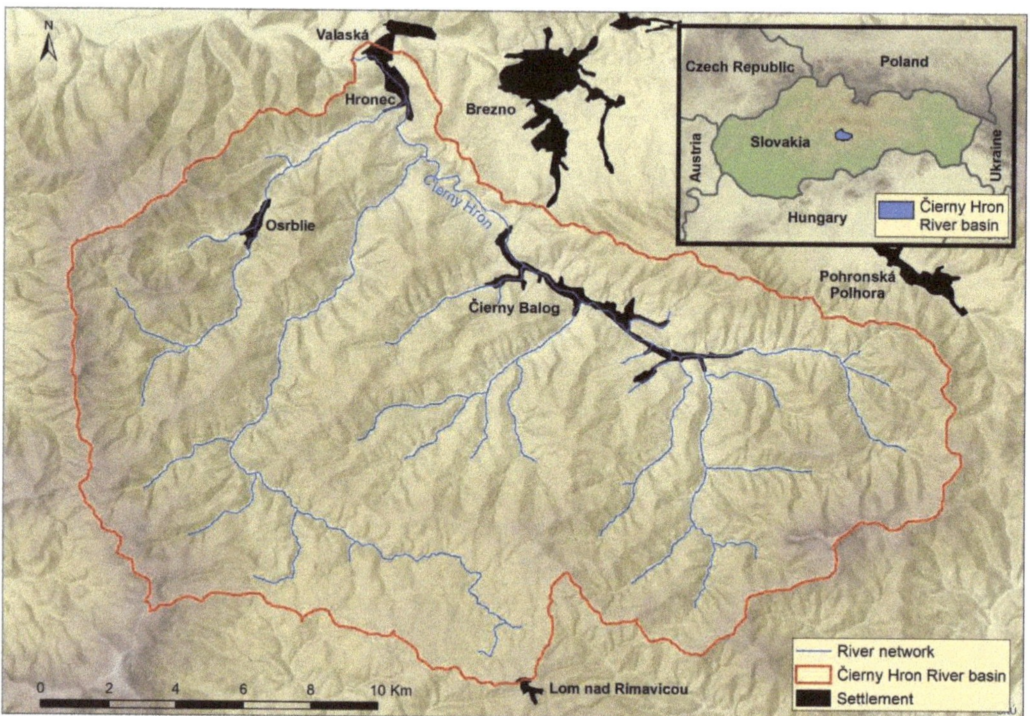

Figure 1. Study area.

2.3. Determination of the Precipitation Amount Causing Floods with Different Land Uses

Since flood protection ES is a service protecting against natural hazards, it was necessary to determine the size of the threat—the extent of the endangered and potentially flooded area and the probability of the precipitation capable of causing floods. Based on the available climate data [55,63,68], it was deduced that for the study area, it is possible to count on the maximum daily amount of precipitation with a recurrence interval of 100 years with a value of 75–80 mm (for the whole river basin). Using the SWAT model, the SCS Curve Number Method [69,70] was applied to model runoff volume in the area and flows in the river channels for four levels of potential daily precipitation, which would evenly affect the entire basin, namely 60, 80, 100, and 120 mm with the current land use. For the same precipitation levels, runoff volume and flows were modelled also for the hypothetical land uses—when the whole river basin (except for existing built-up areas) was used in one way. Five different hypothetical land uses were modelled: forest, permanent grassland, arable land ("row crops/straight row good" in the SCS method), orchard, and built-up area (in the sense of discontinuous urban fabric consisting both of impervious surfaces and pervious green areas such as lawns or gardens). These considered land use types either already exist in the area or are feasible to plan for (i.e., fruit orchards) due to the local natural and socioeconomic conditions. Subsequently, the land use and the amount of precipitation at which the capacity of the river channel is exceeded were modelled in the HEC-RAS program. The corresponding extent of flooding was modelled using HEC-RAS (Figure 2).

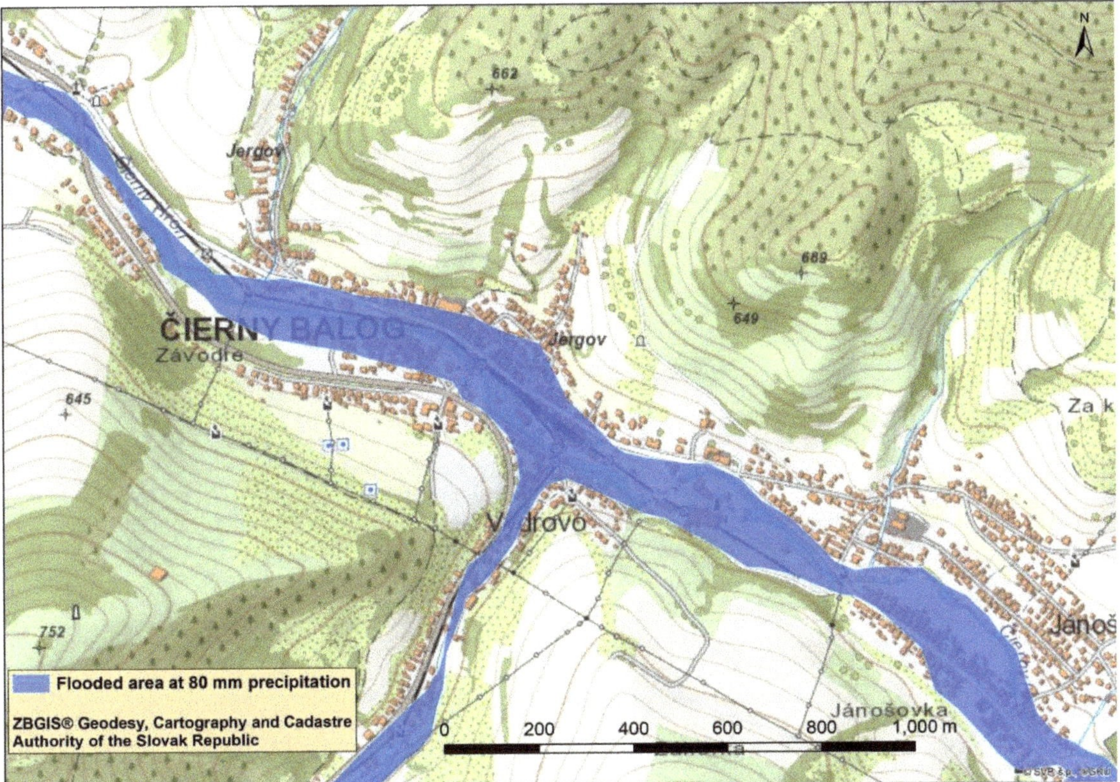

Figure 2. Extent of the modelled flooded area with the current land use and precipitation 80 mm/day in the HEC-RAS model.

2.4. Monetary Valuation of Flood Protection Ecosystem Service

Considering forest as the ecosystem that provides the best flood protection service, it was assumed that the value of its protection is equal to the value of the infrastructure that would be flooded after the precipitation which is already causing a flood (with the current land use). On the other hand, hypothetically, if the whole area was covered by forest (except for existing settlements), this precipitation (runoff) would be safely drained by a river channel and would not cause floods. The cost of damage to the endangered infrastructure (residential buildings, farm and industrial buildings, road and railway network) was calculated according to the methodology for estimating the cost of flood damage caused by floods by the Water Research Institute of Slovakia [71]. This cost was considered as the total cost of the forest ecosystem service in the whole river basin (except for already built-up areas).

However, individual areas in the basin contribute to the total runoff in different proportions depending on the location, soil conditions and slope. Therefore, for each individual area (hydrology response unit—HRU), its share of the total runoff was determined. This also determined its share of the infrastructure protection, which represented the value of the forest ecosystem service in the area of the unit concerned (share of the total ES value of the forest in the whole river basin). Subsequently, the value of the ecosystem service was re-calculated to the cost per ha.

The cost of ecosystem services of other ecosystem types (in the sense of land use types) is lower than that of forest since their retention capacity is lower. The results of runoff modelling for different potential land uses of the whole area (grassland, arable land, orchard, built-up area) show how much higher the water runoff from each HRU was compared to if the HRU was covered by forest. In other words, how much less water the other ecosystem type retains compared to the forest. The cost of the ecosystem service of the given ecosystem is then lowered by the given ratio compared to the cost of the ES provided forest in the given area (Figure 3). The cost of the ecosystem service for each area in the territory of other ecosystems was calculated on the basis of the formula:

$$\text{Cost ES}_{OE} = \text{Cost ES}_{F} \times ((R_{max} - R_{OE})/(R_{max} - R_{F})) \quad (1)$$

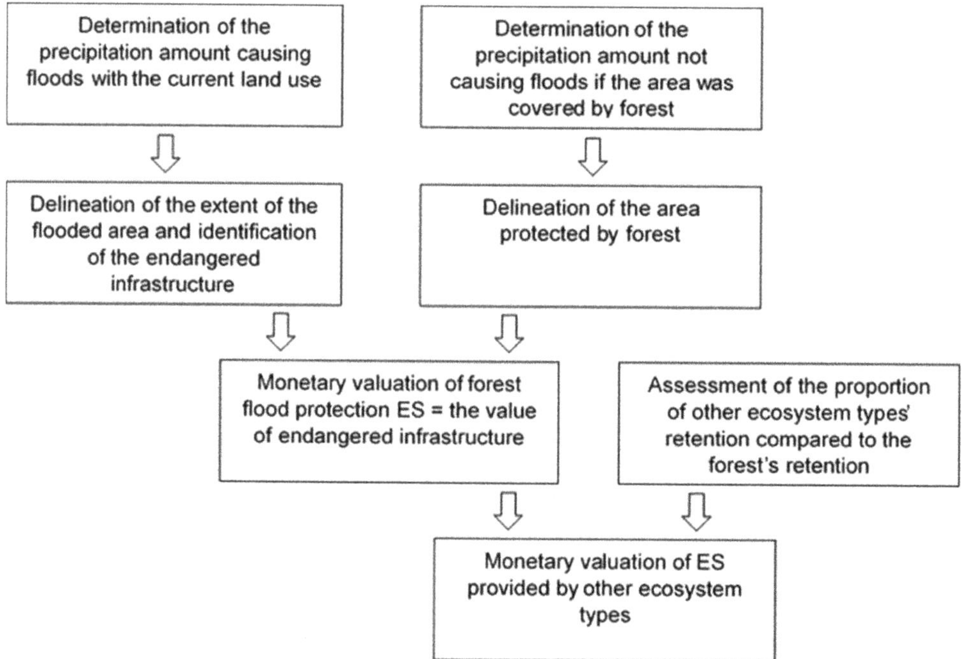

Figure 3. Methodical steps of the assessment.

Cost ES_{OE}—cost of other ecosystem types
Cost ES_{F}—cost of forest ecosystem service
R_{max}—maximum runoff from impervious surface
R_{OE}—runoff from other ecosystem types (different to forest)
R_{F}—runoff from forest

3. Results

Modelling of runoff volume with different precipitation amounts and the current land use in the area and the subsequent modelling of flood threat in the basin (the capacity of the river channel to drain runoff safely) show that 60 mm precipitation per day (without cumulation with precipitation from previous days—antecedent moisture condition (AMC II)), would not cause flooding and the study area is able to deal with it. However, a higher rainfall would already cause a flood. The modelling results also show that if the whole area (except for existing built-up areas) was covered with forest, the flood would not occur, even at 80 mm precipitation. Nevertheless, floods would occur with higher precipitations (100 and 120 mm per day). For this reason, the value of 80 mm precipitation per day became the basis for determining the flood protection ecosystem service. Precipitation of this magnitude occurs in the study area with a probability of approximately once every 100 years [63,68]. Figure 4 shows the ratio of runoff reduction from a rainfall of 80 mm per day for the five considered land use types (types of ecosystems), if they covered the whole territory (except for existing settlements), compared to the maximum runoff (as from completely impervious surface). Based on the modelling results, forest in most of the evaluated area, depending on other natural conditions, retains from 60 to 90% of the theoretical maximum runoff. Orchards are capable of retaining approximately 40–70% on the runoff, and grasslands retain slightly less. They retain about 40–70% in the upper half of the territory, but only between 30 and 40% in the lower half of the territory. Arable land is capable of retaining only up to 20–40% of the maximum runoff, while built-up areas reduce runoff by only 2–5% compared to the completely impervious surface.

Modelling the extent of the flood (Figure 2), with the current land use and precipitation of 80 mm per day, identified the endangered infrastructure (buildings, roads, railways, etc.). The monetary value of the infrastructure was calculated as 20,365,400 euros. This value is the cost of the flood protection service of the forest in the whole territory, if it covers the entire area of the river basin (except for already existing settlements). The cost of a specific area of the forest ecosystem reflects its ability to retain water and it was determined as a share of the whole cost. This share represents a contribution of each HRU (Figure 5) to the total runoff (contribution to the protection of the area), which depends on HRU's location, soil conditions, and slope.

As Figure 6 shows, the cost of this service per ha of the forest is in the range of 1626–1710 euros. Other types of ecosystems are able to retain less runoff water (Figure 4), therefore the cost of their flood protection service reflects this reduced capacity. When comparing water retention in the area for different types of ecosystems compared to the forest retention (Figure 7), orchards retain 72–82% of the amount of water retained by the forest. Grassland retains slightly less, 65–76%, and arable land between 42 and 56%. Built-up areas retain only 5–7% of the amount retained by the forest. These ratios express the share of other ecosystem types in flood protection in a specific place compared to the forest. As this value is lower, they do not fully protect the endangered infrastructure at a precipitation of 80 mm/day as the forests do, therefore the cost for this service is lowered according to this ratio. The resulting costs, calculated per ha of the basin area for each considered ecosystem (land use), are shown in Figure 6.

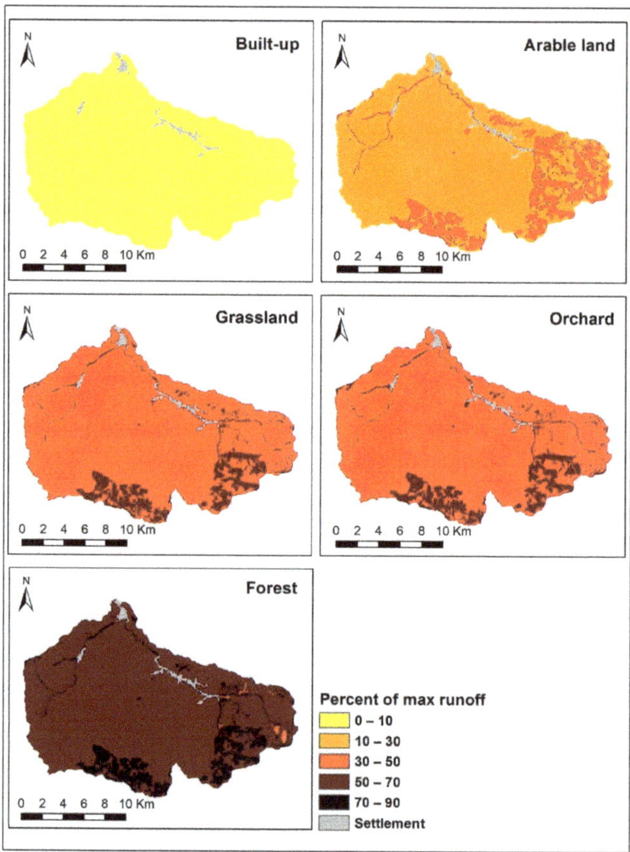

Figure 4. Proportion of maximum runoff retained (from 80 mm precipitation) if the whole area was covered by single land use.

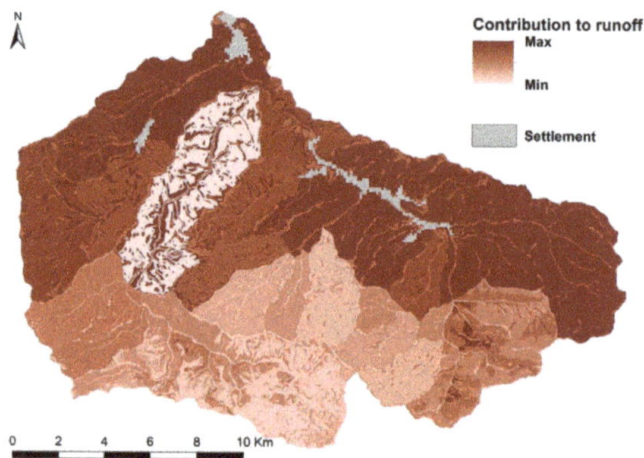

Figure 5. Contribution of the hydrologic response units (HRUs) to the total runoff.

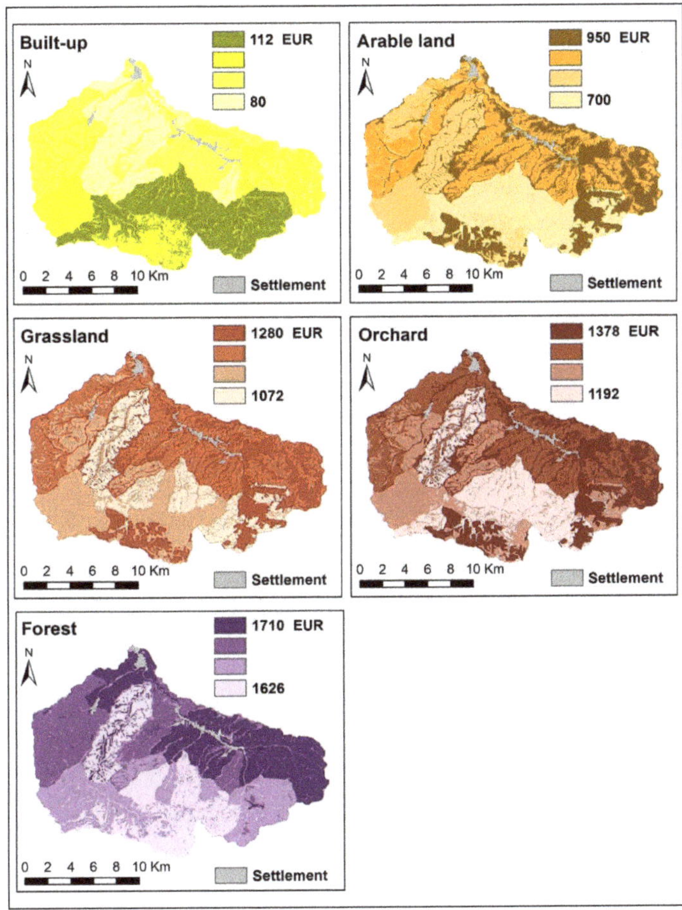

Figure 6. Monetary valuation of flood protection ecosystem service of different ecosystems (land use types).

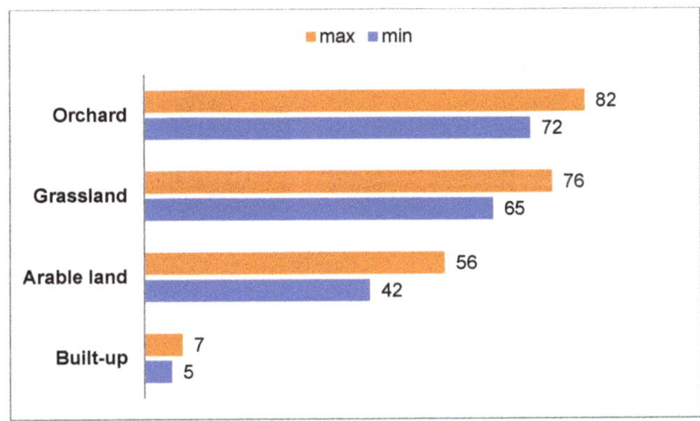

Figure 7. Water retention of different land use types compared to the forest retention (in %).

In the case of orchards, the cost per ha ranges from 1192 to 1378 euros depending on the conditions (location, natural conditions). Grasslands retain a little less rainwater compared to orchards, so the cost for flood protection ES is slightly lower and ranges from 1072 to 1280 euros. The magnitude of flood control on arable land depends on the crop grown and the management performed. Here, for the sake of simplicity, "row crops/straight row good" management (the SCS method) was considered. The calculated cost per ha is in the range of 700–950 euro. Compared to the previous land use types, the ability of residential areas to retain water is considerably reduced, as a large part is built-up and therefore impermeable. All existing settlements within the study area are characterized as discontinuous urban fabric, and the cost of their capability to reduce runoff ranges from 80 to 112 euro per ha. The costs for the flood protection ecosystem service shown in Figure 6 are hypothetical, valid for a specific area in the case of the existence of a given type of ecosystem in a given area. These values can also be used to compare the current situation and possible change to another type of use.

Figure 8 shows the costs of the flood protection ES with the current land use. Despite the fact that almost 80% of the area is covered by forest, and thus most of the area retains 60–90% of the maximum runoff, according to the modelling results, floods would still occur with a precipitation of 80 mm per day, even if there was good forest management. On the other hand, in the case of intense management (clear-cutting, dense road network), floods would occur even with lower precipitation. The cost of the flood protection ES of current ecosystems ranges from 300 euros for arable lands to 1710 euros for forest ecosystems.

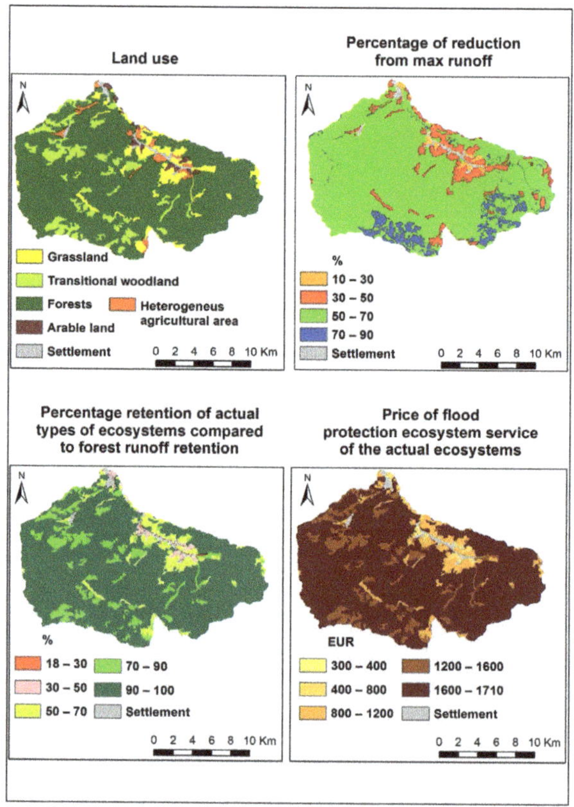

Figure 8. Current land use and the assessment of its flood protection ecosystem service.

Figure 9 presents the cost of the flood protection ES, which society would either gain or lose if the current ecosystem were changed into an alternative one. This comparison makes it possible to identify specific spots in the river basin that are most sensitive to change. In other words, areas where the change in the current land use to one of the five evaluated types would lead to the largest reduction in the performance of the ecosystem service, and to the largest increase in performance. Similarly, it is possible to identify areas whose change to another land use will cause the least loss of performance of the flood protection ES. If the current land use is changed to a forest ecosystem, the cost of ecosystems would remain the same in most areas, as 80% of the area is already covered by forest. Nevertheless, the ES value would increase from 200 to 1010 euros for the remaining 20%. On the other hand, changing to other land uses (besides forest) would reduce the size of the flood protection service in most areas.

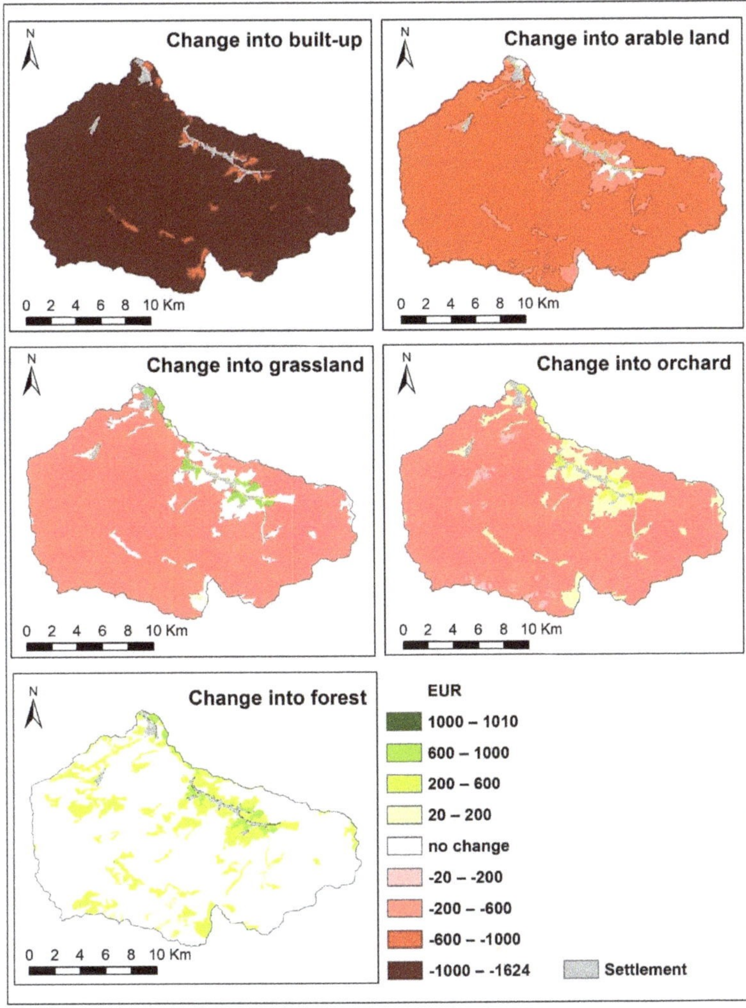

Figure 9. Estimated changes in cost of the flood protection ecosystem services (ES) in the case of land use change.

Changing to an orchard in the areas that are currently covered by forest would reduce the value of the flood protection ES service by 200 to 600 euros per ha, depending on the location and abiotic characteristics of the area. On the other hand, changing to areas that are currently more intensively used to an orchard, especially in the immediate vicinity of settlements, would increase the value of the ecosystem service by up to 600 euros per ha. Deforestation and the consequent change into grasslands decrease the value of the ecosystem service by 200 to 600 euros per ha. However, changing current arable land into grasslands would increase the service value by up to 600 euros per ha. Changing the current land use into arable land would reduce the size of the service in 95% of the territory, and in monetary terms it would mean a reduction of 200 to 1000 euros per ha. Expansion of settlements would reduce the value of the ecosystem service throughout the territory, depending on the location and natural conditions, by 620 to 1630 euros per ha.

4. Discussion

Land use and land management influence the land system's properties, processes and components, which are the basis of service provision. A change in land use management will therefore cause a change in service supply, not only for specific services but for the complete bundle of services provided by that ecosystem [11]. Our spatial approach followed this principle and it presented a way to quantify these potential changes in use and their impact of the flood protection ES provisioning. However, [72] called for caution in evaluating the results of modelling hydrological processes in relation to land use change, especially the sensitivity of the relationship between land use change and flood conditions with increasing return period of the simulated peak flow. In addition, [73] noted that the influence of land use conditions on storm runoff generation is only relevant for convective storms with high precipitation intensities, in contrast with long-lasting advective storms with low rain intensities.

According to [31], ecosystems affect water balance mainly through two processes: interception and infiltration. Interception depends on the structure of the ecosystem above ground (land cover), while infiltration is strongly determined by the soil properties. Surface runoff, which is the main factor for flood formation, also depends on abiotic factors such as bedrocks and topography. Ecosystems (i.e., forests) redirect or absorb parts of the incoming water (from rainfall), reducing the surface runoff and consequently the amount of river discharge. This ecosystem service plays its role before flood occurrence, and in some cases, it can even prevent it. Hence, flood protection ecosystem service assessments should conform to the biophysical characteristics and the likelihood of a flood in a particular area. [31,48] assessed the capacities of different ecosystems to regulate floods through investigations of water retention functions of the vegetation and soil cover. They applied the catchment-based hydrologic GIS model AGWA (and its constituent models KINEROS and SWAT) to express the capacity of the flood protection ES of individual types of land cover using biophysical methods (taking into account infiltration, surface runoff and peak flow). This biophysical assessment was then complemented with a comparison of regional supply–demand balances, with demand expressing the degree of vulnerability defined as "the characteristics and circumstances of a community, system or asset that make it susceptible to the damaging effects of a hazard".

In the presented approach, the extent of the area and the amount of precipitation at which forest would be able to completely protect the area was determined. These results were applied in the estimation of the cost of the ecosystem service. For the practical use of the ES concept, it is necessary to express the value of the ecosystem service in a specific area (i.e., at the habitat level or topical dimension) with its specific properties with respect to the effect of surrounding ecosystems (choral dimension or landscape patterns and spatial relationships of ecosystems) and how this value changes when changing the use of area and/or changing any of the surrounding ecosystems [74,75]. The applied approach follows the SCS curve number method [61,70], which is based on the assumption that the forest performs flood protection (retention) best. As noted by [76,77], the impact of forest and increasing afforestation on hydrological regulations may not be clear, especially in areas with a lack of precipitation and in river basins above 1000 km^2, where there is a great variability of factors (climatic conditions, urbanization, dam construction). However, this is not the case in the study area. In the Carpathian Mountains, many authors have confirmed that forests perform this service best and they are capable of retaining rainfall up to 50 mm [78–80]. This indicates that higher rainfalls will not be fully infiltrated and cause runoff. However, as shown by the modelled results, precipitations up to 80 mm will be drained safely by the existing river channels in the study area. On the other hand, the capacity of forests to provide ecosystem services strongly depends on the actual forest management [16], especially the size of the clearing area and the density of forest roads that could accelerate surface runoff. Nevertheless, none of the possible natural threats or anthropogenic pressures, i.e., negative impact of the intensive forest use, were included in the presented assessment. This might be the reason that, despite the high forest coverage, the study area has experienced numerous floods, resulting either from long-term precipitation or melting of snow, and flash floods triggered by intense summer precipitation in the past [54].

However, the impact of vegetation on water flow is very complex. For example, [81] presented a new procedure to assess drag forces and plant hydrodynamic bending as a function of the stem basal diameter and modulus of elasticity. They also presented other possibilities of calculating the surface roughness of the river channel (vegetation friction forces) for water flow depending on vegetation type (woody plants), its density (frontal projected area) or their elasticity. Other authors [82,83] used more detailed calculation models based on field measurements of flow velocity and morphological vegetation parameters (main morphometrical vegetation features) to determine global water flow resistance (i.e., stem diameters and heights, and bed surface density). [84] proposed an experimental methodology and their results demonstrate the reliability of the Keulegan equation in predicting the flow resistance. Based on the obtained results, a model to evaluate the Nikuradse equivalent of sand-grain roughness starting from the vegetation height and density was proposed and tested. Although these methods are more accurate, a simpler expression using Manning's coefficient was applied in our approach, as it is required as an input by HEC-RAS. More precise methods may be used in the future to refine the modelling of the water flow in the river channel and the possible over bank spills. However, our approach does not aim to refine water flow models, but it focuses on using existing models to determine the value of flood protection ES. Obtaining more precise results on flood risks by using alternative modelling is welcomed but it would not change the proposed methodic steps for determining the ES.

Since regulation services produce or maintain desirable environmental conditions, societal demand should be expressed as the amount of regulation needed to meet a desired end condition [16]. In evaluating disasters and natural hazards, many authors emphasize the need to focus not only on determining the location (spatial distribution), intensity or period of recurrence of natural hazards, but also on determining vulnerability, which depends on the number of people, infrastructure, or ability to deal with the event psychologically or economically, etc. An excellent example of economic risk evaluation in relation to flood hazards was presented by [85,86]. Their basic concept is similar to our case study—to determine the extent of the threat of economic damage in floods with different return periods using hydrological modeling. Their determination of the extent applies probability methods using intensity–duration–frequency (IDF) curves of rainfall height for a return period (derived for the Apulia Region in Southern Italy), by expressing the probability of flood hazard occurrence as also applied by [87], using market costs (unit market values) in determining the economic value and other methods of quantitative risk assessment in terms of monetary values. Our approach focuses more on deriving the value of the ecosystem service for a specific ecosystem. In many ways, our methodic steps are simplified compared to these cited articles and our approach undoubtedly has some constraints related to many drivers, i.e., there are also newer methods in hydrological modelling [88,89]. Despite these limitations, our method makes it possible to express the value of the flood protection ES, and these more precise procedures will serve as inspiration for the further improvement of our approach.

In order to analyse trade-offs between competing ecosystem services, [90] compared the benefits of anti-erosion regulation service regulation water erosion of individual ecosystems versus benefits from crop production of these ecosystems. They evaluated the change in the resulting benefits for the whole area after proposing decreasing intensity of land use (on steep slopes, marginal sites, or habitat connectivity support) to 7% of the territory. It is a very beneficial concept that can be used in argumentation with stakeholders when discussing environmental management of the landscape [91,92]. Even more so, when managerial decisions regarding competing trade-off are based on a comparison of financial benefits or losses.

5. Conclusions

Despite flood protection being considered one of the crucial ecosystem services that builds on an extensive and robust hydrological research, there are still aspects that need further development. These aspects reflect the complex character of hydrological processes both affecting and affected by several economic sectors and a need for the implementation of flood assessment results to land management and planning. This is even more important in land management systems that deal with several formally separated sectoral management policies (spatial planning, agriculture, forestry, water management, and the environment). These sectors have a legal competency over the same territory; however, their managerial and economic interests often diverge, thus successful decision making processes require a common denominator in the form of a monetary valuation of competing trade-off.

In this paper, this challenge was addressed by combining hydrological modelling methods, the method of non-market monetary value for estimating avoided damage costs, and flood risk assessment methods into a rather simple, yet methodically robust, approach. The innovation of the method lies in the fact that on the basis of existing available hydrological models (for which relatively easily accessible data such as DEM, soil map, climatic data, or river channel profiles are required as inputs), it enables repeated operational determination of the extent of endangered property with the current land use, and when changing land use. The available hydrological models can be replaced by newer and more accurate models, depending on more detailed and accurate data being made available (i.e., LIDAR). At the same time, this method also makes it possible to determine the highest level of protection by natural ecosystems (i.e., forests in Slovakia) and, on the basis of this, comparison to determine the monetary value of the ecosystem service (and not just express it in biophysical units) can be performed.

The approach consists of independent and open modules enabling further fine-tuning in order to fit any local natural conditions or available underpinning data. A further development of the method could be the inclusion of forest management intensity or forest road networks as surface runoff accelerators. Although these phenomena were not addressed in the current state of our research presented in this article, the impact of forest management can be applied as a water retention parameter in the follow-up research. Monetary valuation based on damage avoided costs is only one of several feasible methods of estimating flood protection cost. This method was applied as it is an officially recognized method by the national authority. However, flood-related costs also cover prevention costs (construction or maintenance of water channel) and emergency costs or other site- or society-specific costs that could be applied in the assessment in future.

Climate change is leading to an increase in the frequency of extreme (very intense rainfall) storms, which alternate with ever-increasing periods of drought. Therefore, it is important to determine the capacity of ecosystems to retain water, both as protection against floods and, on the other hand, for mitigating drought in the periods between precipitation events. If this existing capacity is not enough in light of the forthcoming changes, it will then be necessary to propose changes in land use and various possible biotechnical measures to ensure sufficient capacity.

Land use management can significantly reduce flood flows in catchments, particularly within the localities where most of the runoff is generated. River and water resource managers or decision-makers should concentrate on enhancing the ability of ecosystems to infiltrate and retain precipitation. The optimal landscape structure, suitable management of natural resources, rational management of current and planned human activities, revitalization and re-naturalization of the landscape and river channels should be of interest for decision-making authorities. The proposed method allows the identification of areas (HRUs) that are more important regarding water retention due to their site conditions, thus deserving higher attention. This is particularly important in land management since it stresses the spatial relations of natural processes in landscape, which are often neglected in other sectoral policies. In addition, it has a strong potential also for building and/or maintaining green and blue infrastructure in the landscape. Information on the monetary values of flood protection provided by the existing and by the planned land uses allows competent and well-based decision-making processes (i.e., suitable forest management vs. reconstruction of channels).

Author Contributions: Conceptualization, I.G. and B.O.; methodology, I.G., B.O., Z.G., T.L.; data curation, field mapping, hydrological modelling and spatial analysis I.G. and Z.G.; writing—original draft preparation, I.G. and B.O.; writing—review and editing, Z.G. and T.L.; visualization, I.G. and B.O.; funding acquisition and project administration, T.L. All authors have read and agreed to the published version of the manuscript.

Funding: This research was funded by the Scientific Grant Agency of the Ministry of Education of the Slovak Republic, grant number 1/0104/19 "Wood pasture ecosystems of Slovakia". The APC was funded by the Scientific Grant Agency of the Ministry of Education of the Slovak Republic, grant number 1/0104/19 "Wood pasture ecosystems of Slovakia".

Institutional Review Board Statement: Not applicable.

Informed Consent Statement: Not applicable.

Data Availability Statement: Publicly available datasets were analyzed in this study. All used data sources are described in Materials and Methods and listed in References.

Acknowledgments: The authors wish to express their gratitude to Richard Charles Scott, B.Ed. for language editing of this manuscript.

Conflicts of Interest: The authors declare no conflict of interest. The funders had no role in the design of the study; in the collection, analyses, or interpretation of data; in the writing of the manuscript, or in the decision to publish the results.

References

1. Millennium Ecosystem Assessment. *Ecosystems and Human Well-Being: A Framework for Assessment*; Island Press: Washington, DC, USA, 2005.
2. van Oudenhoven, A.P.E.; Petz, K.; Alkemade, R.; Hein, L.; de Groot, R.S. Framework for systematic indicator selection to assess effects of land management on ecosystem services. *Ecol. Indic.* **2012**, *21*, 110–122. [CrossRef]
3. Haines-Young, R.H.; Potschin, M.P. The links between biodiversity, ecosystem services and human well-being. In *Ecosystem Ecology: A New Synthesis. BES Ecological Reviews Series*; Raffaelli, D., Frid, C., Eds.; CUP: Cambridge, UK, 2010; pp. 110–139.
4. Kandziora, M.; Burkhard, B.; Müller, F. Interaction of ecosystem properties, ecosystem integrity and ecosystem service indicators—A theoretical matrix exercise. *Ecol. Indic.* **2013**, *28*, 54–78. [CrossRef]
5. Stürck, J.; Poortinga, A.; Verburg, P.H. Mapping ecosystem services: The supply and demand of floodregulation services in Europe. *Ecol. Indic.* **2014**, *38*, 198–211. [CrossRef]
6. Fisher, B.; Turner, R.K.; Morling, P. Defining and classifying ecosystem services for decision making. *Ecol. Econ.* **2009**, *68*, 643–653. [CrossRef]
7. Syrbe, R.U.; Walz, U. Spatial indicators for the assessment of ecosystem services: Providing, benefiting and connecting areas and landscape metrics. *Ecol. Indic.* **2012**, *21*, 80–88. [CrossRef]
8. Burkhard, B.; Kandziora, M.; Hou, Y.; Müller, F. Ecosystem Service Potentials, Flows and Demands—Concepts for Spatial Localisation, Indication and Quantification. *Landsc. Online* **2014**, *34*, 1–32. [CrossRef]

9. Blaschke, T. The role of the spatial dimension within the framework of sustainable landscapes and natural capital. *Landsc. Urban Plan.* **2006**, *75*, 198–226. [CrossRef]
10. Pretty, J.; Brett, C.; Gee, D.; Hine, R.; Mason, C.; Morison, J.; Raven, H.; MD Rayment, M.D.; van der Bijlet, G. An assessment of the total external costs of UK agriculture. *Agric. Syst.* **2000**, *65*, 113–136. [CrossRef]
11. de Groot, R.S.; Alkemade, R.; Braat, L.; Hein, L.; Willemen, L. Challenges in integrating the concept of ecosystem services and values in landscape planning, management and decision making. *Ecol. Complex.* **2010**, *7*, 260–272. [CrossRef]
12. Bastian, O.; Grunewald, K.; Syrbe, R.U.; Walz, U.; Wende, W. Landscape services: The concept and its practical relevance. *Landsc. Ecol.* **2014**, *29*, 1463–1479. [CrossRef]
13. Eade, J.D.O.; Moran, D. Spatial Economic Valuation: Benefits Transfer using Geographical Information Systems. *J. Environ. Manage.* **1996**, *48*, 97–110. [CrossRef]
14. Bateman, I.J.; Jones, A.P.; Lovett, A.A.; Lake, I.R.; Day, B.H. Applying geographical information systems (GIS) to environmental and resource economics. *Environ. Resour. Econ.* **2002**, *22*, 219–269. [CrossRef]
15. Nedkov, S.; Boyanova, K.; Burkhard, B. Quantifying, Modelling and Mapping Ecosystem Services in Watersheds. In *Ecosystem Services and River Basin Ecohydrology*; Chicharo, L., Müller, F., Fohrer, N., Eds.; Springer: Dordrecht, The Netherlands, 2015.
16. Villamagna, A.M.; Angermeier, P.L.; Bennett, E.M. Capacity, pressure, demand, and flow: A conceptual framework for analyzing ecosystem service provision and delivery. *Ecol. Complex.* **2013**, *15*, 114–121. [CrossRef]
17. Vargas, L.; Willemen, L.; Hein, L. Assessing the Capacity of Ecosystems to Supply Ecosystem Services Using Remote Sensing and An Ecosystem Accounting Approach. *Environ. Manage.* **2019**, *63*, 1–15. [CrossRef]
18. Barredo, J. Major flood disasters in Europe: 1950–2005. *Nat. Hazards* **2007**, *42*, 125–148. [CrossRef]
19. European Environment Agency. *River Floods. Indicator Assessment/Data and Maps*; Publications Office of the European Union: Luxembourg, 2019.
20. Stromberg, D. Natural Disasters, Economic Development, and Humanitarian Aid. *J. Econ. Perspect.* **2007**, *21*, 199–222. [CrossRef]
21. Smith, K.; Petley, D.N. *Environmental Hazards. Assessing Risk and Reducing Disaster*, 5th ed.; Routledge Taylor & Francis Group: New York, NY, USA, 2009.
22. Kousky, C.; Walls, M. Floodplain conservation as a flood mitigation strategy: Examining costs and benefits. *Ecol. Econ.* **2014**, *104*, 119–128. [CrossRef]
23. Ciscar, J.C.; Iglesias, A.; Feyen, L.; Szabó, L.; Van Regemorter, D.; Amelung, B.; Nicholls, R.; Watkiss, P.; Christensen, O.B.; Dankers, R.; et al. Physical and economic consequences of climate change in Europe. *Proc. Natl. Acad. Sci. USA* **2011**, *108*, 2678–2683. [CrossRef]
24. Jongman, B.; Ward, P.J.; Aerts, J.C.J.H. Global exposure to river and coastal flooding: Long term trends and changes. *Glob. Environ. Chang.* **2012**, *22*, 823–835. [CrossRef]
25. European Environment Agency. *Economic Losses from Climate-Related Extremes in Europe Publications. Indicator Assessment/Data and Maps*; Office of the European Union: Luxembourg, 2019.
26. Tallis, H.; Mooney, H.; Andelman, S.; Balvanera, P.; Cramer, W.; Karp, D.; Polasky, S.; Reyers, B.; Ricketts, T.; Running, S.; et al. A Global System for Monitoring Ecosystem Service Change. *BioScience* **2012**, *62*, 977–986. [CrossRef]
27. Goulder, L.; Kennedy, D. Valuing ecosystem services: Philosophical bases and empirical methods. In *Nature's Services: Societal Dependence on Natural Ecosystems*; Daily, G.C., Ed.; Island Press: Washington, DC, USA, 1997.
28. Costanza, R.; de Groot, R.S.; Braat, L.; Kubiszewski, I.; Fioramonti, L.; Sutton, P.; Farber, S.; Grasso, M. Twenty years of ecosystem services: How far have we come and how far do we still need to go? *Ecosyst. Serv.* **2017**, *28*, 1–16. [CrossRef]
29. de Groot, R.S. Function-analysis and valuation as a tool to assess land use conflicts in planning for sustainable, multi-functional landscapes. *Landsc. Urban Plan.* **2006**, *75*, 175–186. [CrossRef]
30. Burkhard, B.; Maes, J. (Eds.) *Mapping Ecosystem Services*; Pensoft Publishers: Sofia, Bulgaria, 2017.
31. Nedkov, S.; Burkhard, B. Flood regulating ecosystem services—Mapping supply and demand, in the Etropole municipality, Bulgaria. *Ecol. Indic.* **2012**, *21*, 67–79. [CrossRef]
32. Boyanova, K.; Nedkov, S.; Burkhard, B. Quantification and Mapping of Flood Regulating Ecosystem Services in Different Watersheds—Case Studies in Bulgaria and Arizona, USA. In *Thematic Cartography for the Society. Lecture Notes in Geoinformation and Cartography*; Bandrova, T., Konecny, M., Zlatanova, S., Eds.; Springer: Cham, Germany, 2014; pp. 237–255.
33. Posthumus, H.; Rouquette, J.R.; Morris, J.; Gowing, D.J.G.; Hess, T.M. 2010. A framework for the assessment of ecosystem goods and services; a case study on lowland floodplains in England. *Ecol. Econ.* **2010**, *69*, 1510–1523. [CrossRef]
34. Turner, R.K.; Pearce, D.; Bateman, I. *Environmental Economics: An Elementary Introduction*; Harvester Wheatsheaf: New York, NY, USA, 1994.
35. Farber, S.C.; Constanza, R.; Wilson, M.A. Economic and ecological concepts for valuing ecosystem services. *Ecol. Econ.* **2002**, *41*, 375–392. [CrossRef]
36. Christie, M.; Fazey, I.; Cooper, R.; Hyde, T.; Deri, A.; Hughes, L.; Bush, G.; Brander, L.; Nahman, A.; de Lange, W. *An Evaluation of Economic and Non-Economic Techniques for Assessing the Importance of Biodiversity to People in Developing Countries*; Defra: London, UK, 2008.
37. Freeman III, A.M.; Herriges, J.A.; Kling, C.L. *The Measurement of Environmental and Resource Values: Theory and Methods*, 3rd ed.; RFF PRESS, Taylor & Francis: New York, NY, USA, 2014.

38. de Groot, R.S.; Brander, L.; van der Ploeg, S.; Costanza, R.; Bernand, F.; Braat, L.; Christie, M.; Crossman, N.; Ghermandi, A.; Hein, L.; et al. Global estimates of the value of ecosystems and their services in monetary units. *Ecosyst. Serv.* **2012**, *1*, 50–61. [CrossRef]
39. Crossman, N.D.; Nedkov, S.; Brander, L. Discussion Paper 7: Water Flow Regulation for Mitigating River and Coastal Flooding. Paper Submitted to the Expert Meeting on Advancing the Measurement of Ecosystem Services for Ecosystem Accounting, New York, 22–24 January 2019 and Subsequently Revised. Version of 1 April 2019. Available online: https://seea.un.org/events/expert-meeting-advancing-measurement-ecosystem-services-ecosystem-accounting (accessed on 15 September 2020).
40. Kramer, R.A.; Richter, D.D.; Pattanayak, S.; Sharma, N.P. Ecological and Economic Analysis of Watershed Protection in Eastern Madagascar. *J Environ. Manage.* **1997**, *49*, 277–295. [CrossRef]
41. Grygoruk, M.; Mirosław-Świątek, D.; Chrzanowska, W.; Ignar, S. How much for water? Economic assessment and mapping of floodplain water storage as a catchment-scale ecosystem service of Wetlands. *Water* **2013**, *5*, 1760–1779. [CrossRef]
42. Barth, N.-C.; Döll, P. Assessing the ecosystem service flood protection of a riparian forest by applying a cascade approach. *Ecosyst. Serv.* **2016**, *21*, 39–52.
43. Zhai, G.; Sato, T.; Fukuzono, T.; Ikeda, S.; Yoshida, K. Willingness to pay for flood risk reduction and its determinats in Japan. *J. Am. Water Resour. Assoc.* **2006**, *42*, 927–940. [CrossRef]
44. Entorf, H.; Jensen, A. Willingness-to-pay for hazard safety—A case study on the valuation of flood risk reduction in Germany. *Saf. Sci.* **2020**, *128*, 104657. [CrossRef]
45. O'Connell, P.; Ewen, J.G.; O'Donnell, G.; Quinn, P. Is there a link between agricultural land-use management and flooding? *Hydrol. Earth Syst. Sci.* **2007**, *11*, 96–107. [CrossRef]
46. United Nations Department of Humanitarian Affairs. *Internationally Agreed Glossary of Basic Terms Related to Disaster Management*; DHA/93/36; UN: Geneva, Switzerland, 1992.
47. Department for Environment, Food and Rural Affairs. *Guidelines for Environmental Risk Assessment and Management*; DEFRA: London, UK, 2000.
48. Birkmann, J. Measuring vulnerability to promote disaster-resilient societies: Conceptual frameworks and definitions. *Inst. Environ. Human Secur. J.* **2006**, *5*, 7–54.
49. Pamungkas, A.; Bekessy, S.A.; Lane, R. Vulnerability Modelling to Improve Assessment Process on Community Vulnerability. *Procedia Soc. Behav. Sci.* **2014**, *135*, 159–166. [CrossRef]
50. Dunford, R.W.; Harrison, P.A.; Bagstad, K.J. Computer modelling for ecosystem service assessment. In *Mapping Ecosystem Services*; Burkhard, B., Maes, J., Eds.; Pensoft Publishers: Sofia, Bulgaria, 2017; pp. 126–137.
51. Fohrer, N.; Haverkamp, S.; Frede, H.G. Assessment of the effects of land use patterns on hydrologic landscape functions: Development of sustainable land use concepts for low mountain range areas. *Hydrol. Process.* **2005**, *19*, 659–672. [CrossRef]
52. Kubinský, D.; Weis, K.; Fuska, J.; Lehotský, M.; Petrovič, F. Changes in retention characteristics of 9 historical artificial water reservoirs near Banska Stiavnica, Slovakia. *Open Geosci.* **2015**, *7*, 880–887.
53. Wałęga, A.; Młyński, D.; Wojkowski, J.; Radecki-Pawlik, A.; Lepeška, T. New Empirical Model Using Landscape Hydric Potential Method to Estimate Median Peak Discharges in Mountain Ungauged Catchments. *Water* **2020**, *12*, 983. [CrossRef]
54. Ministerstvo Životného Prostredia SR. *Predbežné Hodnotenie Povodňového Rizika v Čiastkovom Povodí Hrona—Aktualizácia 2018*; MŽP SR, SVP SR: Bratislava, Slovakia, 2018; Available online: https://www.minzp.sk/files/sekcia-vod/hodnotenie-rizika-2018/hron/phpr-hron.pdf (accessed on 20 April 2020).
55. Climatic Atlas of Slovakia. Available online: http://klimat.shmu.sk/kas/ (accessed on 20 April 2020).
56. SWAT Model. Available online: https://swat.tamu.edu (accessed on 14 March 2020).
57. HEC-RAS Model (US Army Corps of Engineers, Hydrologic Engineering Center, Davis). Available online: www.hec.usace.army.mil (accessed on 18 March 2020).
58. Geodesy, Cartography and Cadastre Authority of the Slovak Republic. Geoportal. Available online: https://www.geoportal.sk/sk/zbgis_smd/na-stiahnutie/ (accessed on 20 March 2020).
59. National Forest Centre, Slovakia. Forest GIS. Available online: http://gis.nlcsk.org/lgis/ (accessed on 20 March 2020).
60. Soil Science and Conservation Research Institute, Slovakia. Soil Maps. Available online: http://www.podnemapy.sk (accessed on 21 March 2020).
61. Neitsch, S.L.; Arnold, J.G.; Kiniry, J.R.; Williams, J.R. *Soil and Water Assessment Tool Theoretical Documentation, Technical Report No. 406*; Version 2009; Grassland, Soil and Water Research Laboratory—Agricultural Research Service Blackland Research Center—Texas AgriLife Research, Texas Water Resources Institute: Forney, TX, USA, 2011.
62. Essenfelder, A.H. *SWAT Weather Database: A Quick Guide*; Version: V.0.16.06; 2016; Available online: https://www.google.com.hk/url?sa=t&rct=j&q=&esrc=s&source=web&cd=&ved=2ahUKEwjHm_PM_pzuAhURK6YKHS8bC1AQFjABegQIAhAC&url=https%3A%2F%2Fwww.researchgate.net%2Fprofile%2FArthur_Hrast_Essenfelder%2Fpublication%2F294535100_SWAT_Weather_Database%2Fdata%2F5756f68e08aef6cbe35f4e5b%2FWeatherDatabase-QuickGuide.pdf&usg=AOvVaw1R_TylSnZhwKtGjYkWZu63 (accessed on 22 March 2020).
63. Slovak Hydrometeorological Institute. Available online: http://www.shmu.sk/en (accessed on 22 March 2020).
64. Slovak Road Administration. Available online: www.ssc.sk/en/home.ssc (accessed on 22 March 2020).
65. Slovak Environment Agency. Available online: www.sazp.sk/en (accessed on 22 March 2020).
66. OpenStreetMap. Available online: www.openstreetmap.org (accessed on 22 March 2020).

67. Slovak Water Management Enterprise. Available online: www.svp.sk/en (accessed on 22 March 2020).
68. Gaál, L.; Lapin, M.; Faško, P. Maximálne viacdenné úhrny zrážok na Slovensku. In *Extrémy Počasí a Podnebí, Proceedings of Seminar Extremes of Weather and Climate, Brno, Czech Republic, 11 March 2004*; Rožnovský, J., Litschmann, T., Eds.; Český Hydrometeorologický Ústav: Brno, Czech Republic, 2004; ISBN 80-86690-12-1.
69. Soil Conservation Service. *National Engineering Handbook, Section 4, Hydrology*; Department of Agriculture: Washington, DC, USA, 1964.
70. United States Department of Agriculture. *Module 104: Runoff Curve Number Computations. Study Guide. Hydrology Training Series*; United States Department of Agriculture: Washington, DC, USA, 1989.
71. Water Research Institute of Slovakia. Available online: http://www.vuvh.sk (accessed on 25 March 2020).
72. Brath, A.; Montanari, A.; Moretti, G. Assessing the effect on flood frequency of land use change via hydrological simulation (with uncertainty). *J. Hydrol.* **2006**, *324*, 141–153. [CrossRef]
73. Niehoff, D.; Fritsch, U.; Bronstert, A. Land-use impacts on storm runoff generation: Scenarios of land-use change and simulation of hydrological response in a meso-scale catchment in SW-Germany. *J. Hydrol.* **2002**, *267*, 80–93. [CrossRef]
74. Boyd, J.; Banzhaf, S. What are ecosystem services? *Ecol. Econ.* **2007**, *63*, 616–626. [CrossRef]
75. Nelson, E.; Mendoza, G.; Regetz, J.; Polasky, S.; Tallis, H.; Cameron, D.R.; Chan, K.M.A.; Daily, G.C.; Goldstein, J.; Kareiva, P.M.; et al. Modeling multiple ecosystem services, biodiversity conservation, commodity production, and tradeoffs at landscape scales. *Front. Ecol. Environ.* **2009**, *7*, 4–11. [CrossRef]
76. Zhang, M.F.; Liu, N.; Harper, R.; Li, Q.; Liu, K.; Wei, X.; Ning, H.; Hou, Y.; Liu, S. A global review on hydrological responses to forest change across multiple spatial scales: Importance of scale, climate, forest type and hydrological regime. *J. Hydrol.* **2017**, *546*, 44–59. [CrossRef]
77. Farooqi, T.J.A.; Li, X.; Yu, Z.; Liu, S.; Sun, O.J. Reconciliation of research on forest carbon sequestration and water conservation. *J. For. Res.* 2020. [CrossRef]
78. Minďáš, J.; Škvarenina, J.; Střelcová, K. Význam lesa v hydrologickom režime krajiny. *Zivotn. Prostr.* **2001**, *3*, 146–150.
79. Minďáš, J.; Čaboun, V. *2002: Influence of Vegetation on Catchment Runoff*; Final Report of Project VTP 27-64 E0203; LVÚ: Zvolen, Slovakia, 2002.
80. Bíba, M.; Oceánska, Z.; Vícha, Z.; Jařabáč, M. Forest-hydrological research in small experimental catchments in the Beskydy Mountains. *J. Hydrol. Hydromech.* **2006**, *54*, 113–122.
81. Pasquino, V.; Saulino, L.; Pelosi, A.; Allevato, E.; Rita, A.; Todaro, L.; Saracino, A.; Chirico, G.B. Hydrodynamic behaviour of European black poplar (*Populus nigra* L.) under coppice management along Mediterranean river ecosystems. *River Res. Appl.* **2018**, *34*, 586–594. [CrossRef]
82. Douglas, J.; Gasiorek, J.; Swaffield, J.; Jack, L. *Fluid Mechanics*, 5th ed.; Harlow: Pearson, UK, 2005.
83. Lama, G.F.C.; Errico, A.; Francalanci, S.; Solari, L.; Preti, F.; Chirico, G.B. Evaluation of flow resistance models based on field experiments in a partly vegetated reclamation channel. *Geosciences* **2020**, *10*, 47. [CrossRef]
84. Gualtieri, P.; Felice, S.D.; Pasquino, V.; Doria, G.P. Use of conventional flow resistance equations and a model for the Nikuradse roughness in vegetated flows at high submergence. *J. Hydro. Hydromech.* **2018**, *66*, 107–120. [CrossRef]
85. Pellicani, R.; Parisi, A.; Iemmolo, G.; Apollonio, C. Economic Risk Evaluation in Urban Flooding and Instability-Prone Areas: The Case Study of San Giovanni Rotondo (Southern Italy). *Geosciences* **2018**, *8*, 112. [CrossRef]
86. Apollonio, C.; Bruno, M.F.; Iemmolo, G.; Molfetta, M.G.; Pellicani, R. Flood Risk Evaluation in Ungauged Coastal Areas: The Case Study of Ippocampo (Southern Italy). *Water* **2020**, *12*, 1466. [CrossRef]
87. Lamb, R.; Keef, C.; Tawn, J.A.; Laeger, S.; Meadowcroft, I.; Surendran, S.; Dunning, P.; Batstone, C. A new method to assess the risk of local and widespread flooding on rivers and coasts. *J. Flood Risk Manag.* **2010**, *3*, 323–336. [CrossRef]
88. Grimaldi, S.; Petroselli, A.; Arcangeletti, E.; Nardi, F. Flood mapping in ungauged basins using fully continuous hydrologic–hydraulic modeling. *J. Hydrol.* **2013**, *487*, 39–47. [CrossRef]
89. Teng, J.; Jakeman, A.; Vaze, J.; Croke, B.F.; Dutta, D.; Kim, S. Flood inundation modelling: A review of methods, recent advances and uncertainty analysis. *Environ. Model. Softw.* **2017**, *90*, 201–216. [CrossRef]
90. Bastian, O.; Syrbe, R.U.; Rosenberg, M.; Rahe, D.; Grunewald, K. The five pillar EPPS framework for quantifying, mapping and managing ecosystem services. *Ecosyst. Serv.* **2013**, *4*, 15–24. [CrossRef]
91. de Groot, R.S. *Functions of Nature: Evaluation of Nature in Environmental Planning, Management and Decision Making*; Wolters-Noordhoff BV: Groningen, The Netherlands, 1992.
92. Izakovičová, Z.; Špulerová, J.; Petrovič, F. Integrated Approach to Sustainable Land Use Management. *Environments* **2018**, *5*, 37.

MDPI
St. Alban-Anlage 66
4052 Basel
Switzerland
Tel. +41 61 683 77 34
Fax +41 61 302 89 18
www.mdpi.com

Water Editorial Office
E-mail: water@mdpi.com
www.mdpi.com/journal/water

www.ingramcontent.com/pod-product-compliance
Lightning Source LLC
LaVergne TN
LVHW070930170726
843515LV00005B/908
* 9 7 8 3 0 3 6 5 5 2 0 8 8 *